20th Century

Revolutions in Technology

20TH CENTURY

REVOLUTIONS IN TECHNOLOGY

Edward Nathan Singer

NOVA SCIENCE PUBLISHERS, INC.
Commack, NY
1998

Editorial Production: Susan Boriotti
Office Manager: Annette Hellinger
Graphics: Frank Grucci and John T'Lustachowski
Information Editor: Tatiana Shohov
Book Production: Donna Dennis, Patrick Devin, Christine Mathosian, Tammy Sauter and Diane Sharp
Circulation: Maryanne Schmidt
Marketing/Sales: Cathy DeGregory

Library of Congress Cataloging-in-Publication Data available upon request

ISBN 1-56072-432-3

6080 Jericho Turnpike, Suite 207
Commack, New York 11725
Tele. 516-499-3103 Fax 516-499-3146
e-mail: Novascil@aol.com
Novascience@earthlink.net
Web Site: http://www.nexusworld.com/nova

Printed in the United States of America

Contents

PREFACE

A French politician once said, "War is too important to be left in the hands of Generals." Technology is too important to be left solely in the hands of scientists and engineers. The technological revolutions of the 20th century have brought both great benefits and great environmental and other problems. Humanity is riding a wild stallion called technology and if we do not control it then it will throw and destroy us. However, we can use new forms of technology to rein in the wild stallion and ride safely into the 21st century and beyond.

This book reviews the great technological revolutions of the 20th century and points out the problems they created and some possible solutions. It also attempts to differentiate the real problems from those that up to date have no proven basis. This is vital if we want to keep the benefits without destroying the technology.

The scientists and engineers who shaped the great technological revolutions of the 20th century were a reflection of humanity as a whole. They ranged from bigots and racists to those who respected and loved all humanity. Their stories are an integral part of the world they created.

This book is intended for the general readers who are interested in understanding the technology of the 20th century that affects their lives. This includes both people who have no science background as well as those who do. A short summary of each chapter which follows illustrates the concepts of this book.

CHAPTER 1

APPLYING ELECTRICITY TO EVERYTHING FROM AIR CONDITIONERS TO TOOTHBRUSHES

Electricity forms the basis of many of the 20th Century revolutions in technology. The roots of the development of electricity and magnetism are chronicled as well as the lives of some of the early geniuses and eccentrics.

Benjamin Franklin determined that the direction of electric current was from plus to minus. Many years later it was found that the electric current in a conductor is carried by electrons which flow from minus to plus. The U.S. Army solved this contradiction by issuing a military directive in World War II .

Some of the applications of electricity produced very serious problems that could affect life on earth. For example, the use of CFCs as the refrigerant in electric refrigerators and air conditioners led to partial destruction of protective ozone layer. This can cause a drastic increase in skin cancers. Some of the efforts to use new refrigerants and technologies to reduce the danger are examined.

Another danger is the pollution of the atmosphere by the generation of electric power. Some solutions such as the fluidization of sulfur bearing coal are described.

A third perceived danger is the effect of EMF (Electromagnetic Fields) on people exposed to power lines and devices such as electric blankets. The pros and cons of the low frequency EMF problem are examined together with the latest scientific investigations.

CHAPTER 2

USING THE ELECTROMAGNETIC WAVE TO HEAR AND SEE AT A DISTANCE

The development of radio and television are traced from the theoretical work on electromagnetic fields by Maxwell and the experimental verification by Hertz in the 19th Century to the scientists and inventors who made radio, television, and radar the everyday miracles of the 20th century.

Radio has become an integral part of 20th century life. The cellular radio telephone industry is rapidly spreading throughout the world. There have been lawsuits about brain cancer from the use of portable cellular radio telephones. There have been scares about male sterility from microwaves used in radar and communication relays. These are evaluated as well as studies that try to put them in perspective. Existing standards for limiting electromagnetic radiation are listed.

CHAPTER 3

HOW MANY TRANSISTORS CAN DANCE ON THE HEAD OF A PIN?

The roots of the chip are traced from the weird world of quantum mechanics to the transistor to the integrated circuit to the millions of electronic circuits on a substrate of silicon the size of a thumbnail.

The environmental dangers of manufacturing chips are enumerated. Technologies are described that can minimize these dangers.

CHAPTER 4

ADDING AN EXTRA BRAIN

The chip made possible the rapid expansion of the computer to an integral part of business, education, communications, music, art, and many other fields. The computer with its rapidly developing chip has changed every facet of life in the last quarter of the 20th century.

The roots of the computer are examined together with some of the people who developed this technology. Computer personalities are described including homosexuals who suffered for their sexual orientation.

The computer developed slowly until the explosion around 1981 when personal computers took off. Computers became faster and faster and smaller and smaller. The computer's size went from rooms of equipment to a computer that can be held in the palm of a hand.

The computer has problems some of which are basic and inherent and others which can be addressed by new technology. These are evaluated in detail together with some solutions that are available.

CHAPTER 5

LET THERE BE PURE LIGHT

The roots of the laser are traced from Einstein's theoretical work in the early 20th century through the maser and the early work in the United States and Russia. The bitter patent fights together with some of the personalities involved are chronicled.

The basic operation of the laser is described together with various applications from surgery to communications. The laser technology is also now being used with CDs to record music and information.

The dark side of the laser, the dazer, can be used by the military to blind enemy soldiers. The dazer was deployed but not used by the U.S. in the Persian Gulf War.

CHAPTER 6

DEVELOPING RECORDING TECHNOLOGY

This first covers the development of motion pictures from the silent movies of the late 19th century and the early 20th century to the talkies and then the introduction of color.

The phonograph is also traced from its invention of the recording cylinder in the late 19th century followed by the development of a flat disk or record which became popular around 1915 with the 78 RPM records. The improvements in records are described in the first half of the 20th century.

The audio magnetic tape is traced from 1927 to the development of cassettes which by the early 1980's was exceeding the production of records. Records in turn were superseded by CDs.

The CDs use the laser to digitally record as much as 72 minutes of music. The CDs are also used with computers to record large amounts of information.

Video types of recording are also described including the VCR tape machine and the laser video disk.

Some of the problems with recording devices are discussed including non-lasting recording both in early movies and in VCRs.

The rapid obsolescence of recording technologies and collections are discussed . These include 45 RPM records and eight track cassettes.

CHAPTER 7

REVOLUTION OF THE ARTIFICIAL MOON

The satellite and its uses have helped change the world in the second half of the 20th century. This includes communications, weather data, material surveys, and navigation.

The history of satellites is traced from the first rocket experiments in the United States, Germany, and Russia. The competition between the two super powers after World War II produced great forward strides such as Sputnik, the first artificial moon.

This was followed by a series of more and more advanced rockets and satellites from Gemini through Apollo. The culmination was putting men on the moon.

The development of space stations leading to international cooperation in the 21st century is discussed..

CHAPTER 8

MOVING TOWARDS UNIVERSAL COMMUNICATIONS

Future Public Land Mobile Telecommunication System (FPLMTS) is the official name designated by the International Telecommunications Union of the United Nations. FPLMTS is intended to become a worldwide personal communication network offering all of the services of a telephone network including voice, facsimile, and data. FPLMTS includes the functions of cordless telephony, wireless pay phones, private branch exchanges, and rural radio and telephone exchanges among terminals on land, on sea and in the air. Calls within the mobile system would be routed to and from the existing public switched telephone networks through satellites. Personal and mobile stations will be able to connect into the system by radio at any time and from any place in the world. This system is expected to begin operation before the end of the 20th century.

This chapter will include the development of digital telephone networks, facsimile, digitized radio voice, E-Mail, the Internet, World Wide Web, and special satellite networks.

Satellite systems for FPLMTS are being developed such as little LEO, big LEO, and GSO. The Iridium and the Tritium proposals for this technology are described.

CHAPTER 9

LEAVING THE HORSES BEHIND

The 20th century saw the development of the automobile from its beginnings in the late 19th century. The automobile, especially the one using the internal combustion engine, produced vast amounts of pollution. The technologies to reduce the pollution are evaluated from catalytic converters, oxygenated fuel, and finally electric cars.

Automobile crashes caused many deaths in the 20th century. The mortality rate often exceeded the number of people killed in a war. Technologies to make the car safer are examined.

CHAPTER 10

ON THE WINGS OF EAGLES

Humanity from the early Greek times through Leonardo da Vinci dreamed of people flying above the earth. The 20th century saw the first airplane in 1903 at the beginning and the Boeing 777 in 1995 towards the end of the century. This latter was the first civilian aircraft designed by computer.

The development of the airplane from the crude piston engine of the Wright Brothers first plane to the sophisticated fan jet aircraft of the end of the 20th century is described.

Some of the factors contributing to crashes are examined together with technologies that would minimize accidents.

CHAPTER 11

MARRYING TECHNOLOGY TO MEDICINE

In the 20th century there were a number of great revolutions in medicine. One was putting microscopic organisms to work in the war on infections. Antibiotics like penicillin when used too often generated resistant bacteria. It was found that antibiotics had to be changed from time to time to avoid this problem.

At the end of the 19th century new diagnostic technology began with the discovery of X-rays. However, it was discovered in the 20th century that too much exposure to X-rays could cause cancer. It became a matter of a cost-benefit ratio. Soon other diagnostic methods were discovered such as PET (Positron Emission Tomography), MRI (Magnetic Resonance Imaging) and SQUID (Superconducting Quantum Interfering Device). These are depicted in this chapter with the advantages and disadvantages of each.

Another new medical development in the 20th Century was the trading of new parts for old in the human body. A number of these technologies are evaluated with a comparison of the advantages and disadvantages of each.

The 20th century saw the development of a great many new drugs. Many of these have serious side effects and a cost/benefit analysis has to be made.

A great revolution in the last decade of the 20th century is minimally invasive surgery in which the surgeon looks not at the patient but at a video screen. A tiny TV camera equipped with fiber optics is inserted into the body either through an orifice or a puncture hole. A small surgical instrument is also
inserted and the surgeon performs the operation guided by what he sees on a TV screen.

This can result in a patient recovering much quicker. A three dimensional laparoscope system is described. However, there are problems that must be solved.

One of the problems of laparoscopic surgery is the training of surgeons in the new technique. A possible solution is the use of simulators similar to flight simulators in the aviation industry.

CHAPTER 12

PLAYING GOD WITH GENETIC ENGINEERING

The discovery of the structure of DNA in the 20th century led to the insertion of human genes into animals to produce beneficial compounds for humans. An example of this type of genetic engineering is the insertion of a human gene in a cow to produce large amounts of human proteins in the cow's milk. The human proteins are used to fight human diseases.

However, this technology could be used for other less than benign results. There has been a great deal of concern about genetic engineering.

This chapter discusses the origin and development of genetic engineering and the fears of some people and the hopes of others.

CHAPTER 13

FOOD DEVELOPMENTS IN THE 20TH CENTURY

Three great revolutions in food have occurred in the 20th century. The first is in the growing of food. The green revolution using special seeds, fertilizers, pesticides, and mechanization increased food production dramatically but brought its own problems. The specialization of specific hybrid seeds increased the possibility of a particular food plant being wiped out by disease. The use of fertilizers and pesticides had its own dangers. Pesticides posed problems such as increasing cancer. Possible solutions are discussed.

The second revolution is preparing food in new ways. These include freezing, canning, microwave cooking, the use of chemical additives for preservation and irradiation of food. Some of these have created their own problems. These are discussed and possible solutions are presented.

The third great revolution is the controlling of reproduction in animals by artificial insemination leading to factory-like production of animal food.

CHAPTER 14

USING ARTIFICIAL MATERIALS

One great change was the use of plastics which has filled garbage dumps all over the world with material that does not decompose. Several solutions are discussed.

Another revolution was the use of artificial fibers for clothing such as rayon and nylon partially replacing the natural ones such as silk and cotton.

A third revolution is the use of special materials such as PCP, DDT, CFCs, and many others that cause very serious problems to humans, animals and the environment in general.

CHAPTER 15

EXTRACTING ENERGY FROM THE ATOM

Atomic fission which was the first method of extracting energy from the atom has caused enormous problems. One of these is how to dispose the radioactive waste produced by atomic electric power reactors. A possible future technology that may help is discussed.

Another problem with atomic fission is that some of the radioactive wastes can be converted to plutonium, a deadly poison not found in nature. France is now exporting shiploads of plutonium to Japan. This poses a tremendous problem for the world.

A method of extracting energy now being studied is fusion. There is a discussion among scientists about the fuel to be used. One method will result in some radioactivity even though it will be lower than fission. The other method proposed is considerably more expensive but uses a different fuel that will result in practically no radiation. This will probably be a decision made at the end of the 20th or early 21st century.

EPILOGUE

In the 20th century, technology changed everything except the way people treat each other. The internal hatreds of humanity becomes much more dangerous in the world of atom bombs and genetic engineering. The teaching of tolerance in the future will be extremely necessary for survival.

The challenge of meeting a future asteroid or comet threat is described which will demand the cooperation of all people, without which humanity could be wiped out. A cooperative humanity with a minimum of internal hate can meet the challenge using the basic technology developed in the 20th century.

ACKNOWLEDGMENTS

This book would not be possible without the assistance and suggestions of my wife, Hilda Gofstein Singer.

I also wish to thank the staff of a number of universities in the United States and abroad for their help. Among them are Columbia University, Polytechnic University, Princeton University, Rutgers University, the University of California at Irvine, the University of Edinburgh and the University of Minnesota.

I wish to thank Mr. John Fiorentino for his assistance

ABOUT THE AUTHOR

Bachelor of Science Degree in natural sciences from the College of the City of New York

Masters Degree in Electrical Engineering from Polytechnic Institute of Brooklyn (now Polytechnic University)

Forty years of technical experience as an engineer.

Taught Physics for three years at a Community College.

Five patents in electronics and allied fields

Wrote many journal articles which were published in various technical journals and magazines including I.R.E. (now I.E.E.E.) publications, Electronics and others.

Received the 1984 National GEICO Public Service Award for work in radio communications for the New York City Fire Department.

Professional Engineer registered in New York State

Senior member of the I.E.E.E.

Fellow of the Radio Club of America

Member of Sigma Xi (the Scientific Research Society)

Member of the New York Academy of Sciences

Author of a book, "Land Mobile Radio Systems," published by Prentice-Hall in 1989. A second edition was published in 1994.

Listed in "Who's Who in Science and Technology," published by Marquis 1992.

CHAPTER 1

APPLYING ELECTRICITY TO EVERYTHING FROM TOOTHBRUSHES TO AIR CONDITIONERS

1.1 THE ROOTS OF 20TH CENTURY ELECTRICAL TECHNOLOGY

Static electricity was known in the ancient world. The Greeks used rods of amber rubbed by a cloth to attract bits of straw. William Gilbert (1544-1603), a British scientist, discovered that many other materials such as glass, rock crystal and sulfur were also sources of static electricity when rubbed. By the early 1700's generators using a friction pad against a revolving globe of glass produced static electricity.

Around 1740 Pieter van Musschenbroeck (1692-1761) of Leyden, Holland, used a special glass jar to store static electricity. He coated the inside and outside of the Leyden jar with a metal foil. The static charge was applied to the inside of the foil. This device could accumulate static charge over a period of time and produced very large shocks of electricity. This was demonstrated in 1750 by a group of monks who held hands and formed a 900-foot circle in front of the King of France. Two adjacent monks unclasped their hands and touched a charged Leyden jar, one on the inside and one on the outside foil. The great circle of monks sprang up into the air with robes flying in all directions. The King burst into laughter as the monks tumbled around in confusion.

In the American colonies, Benjamin Franklin (1706-1790) was fascinated by the progress of the study of electricity in Europe. In 1747, Franklin proposed that electric fire (as he called electricity) is a common element existing in all bodies. If a body had an excess, it was called plus. If a body had a deficiency, it was called minus. Electric current flowed from an excess to a deficiency or from plus to minus.

Franklin's single fluid theory of electricity stated that electric current flowed from an excess which he called plus to a deficiency or minus. He worked with a glass ball which rotated against a friction pad. The glass ball became charged with electricity which could be stored in a Leyden jar. Franklin assumed that when the glass pad was rubbed an excess charge appeared on the glass and he called that plus. This started a long history of accepting that the flow of electric current was from plus or positive (as determined by

Franklin) to minus or negative. Many rules were established based on the flow of electricity from positive to negative.

Many years later in 1897 it was found that electric current in a conductor was actually the flow of negatively charged particles called electrons. The electrons flow from negative to positive. Electrical engineering schools continued teaching that electricity flowed from positive to negative in order not to change well-established rules.

The above was forcibly brought to the attention of the author in 1942 while he was a Staff Sergeant teaching at the Anti-Aircraft Radar School at Camp Davis, North Carolina. The students were Second Lieutenants just graduated from electrical engineering schools. When it was explained that an electrical current in a conductor or a vacuum was a flow of electrons which proceeded from minus to plus, an uproar ensued among the students who then complained to the Commanding Officer of the school. The next day an official military order appeared on the bulletin board signed by the Brigadier General. It stated that, "Henceforth and forever, electrons will flow from negative to positive." It also stated that due to previous commitments, electric current would flow from positive to negative. Thus Benjamin Franklin had the last word.

Benjamin Franklin became famous by showing that lightning was the same as static electricity in Leyden jars. In 1752, he flew a kite in a lightning storm. The kite had a sharp pointed wire rising a foot above it. A long piece of kite twine was tied to a silk ribbon. The silk ribbon was indoors and had to be kept dry. A key was tied at the junction of the twine and silk ribbon. During a storm an electric charge proceeded down the wet twine to the key. Benjamin Franklin stayed in a shelter taking care that the twine did not touch the frame of the door or the window. The key was then used to charge a Leyden jar just like any static electric source. The experiment was very dangerous and the next person to perform it was electrocuted.

Benjamin Franklin became well known even in Europe for his work in electricity and other matters in his old age, especially among young women. During the American Revolution, he served as the Continental Congress's Ambassador to France. In Paris young ladies stood in line for a chance to sit on his lap and caress the few hairs on his bald head and whisper sweet nothings in his ear.

Franklin never legally married, but he had a common-law wife, Mrs. Deborah Read, whose husband had disappeared at sea. Franklin had her agree to rear his illegitimate son, William Franklin, although she was not the mother. He continued to have many mistresses even in his late seventies. He boasted in his *Autobiography* that he resolved to cure himself of his obsessive girl chasing by bedding only the minimal number of women necessary to maintain his "health." He thought all women were wonderful, young or old, lovely or homely. He flattered every woman he met and they were all wild over him. He died in 1790 at the age of 84.

In 1800, a great change in the development of electricity occurred. A steady source of electric current was constructed by Alessandro Volta (1745-1827), a professor at the University of Pavia in Italy. He studied electricity for thirty years culminating in the development of the first battery. This consisted of a vertical pile of alternate zinc and silver disks separated by paper strips soaked in brine. An electric current was generated which unlike static electricity was continuous. For the first time a steady electric current was available for experiments. The voltaic pile was constructed all over Europe. In England the new steady electric current was used to decompose water into its constituent gases. The voltaic pile was used in universities to demonstrate electricity. One of these was at the University of Copenhagen in Denmark where Hans Christian Oersted (1777-1851) was a professor of natural philosophy.

Oersted in 1819 was giving a lecture using the voltaic pile to demonstrate an electric current. A magnetic needle compass was lying near the wire carrying the current. He noticed that when the electric current was turned on, the needle of the compass moved. Oersted had discovered the relationship between electricity and magnetism that altered the course of technology. He experimented for months and found that the magnetic compass pointed in a direction perpendicular to the wire. When Oersted reversed the flow of current, the compass needle turned 180 degrees. In 1820, he published a paper stating that he had discovered that a magnetic field had been generated by the flow of an electric current. He concluded that the magnetic effect was circular around the wire.

This was the first time that electricity and magnetism were proved to be interconnected. Magnetism like static electricity was known by the ancient Greeks. They knew that a massive lodestone in Magnesia, Asia Minor, attracted the iron tips of shepherd's staffs. The term magnetism was derived from this location. Later it was learned that an iron needle could be magnetized by a lodestone. This led to the invention of the magnetic compass. Columbus used the magnetic compass to navigate his way to the New World. The magnetic compass was perfected by Charles Augustus Coulomb (1736-1806). By Oersted's time the magnetic compass was a sensitive instrument that could be used to demonstrate the relationship between electricity and magnetism.

Oersted's discovery aroused great interest among the scientists of the day including Andre Marie Ampere in Paris and Michael Faraday in London. Ampere (1775-1836) within two weeks of the Oersted report showed that a coil of wire carrying a current developed magnetic poles and acted as a bar magnet without the presence of iron. He named this device a solenoid. Ampere also developed a mathematical formula showing a quantitative relation connecting the current and the magnetic field and the length of the conductor.

Michael Faraday (1791-1867) was greatly interested in these developments and it was Faraday who later invented the principle of the electrical motor and electrical generator. These became the foundations of the great technological revolutions of the 20th century.

Faraday was born into a poor family and received a very minimal elementary education. He was apprenticed at the age of fourteen to a bookbinder in London. Faraday was trained to work rapidly and accurately with his hands which served him in good stead when he later became a hands-on scientist. He read many books that were brought in for binding including the Encyclopedia Britannica. Here he read about electricity and other interesting scientific topics. He bound books on chemistry and read them all.

Michael Faraday educated himself in other ways than reading books in his shop. He attended science lectures given every two weeks under the auspices of the City Philosophical Society at the home of Mr. John Tatum. Tatum had a science library and a laboratory in his house. He illustrated his lectures with experiments, and demonstrated the voltaic pile developed some ten years before. These lectures which were given from 1810-1811 served as a science education for the young apprentice.

Michael Faraday realized that his writing of the English language was very poor. He induced a friend at the City Philosophical Society to spend two hours a week helping him improve his spelling, grammar, and punctuation. This lasted seven years and Faraday vastly improved his writing skills. He wrote frequent letters to friends to practice newfound skills. Faraday also took drawing lessons from a Frenchman. His acquired writing proficiency together with his skill in drawing was very useful in describing his future scientific experiments.

Faraday set up a small laboratory in the back of the bookshop. Here he built his own voltaic pile and performed experiments in electrochemistry which he described in letters to his friends. In the year 1812, a friend showed one of the letters to his father. The latter was very impressed and obtained an admission ticket for Faraday to attend a series of four lectures by Sir Humphry Davy at the Royal Institution.

Sir Humphry Davy was the most important chemist in England at the time. Faraday prepared notes on the lectures and sent them to the great scientists who were very impressed. An opening as a laboratory assistant to Sir Davy opened shortly afterwards in the Royal Institution. In the spring of 1813, Michael Faraday was hired. A very fruitful relationship started between the twenty-two year old former bookbinder apprentice and the Royal Institution, the center of British science.

Sir Humphry Davy had married a rich widow in 1812 and he decided on a two-year scientific tour of Europe with his bride. They left in 1813 accompanied by Michael Faraday as Sir Humphry Davy's assistant. Lady Davy resented the young man of the lower classes and treated him as a servant. Faraday swallowed his pride and addressed her with great respect for the entire tour of Europe.

The young assistant prepared the chemicals Sir Davy used in his demonstrations all over Europe. He got to meet the greatest scientific minds of the day in France, Italy, Switzerland, and Germany. This was a tremendous educational opportunity for the young man who picked up a working knowledge of French and Italian.

Michael Faraday returned to England with a practical education which enabled him to embark on a career as a scientist. He was appointed Superintendent of the Apparatus of the Laboratory at the Royal Institution. Faraday used the library facilities to upgrade his knowledge of theoretical chemistry. He proceeded to systematically investigate the progress of chemistry from the end of the 18th century. This additional knowledge and training enabled him to publish his first scientific paper on chemistry analysis in 1816. The self-educated young man twenty-five years old was on his way to becoming well known in the London scientific establishment. He now gave lectures on chemistry before the City Philosophical Society. Faraday did not consider that he was a chemist but rather a natural philosopher.

In 1821, Faraday made a study of electricity and repeated the experiments of Oersted, Ampere, and others. He wrote a "Historical Sketch of Electromagnetism Motions" which was published in the Annals of Philosophy. Faraday realized from his study that there was a phenomenon, electromagnetic rotation, which opened up new vistas. He published a paper in October 1821 describing his experiments to convert electricity into mechanical energy. He had discovered the basic principle of the electric motor which became one of the workhorses of the 20th century.

Faraday became famous with his electric motor demonstrations and Sir Humphry Davy became increasingly jealous. The assistant was eclipsing his master and the older man began to resent the younger man. Faraday was proposed as a member of the Royal Society in 1824 but Davy was the only one who voted against him. Faraday was now a member of the most prestigious scientific institution in England. He continued his research into electricity as a scientist on his own.

In 1831, Faraday made the first dynamo which was the forerunner of the 20th century electrical generator. He used a copper disk rotating between the two poles of a horseshoe magnet to generate an electric current. The mechanical energy (the rotation of the copper disk) was converted into electric energy. The electric current was collected by a brush at the axis and a brush on the edge of the rotating disk. The brushes were connected to a galvanometer which measured the current. The rotating conductor (the copper disk) cut through the lines of force of the magnetic field of the horseshoe magnet and an induced current was set up in the conductor. Faraday had established the principle of the electric generator. However, he considered himself a natural philosopher and not an engineer. Therefore, he did not bother to develop a practical generator.

Engineers replaced the single copper disk by a coil of wire called the armature. The coil is mounted on a shaft and rotates with it. The shaft is turned by some means, such as a steam engine or water powered wheel. As the coil rotates, it cuts magnetic lines of force and a current is induced in the coil. A connection is made from slip rings through brushes to the external circuit. The current flows in one direction for one-half rotation of the armature and then flows in the other direction for the other half rotation creating an Alternating Current.

The Alternating Current was later changed to DC or Direct Current by changing the current direction every half revolution. This was done by a device called a commutator which in its simplest form is a split-rotating ring. This in effect reverses the current every half cycle producing a Direct Current. Using a number of armature coils and a multi-segmented commutator produced a more even Direct Current. The idea of the commutator was also used to make a DC motor. A battery is used and the shaft rotates doing mechanical work.

Many engineers took the basic model of the generator and over the years converted it into the modern electrical generator of the 20th century. Originally clusters of permanent horseshoe magnets provided the magnetic field. Later electromagnets supplied by a current from a separate machine were used. Werner von Siemens of Germany in 1867 discovered the principle of self-excitation using current from the DC generator itself to excite the magnetic field winding. This eliminated the separate machine from the electromagnets. Siemens and Halske, an engineering firm, developed in 1872 an efficient new kind of armature winding which was widely adopted. This type of winding exposed the maximum length of the conductor wire in the armature perpendicular to the magnetic field. The armature coils were wound on iron cores to get a large number of lines of force to pass through the coils. The field magnets were contoured to the armature to maximize the effect of the magnetic field. The air-gap between the armature and field were made as small as possible.

The new type of generator incorporating the improvements was soon put to work. In 1879 Siemens and Halske demonstrated the first practical electric railway which originally operated on a line at the Berlin Trade Exhibition. The next year the first electric elevator was exhibited by Siemens and Halske at the Mannheim exhibition. One further major decision was needed before Faraday's invention was ready for the technological revolution of the 20th century. This was the choice of DC or AC for distribution of electricity from the electrical generator to homes and factories. This led to the so-called current wars between the proponents of DC and AC.

The greatest advocate for DC was Thomas Alva Edison (1847-1931). He had invented a useful incandescent lamp in 1879. This made a central electrical generator practical for supplying electricity to homes and factories for lighting. In the next few years, Edison developed the components for a DC electrical system. These were a dynamo to generate DC power, a chemical meter to measure the amount of electricity used, fuses, switches, and resistance boxes. On September 4, 1882, he started up his first central power station at 257 Pearl Street in New York City. This supplied electric light for a one square mile area of lower Manhattan.

Edison's formal education was very limited and was supplemented by four years of instruction by his mother at home. At the age of 12, he rode the trains to sell newspapers and food and to pick up odd jobs. He became interested in chemistry and electricity and performed experiments at home and on the train. At age 12 he began to grow deaf to the point that he could not hear above a shout. He shut himself off from others and started a vast self-directed program of reading. At age 15 he read Newton's *Principia* which gave him a profound distaste for mathematics. He also read the two volumes of Faraday's *Experimental Researches* and he was encouraged by Faraday's achievements without a formal education in mathematics.

Edison employed engineers and scientists who made up for his lack of formal education. Examples were Francis Upton with an M.A. in science from Princeton and William J. Hammer, an engineer.

Thomas Edison went on to invent many of the electrically operated devices which changed the 20th century. These included the phonograph, the movie projector, the radio vacuum tube, the electric light and many others. He was issued 1,093 patents which is still a record. It is ironic that this man who had no formal education to speak of received a Bachelor of Science degree in 1992, 61 years after his death. The institution granting the degree was the Thomas Edison State College in Trenton, New Jersey. The degree was announced by the President of the college as an earned one and not an honorary degree.

In 1887, 120 central DC stations of the type developed by Edison were in use. But the DC system had a major flaw. The distance that DC could be transmitted economically was limited by the current flow. Edison stations transmitted Direct Current at low voltages of either 120 volts or 240 volts. Heat loss in the transmission line was equal to the square of the current times the resistance of the line. Practically, this meant that the DC central stations could not be located more than a mile from the customers. Also carbon arc lamps encouraged AC because the two carbon rods wore unevenly when connected across a DC source.

The advantage of an AC system was that the central station could be located many miles away from its customers. In AC systems, the voltage could be stepped up by transformers at the power stations to very high voltages in the order of 30,000 volts or higher. This meant that the current in the transmission line would be very small resulting in low losses. A step down transformer near the customer's house stepped the voltage down to safe levels. Direct Current systems could not operate transformers and were thus limited in the distance they could operate.

The primary proponent for AC systems was George Westinghouse. He realized that three components were needed to make an AC system practical. There was an efficient transformer, a meter that would measure the AC power the customer used, and a motor that would operate on Alternating Current. Westinghouse in 1885 bought an option on the European patents for the Gaulard and Gibbs transformer and contracted with William Stanley to improve it. Within a year Stanley had a practical transformer that he demonstrated in an experimental alternating central station in Great Barrington, MA. Stanley's transformer meant that an alternating voltage could be raised (stepped up) for efficient transmission over long distances and then lowered (stepped down) for safe use at the customer site. Westinghouse's first commercial AC plant opened in Buffalo in 1886. A flat rate per lamp was charged since an AC meter was not available.

In 1888, Westinghouse's chief engineer, Oliver Shallenberger, invented a practical AC power meter. He used a rotating metal disc to convert electrical energy into mechanical motion.

However, without a practical AC motor the applications of Alternating Currents were limited to lighting. On May 1, 1888, Nikola Tesla (1856-1943), an immigrant from Croatia, received patents for his design of an AC induction motor which later became one of the electrical workhorses of the 20th century. He used the idea of a rotating magnetic field produced by two or more Alternating Currents out of phase. An example is a three-phase Alternating Current applied to three sets of wiring in the stator of the motor. Each phase of the AC is separated by 120 degrees. Similarly each set of wiring in the stator is geometrically separated by 120 degrees.

This combination produces a rotating magnetic field in the stator of the motor. The rotating magnetic field of the stator induces a current in a rotor which produces its own rotating magnetic field. The magnetic field of the rotor follows the rotating field of the stator causing the rotor to rotate. It should be noted that the induction motor does not use a commutator and brushes that cause sparking and brush wear common in DC motors.

Westinghouse bought Tesla's patents in July 1888 and hired Tesla to develop the motor. All of the basic components of a practical AC system in the United States were now present. At the same time Europe was working on a practical AC system. In August 1891, a demonstration of Alternating Current was made in Germany.

Electricity from the Lauffen water powered generator was sent to Frankfurt, 108 miles away. It operated incandescent lamps and one 100 horsepower motor and used transformers at both ends of the system.

The Edison DC interests now focused the battle in the current wars on the issue of safety. Harold Brown, a New York electrical engineer, compared the safety of low DC voltage with high voltage AC using animals. He showed that the low DC voltage did not kill the animals while the high voltage AC electrocuted them. Brown's demonstrations did not discourage the use of AC but did convince New York State to choose AC for the first legal execution by electrocution. William Kemmler on August 6, 1890 was strapped into an electric chair that Brown had used to kill small animals but was not enough for a quick execution of a human. The half-dead Kemmler had to be given another jolt before he was pronounced dead. The Edison Company consolidated with Thomson-Houston on February 17, 1892 into the General Electric Company. G.E. adopted the AC format and the current wars were essentially over.

In 1893, Westinghouse won the contract to light the Chicago World's Fair of that year. AC was used to light 8,000 arc lights and 130,000 incandescent lights. Westinghouse also demonstrated AC motors. Two years later, in 1895, the first transmission of high voltage Alternating Current from the hydroelectric generating plant at Niagara Falls reached Buffalo, some 20 miles away. The use of AC was now established and electricity was ready to be widely applied in the 20th century.

The engineer who put the production of Alternating Current on a scientific basis was Charles Proteus Steinmetz, a German Jewish immigrant, who was an employee of General Electric. He changed the design of transformers used in AC from trial and error to one based on mathematics.

Steinmetz was born in Breslau, Germany, now Wroclaw, Poland, on April 9, 1865. His given name was Karl August Rudolf. He was very small and crippled with a twisted body like his father and grandfather. Steinmetz began his college education at the University of Breslau where he studied mathematics for five years. While at the University, he became a dedicated socialist. Steinmetz was placed under police surveillance in 1887 and just before he was to receive his Ph.D. he fled to Zurich where he studied mechanical engineering.

In 1889 Steinmetz immigrated into the United States. He became an American citizen in 1894 and took the name Charles Proteus Steinmetz. In the United States,

Steinmetz was employed as a draftsman by Rudolf Eickemeyer, an inventor and pioneer in electrical patents. In 1893, the newly founded General Electric Company bought Eickemeyer's patents along with the services of Steinmetz. The immigrant had already written two long papers on the mathematical law of magnetic hysteresis which became the basis of the scientific design of transformers.

In 1897 Steinmetz was the co-author of a textbook on electricity, "Theory and Calculation of Alternating Current Phenomena." This book played a decisive role in the turn-of-the-century debate between Alternating and Direct Current technologies. The mathematical techniques he described are still used by electrical engineers at the end of the 20th century.

Steinmetz was made a consulting engineer at General Electric after 1895. In the next 28 years he produced 195 patents in electrical energy. He was one of the earliest advocates of atmospheric pollution control, research on solar energy conversion, nationwide electrical networks, electrification of railways, and electric automobiles.

Steinmetz wrote ten technical textbooks which were widely used in electrical engineering colleges. He served on the faculty of Union College where he created the Electrical Engineering Department. His honors were many and he served as both the President of the American Institute of Electrical Engineers (1901) and the Illuminating Engineering Society (1915).

Steinmetz retained a lifelong interest in socialism and after the election of a socialist city government in Schenectady, New York; he served in that administration with distinction. Charles Proteus Steinmetz died in Schenectady on October 26, 1923 after making a very distinct contribution to electrical energy in the 20th century.

While central power stations were being developed, new batteries were also invented. In 1860 Gaston Plante produced the first chargeable battery. He immersed lead plates in a solution of sulfuric acid. This storage battery is similar to that used in cars throughout the 20th century. In 1880 George Leclanche (1839-1882), a French engineer, developed the first dry cell. The container and the negative electrode were made of zinc. The electrolyte inside was made of ground carbon, manganese dioxide and sal ammoniac. A carbon rod going through the electrolyte was the positive electrode. Electricity was now portable. Hearing aids, automatic focusing cameras, flash guns, quartz wrist watches, electric calculators, electric toothbrushes, inventions of the 20th century, all depend on portable electricity invented by George Leclanche.

1.2 LIGHTING UP THE WORLD IN THE 20TH CENTURY

The light bulbs of the late 19th and early 20th century were incandescent lights using carbon filaments which glowed in a vacuum when an electric current passed through them. They were very inefficient and a great deal of research was performed in both the United States and Europe to improve the incandescent lamp. Dr. Irving Langmuir of the General Electric Research Laboratory developed a tungsten filament which he installed in a glass bulb filled with nitrogen. The inert gas pressure prevented evaporation of the tungsten filament. The new incandescent lamp became commercial in 1913 and was a much more efficient light than the carbon filament.

It is ironic that Edison soon after made a carbon filament incandescent lamp that gave off light for more than 65 years of almost continuous service. There is a carbon filament lamp in Edison's home in Fort Meyers, Florida (now a museum) which has given off light since 1927, 10 hours a day, 6 days a week. This type of electric light became a sort of heirloom in some countries. However, the carbon filament lamp in most parts of the world was soon replaced by the more efficient tungsten filament lamp.

A new type of lighting from electrical discharge lamps was also being developed in the early part of the 20th century. Around 1850 Heinrich Geissler discovered that an electric discharge through a low-pressure gas gives off light of a spectrum characteristic of that gas. The first important commercial development in electric discharge lighting was the neon tube. Georges Claude demonstrated his first neon gas sign in 1910. This was the beginning of the neon sign industry which became widespread in the cities of the 20th century. The principle of electric discharge lamps became widely used in the 20th century in fluorescent lights, high-pressure mercury lamps, sodium lamps, and metal-halide arc lamps.

The above electrical discharge lamps convert electrical energy into light by first accelerating free electrons in the gas in the tube by applying electricity to the end of the tube. The free electrons collide with electrons in the atoms of the gas raising the latter's energy level. The internal energy of the electrons in the gas atom is dissipated as light radiation of a specific frequency. The advantages of electric discharge lamps are high efficiency, long life, and a good maintenance of light output up to the end of life. The disadvantages are the initial expense and the requirement of auxiliary apparatus to regulate lamp power.

One of the most popular 20th century electrical discharge lamps was the fluorescent light. In the 1920's it was discovered by a number of people that a discharge through a mixture of low-pressure mercury vapor and a rare gas (Argon) would convert 60% of the electrical energy into a single invisible ultraviolet frequency. In the late 1930's a fluorescent phosphor was applied to the inside walls of the discharge tube. The phosphor converted the invisible light to visible light. The early fluorescent light required auxiliary control called ballasts which were developed soon after. The fluorescent lamp became a commercial item in the 1940's when its great efficiency was an asset to a nation at war. A fluorescent lamp radiates less than 1/4 as much in the form of heat as an incandescent lamp of the same light. The efficiency of a fluorescent lamp increases rapidly with its length up to about 40 inches and then increases more slowly. The most efficient diameter is 1.5".

The fluorescent lamps are mostly of the hot cathode type. A starter circuit heats up the cathodes. Then the starter circuit opens up and the ballast, a coil wound around an iron core builds up the voltage to leap the gap and starts the discharge through the Argon gas. The discharge is then taken over by mercury. The Argon slows down the movement of mercury ions towards the cathodes and reduces the violence of cathode bombardment.

The ballast has two functions: one as stated above is to give the momentary high voltage required to initiate the electrical discharge through the gas of the fluorescent tube. The other function is to smooth out the flow of power to the fluorescent tube. The type of ballast consisting of a coil and a capacitor is referred to as a magnetic ballast. Another type is the electronic ballast which provides the two functions of a ballast.

Later another electrical discharge light was developed called the halogen lamp. Halogens are a special group of elements, fluorine, chlorine, bromine, and iodine. A halide is a compound that includes a halogen. In the halogen lamp the arc tube contains a rare gas for starting and a charge of mercury plus one or more of the metal halides (usually iodides). In operation all of the mercury is vaporized resulting in a high-pressure arc consisting principally of mercury vapor at several atmospheric pressures. The metal iodides evaporate from the tube walls, the molecules diffusing into the high temperature arc column where they dissociate into metal and halide ions. The metal ions are excited to give off their own characteristic spectral lines. The metals are selected to cover the visible spectrum. The metal iodide lamps give very good luminous efficiency in the order of 50% of the total radiant energy in the visible spectrum.

Another important electrical discharge light is the sodium lamp. There are actually two types: the high-pressure sodium (HPS) and the low-pressure sodium (LPS) lamps. HPS lamps fill the night with a broad spectrum of colors that combine to make a pinkish color similar to incandescent or "white" light. The LPS lamps give off a characteristic yellow light which is used for street and tunnel illumination. LPS has the advantages of lower cost and fewer glares than HPS. Many cities have installed LPS in spite of lobbying by lighting and utilities for HPS.

1.3 Keeping Things Cool

In 1844 Dr. John Gorrie, a Director of the U.S. Marine Hospital at Appalachia, Florida, described in a newspaper the first commercial machine built and used for refrigeration and air conditioning. He wrote, "If air were highly compressed it would heat up by the energy of the compression. If the compressed air were run through metal pipes cooled with water, and if this air, cooled to the water temperature, was expanded to atmospheric pressure again, very low temperatures could be obtained even low enough to freeze water pans in a refrigerator box." He further wrote, "There are advantages to be derived from the generation within any building and this is equally applicable to ships as well." The compression was accomplished by steam since electricity was not available to operate the compression of the air. Dr. Gorrie obtained in 1851 U.S. Patent Number 8080 for his invention.

An important improvement was the use of ether instead of air as the refrigerant. Alexander Catlin Twinning of New Haven received a British patent in 1850 for this and U.S. Patent Number 10,221 in 1853.

Many improvements were made over the years and many ice-making machines were used. By 1894 there were advertisements by 17 manufacturers of ammonia compressors to make ice. They were all driven by steam. Gradually electricity took over and by 1900 Tammany Hall politicians in New York City formed the "Ice Trust" officially known as the American Ice Company. They cornered the market for artificial ice making and the price doubled in twelve months. New York City's Mayor and several Judges were involved in the illegal scheme. The market for household ice ended in 1950.

In 1911 General Electric began the manufacture of the Audiffren home refrigerator using electricity. By 1919 electric power began to dominate refrigerating plant drives and high-speed short-stroke compressors invaded the field. An air-cooled small refrigeration machine was developed and further improvements were made. General Electric began manufacturing in 1926 the G.E. Monitor Top Unit on a large-scale production for household and store use. In 1930 sales of refrigerators, 850,000, exceeded those of iceboxes for the first time. In 1941, 4,000,000 refrigerators were sold.

The basic principle and operation of the electric refrigerator involves the use of a refrigerant which has two phases, one of which is a vapor and the other is liquid. A compressor run by an electric pump using an induction motor compresses the vapor. The hot compressed gas goes to an air-cooled condenser where it is cooled and liquefied. The liquid refrigerant is released into a region of lower pressure in the cooling compartment. The refrigerant takes away heat from this compartment and becomes a vapor again. One popular refrigerant in the earliest refrigerators was ammonia (not household ammonia which is ammonia in water). The refrigerant ammonia is a liquid when pressure is applied and a vapor when the pressure is released. Later other refrigerants were used until the 1930's when "Freon" took over as explained later in section 1.4.

In addition to the development of refrigeration, the application of electricity in the early 20th century led to air conditioning. The father of air conditioning is considered to be Dr. Willis Haviland Carrier (1876-1950). He grew up as the only child in a household of adults on a farm in Erie County, New York State. When he was nine, his arithmetic class reached fractions. Willis could not grasp the subject and his mother noticed his frustration. She told him to go down to the cellar and bring up a pan of apples. She had him cut the apples into halves, quarters, and eighths and add up and subtract the parts. He learned that problems could be broken down to something simpler that would be easier to solve. His mother died when he was eleven. She had taught him to figure out things for himself.

In 1897 Carrier won a four-year scholarship to Cornell University and received the degree of Mechanical Engineering in Electrical Engineering in 1901. He took his studies and applied them to the development of a new air conditioning technology using the availability of electricity. He invented a centrifugal compressor and sought a new non-toxic refrigerant. He selected Dielen, C_2H_2 manufactured in Geneva.

His first "comfort job" sale of the new system was in 1924 in the J.L. Hudson Company department store in Detroit. The store had found that on bargain days the temperature in the basement soared and customers fainted. Three centrifugal machines were installed resulting in an efficient system. This was the first air conditioning installation in a department store. He then designed and installed his machine in the Rivoli Theater in New York City in 1925. Carrier also began building inexpensive air conditioning units for small shops. Carrier's reputation spread and in 1928 he installed his cooling system in the United States House of Representatives and in the Senate in 1929.

1.4 THE OZONE LAYER DEPLETION PROBLEM

In 1930 a new development occurred in the field of refrigeration and air conditioning that would have tremendous repercussions in the last years of the 20th century. Thomas Midgley of Frigidaire developed a new class of refrigerants in 1928 called "Freon." They were the infamous CFCs or chlorofluorocarbons. They became widely used in refrigeration, air conditioning, and also in cleaning solvents, packaging and insulation.

The CFCs are extremely stable compounds of chlorine, fluorine and carbon. They are odorless, colorless, noninflammable, and inexpensive. In the air conditioning and refrigeration industry the most commonly used CFCs in the 20th century were called R-11 and R-12.

In 1973 F. Sherwood Rowland and Mario J. Molina showed that there was a CFC ozone depletion problem. They found there was a serious long term pending disaster. This was demonstrated in chemical experiments at the University of California at Irvine, Figure 1-1. Rowland, Molina, and Paul Crutzen of Germany's Max Planck Institute for Chemistry, were awarded the Nobel Prize for Chemistry in 1995 for their work in demonstrating the ozone depletion problem.

When CFCs are released in the lower atmosphere, they do not break down readily. Although they are heavier than air, they eventually migrate to the earth's stratosphere. There, under continuous bombardment by ultraviolet rays from the sun, CFC molecules eventually break down and release chlorine atoms. The chlorine atoms bond with ozone molecules, breaking them down. One chlorine atom can destroy 100,000 ozone molecules. Sufficient CFC molecules have migrated to the stratosphere, broken down, and bonded with ozone molecules to cause a measurable diminishment of the earth's ozone layer.

The ozone layer is a significant barrier to ultraviolet rays which in quantity are harmful to plant and animal life. As the ozone layer thins, more and more ultraviolet rays are let through to reach the earth's surface causing a substantial increase in skin cancer and threatening the existence of life. The problem is made more serious by the fact that the extremely stable CFC molecule may remain in the earth's atmosphere for up to 150 years before it is eventually broken down. Thus, the release of CFC molecules into the atmosphere in 1990 may be a factor affecting human health into the 22nd century.

Figure 1-1 Dr. F.S. Rowland (Left) and Dr. M.J. Molina, The University of California at Irvine, 1974, courtesy of Dr. Rowland

Intermittent possible replacements for the CFCs are the hydrochlorofluorocarbons or HCFCs. They are similar to CFCs but in addition to chlorine, fluorine and carbon, HCFCs contain hydrogen atoms. HCFCs are not as stable as CFCs and are much more likely to break down in the lower atmosphere. Since most HCFC molecules never reach the stratosphere, they have a lower effect on the ozone layer. HCFCs have only 5% of the ozone depletion potential of CFCs. The most common HCFC is R-22 which is used in residential air conditioning systems and in many commercial air conditioning systems.

The ozone layer depletion problem is an example of riding the wild stallion of technology. If science cannot solve this problem, the wild stallion can cause humanity a great deal of harm. Fortunately, the world community has realized the seriousness of the problem and has taken a series of actions to control the depletion of the ozone layer.

In 1978 CFC use in aerosols were banned in the United States and other countries. The Vienna Convention in 1985 discussed the phasing out of CFCs. In 1987 at the Montreal Protocol most countries agreed to a phaseout of CFCs by 2000. Scientific measurements of a hole in the ozone layer above the Antarctic continent convinced many governments to speed up the process of banning CFCs. In 1990 there were the London Amendments where some countries voluntarily advanced the phaseout to 1997. A phaseout of HCFCs by 2030 was agreed to. In 1992 the United States advanced its phaseout of CFCs to 1995 and pressure began to advance HCFC phaseout to 2005.

The United States Congress passed the Clean Air Act in 1990 which imposed limitations on the use or disposal of refrigerants and refrigerant-containing equipment. The Clean Air Act prohibits intentional venting of refrigerants during servicing and disposal of equipment effective July 1, 1992. Prior to the time that CFCs and HCFCs were known to be harmful to the environment, venting of these refrigerants was the common and accepted practice. Now a qualified service technician must remove and capture refrigerant from systems and equipment whenever service requirements call for evacuation of the refrigerant line-component replacement, equipment replacement, equipment disposal, etc.

In December 1992, a consortium of United States utility companies announced two finalists in a $30 million contest to develop a new refrigerator that will use no CFCs and exceed federal energy efficiency standards. The two companies were Whirlpool and Frigidaire. In August 1993 Whirlpool was the winner, using a refrigerant called HFC-134a, which would not appreciably harm the earth's ozone layer. The new refrigerator uses a computer chip regulated defrost system and better insulation to increase efficiency.

In the meantime, Harry Rosin, a director of a research institute in Dortmund, Germany, set out to build an ozone-friendly cold-storage room. He mixed propane with butane to obtain a new coolant. The main drawback, the possibility of an explosion, was minimized by using very small amounts of coolant. A refrigerator using this coolant went on sale in Germany in 1994.

It appears that new technology will solve the ozone layer problem as far as refrigerators are concerned. Humanity can ride that wild stallion safely into the 21st century.

1.5 KEEPING THINGS CLEAN

For many centuries people near streams washed their clothes by slamming them against rocks. In some parts of the world this was still being done in the 20th century. Then came the washboard and women scrubbed their clothes against metal corrugated surfaces. This was mechanized by using two curved washboards which were moved in opposite directions by the operation of a hand lever. The clothes were pressed between the two curved washboards which operated in a closed box. This device was still being advertised in Montgomery Ward's catalogue for 1927 as "Our Famous and Faithful."

In 1869 a new type of home laundering unit was described in U.S. Patent 94,005. A small four bladed rotor at the bottom of the tub drove the water through the clothes in a cylinder. A hand-operated crank turned the rotor. This machine had to await the invention of the electric induction motor before it became practical. The rotary operation of the machine was especially suitable for the rotary operation of the electric motor.

In the early part of the 20th century home washing machines operating by electricity began to appear. Maytag's first electric model was introduced in 1911 and in 1919 Maytag converted from wooden models to aluminum tubs. Maytag introduced its Gyrafoam in 1922, a bottom-mounted agitator model. The modern washing machine slowly evolved. The clothes were put in an inner basket with holes on the side and bottom. The sudsy water came spurting through the holes on the bottom. A four-bladed rotor at the bottom of the basket was turned at a slow speed by a small induction motor. The rotor drove the sudsy water through the clothes, washing and followed by rinsing. The rotor operated at a fast speed for rinsing. The washing liquid was extracted by centrifugal force through the horizontal holes in the basket. The water was circulated back to the clothesbasket by a pump. Successive washings and rinsing were used with the operator controlling the speed of the rotor. The operator had to also put in the soap at the proper time after the clothes were soaked.

In the early 20th century the use of electrically operated home laundry machines slowly spread. In 1916, 900,000 units were sold in the United States. Nine years later 1.4 million machines were sold despite the depression which gripped the United States in 1929. Around 1939 the automatic electric washing machine began to be developed. World War II temporarily halted the rapid spread of the home electric washing machine. In 1971 the microprocessor was invented and soon after the washing machine became fully automatic. Electricity had transformed a tedious job of washing into one of loading and unloading the clothes from the machine.

Another use of the small induction motor in cleaning was the electric dishwasher. A mechanically operated dishwasher was patented in 1865. A hand-operated crank which threw water against dishes mounted in a tube was described in U.S. Patent 51,000. This too had to wait for the invention of the electric induction motor to become practical. The electric dishwasher appeared in the 1930's and did not become widespread until well after World War II. The electric dishwashing machine transformed another tedious job into one of loading and unloading the machine.

An additional cleaning tool that was electrified in the early 20th century was the electric vacuum cleaner. An electrical vacuum carpet sweeper was described in a U.S. Patent 889,823 on 2 June 1908. This was improved by the electric suction carpet sweeper, U.S. Patent 1,151,731 on 31 August 1915 which became the basis of the Hoover vacuum cleaner.

Another use of electricity in cleaning is the treatment of sewage. Electric motors are used to move the human waste products and to clarify the water component before discharging into rivers and streams. This was illustrated in Baghdad during the Gulf War of 1991. Electricity in that city stopped flowing because of the bombings. The sewage treatment plants could not operate and raw human waste products were discharged into a river that was being used for drinking water.

1.6 KEEPING THINGS HOT

It was found in the 19th century that electricity flowing through a narrow wire could cause the wire to heat up. This phenomenon was used in the 20th century to make life better.

In the 19th century clothes were ironed by heating a flat iron in a fire until it glowed. The iron rapidly cooled while pressing the clothes. Gas was first used to try to solve this problem. A gutta percha tube directly connected the gas iron to a gas outlet in the ceiling. This system was introduced around 1850 but the stove was still used extensively to heat the iron. The electric iron came on the market in 1909 heavily advertised as the Westinghouse Electric iron. The advertisements stated, "Why not kick the stove out and get your wife a Westinghouse Electric iron. Don't you think she would be grateful for it? If you have any doubt about it, you can have an iron on free trial." The electric iron rapidly became used in most homes where electricity was available.

Practical experiments in cooking by heating with electricity began in the 1890's. A small electrified range, an electric broiler, and an electric kettle were exhibited in the Chicago World's Fair of 1893. An electrically cooked banquet was held in 1895 in honor of the Lord Mayor of London. Between 1909 and 1913 many electric ranges were developed.

Heating homes by electricity also used the phenomenon of electricity heating up a thin wire. This had to compete with coal, oil and gas which have generally proved to be more economical in heating than electricity. If the cost of electricity in an area goes up dramatically for any reason, then electric heating may become impractical.

The heating effect of the electric current found an important application in the welding of metals together in the 20th century. When two metal rods, completing an electric current, are touched together, there is considerable resistance at the point of contact so that the ends of the rod will melt and fuse together. In this way a welded joint can be formed.

Electric blankets are another application of using electricity to heat things up. A controversy arose because of the argument that small magnetic fields can cause cancer. The scientific community could not agree on this but the fact that the magnetic field of the blanket was very close to the human body for a relatively long time caused concern. This will be discussed later.

1.7 KEEPING THINGS MOVING WITH ELECTRICITY

In the last two decades of the 19th century, electric trolleys with high voltages carried overhead were introduced in many cities of the world. In the United States the electric trolley was mainly replaced in the first part of the 20th century. One of the few trolley survivors in the United States is the Philadelphia electric streetcar. It started operation on December 15, 1892 and in 1992 the 100th anniversary of electric traction in Philadelphia was celebrated.

Electricity at the end of the 20th century is used to operate trolleys, subways and other rapid transit systems in many parts of the world. The distribution of 600 volts or less uses a third rail. The first subway in New York City opened in 1905. It used 600 volts. Voltages above 600 volts are carried in overhead trolley wires. European railway systems use 50 Hertz, 25,000 volts, AC.

Electrically operated and controlled elevators have made tall buildings possible in the 20th century. The elevators may be operated by either AC or DC motors. Slower elevators use AC induction motors and faster ones are operated by a DC motor. The DC voltage is supplied by a motor generator. The motor generator is operated by a 3-phase AC power supply. The generator of the motor generator supplies a DC voltage which operates the elevator. The first automatic push-button elevator went into operation in 1922.

Moving stairways or escalators are used extensively in department stores, subway stations and airline buildings and in commercial buildings. Escalators are usually operated by induction motors. A 32-inch stairway with a rise of 20 feet requires about 10 horsepower when fully loaded.

1.8 Keeping Time with Electricity

The first electric clock began keeping time in 1918. Electrical watches using quartz crystals operating with batteries came into use after World War II.

1.9 Electricity used to Operate New Technologies

Many of the new technologies of the 20th century are operated by electricity. The applications include computers, radio communications and broadcasting, television, X-rays, MRI, lasers and many others. Electricity has become the lifeblood of civilization in the 20th century.

1.10 Battery Development in the 20th Century

Small rechargeable batteries such as the nickel-cadmium battery were developed to avoid the throwing away of used batteries. These batteries were employed in the 20th century to operate walkie-talkies, portable cellular telephones, and electric toothbrushes. Very small button size silver-oxide batteries were developed for operation of quartz-controlled watches and small calculators. Long-lasting lithium batteries were developed for implantable heart pacemakers. These batteries can last at least ten years.

1.11 ENVIRONMENTAL PROBLEMS IN THE PRODUCTION OF ELECTRICITY

Electricity in the 20th century is produced by steam turbine driven generators. The turbine uses the flow of steam at high pressure to spin a fan-bladed rotor. The steam turbine was developed in England by C.A. Parsons (1854-1931). The steam is produced by boiling water and the source of the heat is the burning of fossil fuels or nuclear energy.

In the United States in 1991* the generation of electric power was as follows:

Coal	54.8%
Hydro	9.9%
Natural Gas	9.3%
Oil	3.8%
Other	0.4%
Nuclear**	21.8%

ACID RAIN

A 1,000 megawatt untreated coal fired electric plant can generate 70,000 tons of sulfur dioxide and 25,000 tons of nitrous oxide per year*. These two gases mix with rainwater to form a rain of sulfuric and nitric acids that ruin lakes, rivers, and forests.

One method of ameliorating this problem is the use of advanced flue-gas desulfurization systems to reduce sulfur dioxide from burning coal. An example is a pressurized fluidized-bed coal combustion demonstration facility on the Ohio River near Brilliant, Ohio. This is the 70 million-watt American Electric Power's TIDD plant which was partly financed by the U.S. Department of Energy under its Clean Coal Technology Program.

---

*U.S. Council for Energy Awareness

**Nuclear power and its problems are discussed in Chapter 15

The TIDD plant burns crushed coal with dolomite which captures more than 90 percent of the sulfur released by the coal. Dolomite is calcium magnesium carbonate, a form of limestone. In this process high-velocity air causes particles of coal and dolomite to float randomly or "fluidize." During combustion the dolomite absorbs sulfur. Combustion takes place at a lower temperature than in a conventional boiler. The lower temperature reduces the emission of nitrogen oxides to about 50 percent of the federal limits in the United States. This particular technology also reduces carbon dioxide emission because of its higher efficiency.

Another technology to reduce acid rain pollutants is the use of scrubbers. Scrubbers are a group of add-on pollution control systems for removing acid rain gases from power plant smokestacks. One type of scrubber system sprays the exhaust with a watery mist containing limestone. This combines with the sulfur dioxide exhaust to form a watery sludge. In Germany and England the waste is used to make sheetrock which is sold to construction companies for erecting walls in homes and offices. Thus the technology is used to alleviate the pollution problem caused by burning fossil fuel to generate electricity and at the same time part of the scrubber expense is paid for. Another type of scrubber is a selective catalytic system to reduce the nitrous oxides, the other gas that causes acid rain. Japan has been the leader in this type of scrubber.

Greenhouse Gases

Burning coal, oil, or natural gas, produces carbon dioxide, one of the greenhouse gases. These gases trap the solar heat reflected from the earth. The trapped heat builds up and increases the temperature. The increased temperature could have catastrophic results. The ice at both poles if melted would raise the level of the oceans drowning most coastal cities. Agriculture in some areas could be destroyed while in cold areas agriculture might be vastly improved.

Most scientists believe that the greenhouse effect is an important problem while some think that the effect will be swamped out by natural changes in the climate.

The Japanese utility company, Tokyo Electrical Power, is searching for new technologies to extract carbon dioxide from power plant smokestacks by chemical absorption. Another method of reducing the production of carbon dioxide is by conservation using electricity more efficiently.

1.12 Low Frequency Magnetic Fields

Before going into the question of cancer and magnetic field, it is desirable to obtain a bird's eye view of the low frequency electric distribution network as well as the electrical equipment in the home and work place.

THE ELECTRICAL DISTRIBUTION NETWORK

Typically electric power in the United States starts with a 20,000 volts (20 kV), 60 Hertz, three phase generator. A step-up transformer raises the voltage in the range from 69 kV to 765 kV (typically 500 kV) and fed to high voltage three phase transmission lines, Figure 1-2. These consist of three wires, all oscillating at 60 Hertz. As the voltage of one wire is peaking, the voltage on one of the others is one-third of a cycle behind and the voltage of the remaining wire is one-third of a cycle ahead.

The high voltage transmission lines transmit the electric power to a substation where a step-down transformer changes the high voltage to an approximate range of from 5-45 kV (typically 30 kV) for the primary distribution transmission lines. These transmit the

Figure 1-2. Two vertical three phase high voltage lines. One 3-wire transmission circuit is on the left and one is on the right of each tower for economic reasons.

power to a local distribution step-down transformer, where the voltage is reduced to 115/230 volts for factories and homes. Commercial and industrial facilities use three-phase power while homes use single-phase power. Utility companies connect equal numbers of houses to each phase of a local distribution network to balance the load across the phases.

The power lines have both electric and magnetic fields but the latter are usually the ones measured. The electric fields are disturbed by people making the measurements and by walls, etc. The magnetic fields from 500 kV transmission lines at 10 meters ranges from approximately 80 to 800* milliGauss (mG). At 100 meters the magnetic field varies from 1 to 10 mG. The primary distribution line magnetic field at 10 meters ranges from 0.8 mG to 8 mG. In the United States the earth's magnetic field is 500 mG but it is not an alternating field like that of electric AC power.

In the United States the unit of measurement is the milliGauss but in Europe it is the microTesla. Sweden has become an important country in studying health effects of magnetic fields and they use microTesla. Multiplying that value by 10 converts microTesla to milliGauss.

Electricity in the Home

Electricity in the home with 115 volts operates heating devices like irons, electric blankets, and a number of electric motors. Small motors of 1/4 horsepower or less usually employ universal motors. This type of motor uses commutators with brushes and is called universal because they can be used with either AC or DC. This includes motors used for small electric drills, food mixers, and vacuum cleaners. Motors greater than 1/4 horsepower like washing machines employ induction motors. These use single-phase 110-volt AC power but obtain a second phase 90 degrees apart by using a capacitor or inductance to start the motors.

Some typical readings** in the home are:

Clothes Washers	7 - 400 mG at 1 inch 0.2 - 0.48 mG at 3 feet	Dishwashers	25.0 mG at 4 inches 7.6 mG at 18 inches
Electric Clothes Dryers	3 - 70 mG at 1 inch	Electric Stoves, One Burner	860.4 mG at 4 inches 20.6 mG at 18 inches
Heating Irons	80 - 300 mG at 3 feet	Clock-Timers	63.8 mG at 4 inches 20.6 mG at 18 inches
Electric Hair Dryers	106 mG at 4 inches	Electric Blankets	3 mG - 50 mG at 1 inch

* Magnetic fields from transmission lines

** From "Today's View of Magnetic Fields," in IEEE Spectrum, December 1994

Magnetic Fields in the Office

Copying Machines	5.5 mG
Fax	6.2 mG
Fluorescent Lights	4.7 mG at 8 feet from floor
Tabletop Fluorescent Lamps	2.0 mG at 1 foot
	0.8 mG at 8 feet
Halogen Lamp	0.6 mG at surface
Computer Monitors	14.4 mG at screen
(Swedish Standards)	1.8 mG at 1 foot
	33 mG at sides
Computer Monitors	60.1 mG at screen
(Not designed to	5.6 mG at foot
Swedish Standards)	376 mG at side measurement
Electrical Closet	At wall,
	15.1 mG at floor

Magnetic Fields in the Workshop

Large machines require 230 volts, 3-phase power. The magnetic fields will vary greatly with the size of the equipment. Some examples are:

Circular Saws	2100 mG - 10,000 mG at 1 inch
	0.2 - 10 mG at 3 feet
Electric Drills	400 - 10,000 mG at 1 inch
	0.8 - 2 mG at 3 feet

1.13 Low Frequency EMF and Cancer?

From 1979 and into the 1990's a number of epidemiology studies seemed to indicate a statistical connection between low frequency electromagnetic fields (EMF) and cancer. The strongest example was a detailed study in Sweden published in 1992. It indicated that children are more likely to develop cancer if they live near high-tension power lines.

The Swedish EMF-Cancer Study

An English edition of this study has been issued by the Swedish National Institute of Occupational Health entitled, "New Evidence of Cancer from Electromagnetic Fields." The investigation was based on half a million people who have lived near high-tension power lines. It covered the effect of fields from 220 and 400 kV lines.

In a normal home in Sweden the magnetic fields rarely exceeds a daily average of 0.1 microTesla (1 mG). The incidence of child leukemia was doubled in homes where the average magnetic field was above 0.2 microTesla (2 mG) and when the average field was above 0.3 microTesla the risk was 3.8 times normal. For children who live within 50 meters of 400 kV power lines, the risk was 2.9 times normal.

For adults, the study indicated elevated risks for acute and chronic myeloid leukemia. A magnetic field of 0.2 microTesla (2 mG) was correlated to a risk 1.7 times normal.

The Fight over EMF-Cancer

Several physicists, such as Dr. Robert Adair and Dr. Bennett of Yale, believed that the energy of low frequency EFM from power lines and electrical appliances was too low to break DNA bonds and cause cancer. They thought that the epidemiology studies were statistically flawed. Other scientists such as Dr. Granger Morgan at Carnegie Mellon University pointed out that cancer could be caused by the effect of EMF on hormone production.

Journalists, both print and TV, took up the controversy and some hinted at a cover-up to protect the electric power companies. This became a very bitter emotional fight at scientific and public meetings. The controversy reached President Clinton when mothers from Omaha, Nebraska, showed him a map of clusters of childhood leukemia within a mile of high voltage electric lines.

A Controlled Laboratory Study

In 1991 in response to the EMF-cancer controversy, the United States Congress appropriated $65 million for a detailed laboratory study under controlled conditions to determine the extent of the problem, if any. Dr. Henry Boorman was appointed the head of the EM program of the National Institute of Environmental Health Sciences. Dr. Boorman hired the best scientists he could find and he contracted excellent scientific institutions to perform very detailed controlled experiments. Work began in 1993.

A contract for $9 million was awarded to the Illinois Institute of Technology to study the effects of low frequency EMF on rodents under controlled conditions. A special laboratory was constructed with variable low frequency EMF. Three thousand mice could be tested at one time. By the spring of 1995 tests were completed in four areas with negative results in all of them. They were fetal abnormalities, cancer, reproduction, and immunology.

There was an additional study at the Pacific Northwest Laboratory by Dr. Jeffrey Safir to see if EMF had a subtle effect on a specific cancer gene. Previously, in New York City, a study had indicated that this was so. However, Dr. Safir's tests showed negative results. Dr. Safir then went to New York City and used the same equipment and tests. However, Dr. Safir's results were still negative. The results of Dr. Safir's work were published on May 4, 1995 in the British Science Journal, *Nature.* Another article in the same issue also described negative results by a team from Cambridge, England.

THE STATISTICS BATTLE

Many scientists and statisticians believe that there has to be a very strong correlation between two factors to show cause and effect. They also believe that the statistics should be backed up by laboratory studies. Finally, an understanding of the operation would clinch the cause and effect relationship. For example, in cigarettes, the correlation between smoking and cancer was found to be from 10 to 20. This was followed up by controlled animal studies which confirmed the correlation.

In the case of EMF, most of the studies showed that the correlation was less than 2. However, the Swedish government's study had a maximum correlation of 3.8. Several scientists such as Dr. John Moulder of the Medical College of Wisconsin, Dr. Patricia Butler, Dean of the School of Public Health U.C. Berkeley, California, and Dr. Robert Adair of the Physics Department of Yale University have attacked this study as statistically flawed.

THE MELATONIN PROBLEM

Melatonin is a hormone produced by a gland in the brain. It inhibits the growth of some cancers. Melatonin production increases at night but it can be suppressed by bright lights. Some laboratory experiments have indicated that magnetic fields suppress the production of melatonin from the pineal gland in the brain. More laboratory studies are necessary to determine if low frequency power fields actually affect the production of melatonin.

RISK EVALUATION

The possible risk of cancer from magnetic fields is very small. The Swedish study estimated that one out of Sweden's 1.5 million children contract leukemia every year because they live or have lived next to the high voltage grid and its high level of electromagnetic fields.

The possible risk must be compared to risks taken to avoid the EMF problem. An example is a child in a day care center situated next to a large power line. If the parents move the child to another day care center, there may be a much greater risk factor. For example, the new day care center may be on the side of a heavily trafficked street. Similarly, moving a child by bus to another day care center or school could raise the risk significantly.

THE POLICY OF PRUDENT AVOIDANCE

Until there are enough definitive studies to settle the problem to most people's satisfaction a policy of prudent avoidance may be useful. This was initially proposed by Dr. Granger Morgan of Carnegie Mellon University. This policy is between doing nothing and the tremendous expense of moving power lines or moving people.

Prudent avoidance includes not using electric blankets all night. Another caution is to move a motor driven electric clock from a bedside table to a dresser across the room farther away from the person's head. Alternately, the motor driven clock could be replaced by a digital clock or a wind-up clock. A more detailed discussion can be found in the Bibliography under Morgan, Granger.

It should be pointed out that details of the policy of prudent avoidance are subject to change as the results of more tests are received. For example, a recent investigation of pregnant women using electric blankets did not show any harmful effects.

THE 1996 NATIONAL RESEARCH COUNCIL REPORT

A report by the National Research Council, *Possible Health Effects of Residential Exposure to Electric and Magnetic Fields*, National Academy Press, Washington, DC, 1996, stated that a review of the scientific literature found no persuasive evidence for cancer caused by weak electromagnetic fields.

The Council is the research arm of the congressional chartered National Academy of Science. It is not certain that this report will finally end the controversy.

BIBLIOGRAPHY

Bennett, William J. — *Biological Effects of Magnetic and Electromagnetic Fields,* Plenum Press, 1995

Bordeau, Sanford P. — *Volts to Hertz -- The Rise of Electricity,* Burgess Publishing Company, 1982

Carnegie Mellon University — "Electric and Magnetic Fields from 60 Hz Electric Power: What do we know about possible health risks?", 1989

Carnegie Mellon University — "Part 1: Measuring Power-Frequency Fields," 1992
"Part 2: What Can We Conclude from Measurements of Power-Frequency Fields," 1993
The above three brochures may be bought from the Department of Engineering and Public Policy, Carnegie Mellon University, Pittsburgh, PA 15213
Attention: EMF Brochure

Cheney, Margaret — *Tesla, Man Out of Time,* Dorset Press, 1981

Floderus, Birgitta — *"New Evidence of Cancer from Electromagnetic Fields"* English Edition. Obtainable from the Swedish Information Service, One Dag Hammarskjold Plaza, New York, NY 10017-2201

Morgan, Granger, et al. — "Controlling Exposure to Transmission Line Electromagnetic Fields," *Public Utilities Fortnightly,* March 17, 1988, pages 49-58

Morgan, Granger, et al. — "Power Frequency Fields: The Regulatory

Dilemma," *Issues in Science and Technology*, Summer 1987, pages 81-91

Perry, Tekla S. "Today's View of Magnetic Fields," *IEEE Spectrum*, December 1994

Ueno, Shoogs *Biological Effects of Magnetic and Electromagnetic Fields,* Plenum Press, 1995

CHAPTER 2

USING THE ELECTROMAGNETIC WAVE TO HEAR AND SEE AT A DISTANCE

2.1 THE ROOTS OF THE ELECTROMAGNETIC WAVE CONCEPT

The mathematical father of the 20th century radio and television was James Clerk Maxwell (1831-79). He was born in Edinburgh, Scotland. It was evident very early that he was extremely gifted. Maxwell at the age of 14 wrote his first scientific paper on a method of drawing oval curves. In his youth Maxwell was acquainted with very distinguished Scottish scientists including the famous Lord Kelvin.

Maxwell was educated at Edinburgh University and Trinity College. On October 1855 he was one of three mathematicians selected for a fellowship at Trinity. He had a distinguished teaching and research career at several United Kingdom colleges culminating in his appointment in 1871 as the first Cavendish Professor of Physics at Cambridge.

He was a great admirer of Faraday and his work in electricity and magnetism. In 1856 he wrote a paper, "On Faraday's Lines of Force." He converted Faraday's ideas and methods into mathematical concepts and formulas. He continued to study electromagnetism and in 1864 he wrote *On a Dynamical Theory of the Electro-Magnetic Field.* He mathematically stated that a varying electrical field in space produced a varying magnetic field. This latter field would in turn produce a varying electrical field and so on. The electric and magnetic fields were perpendicular to each other. The resulting wave traveled in a direction perpendicular to both the electric and magnetic fields with a velocity in space equal to the speed of light.

Maxwell started out with 20 equations which were reduced to 8 equations and later to 4 equations. These four equations are now known universally as Maxwell's equations. He noted that both light and the electromagnetic wave were transverse (vibrating perpendicular to the direction of propagation). Maxwell also observed that the velocity of the electromagnetic wave was the same as light. He concluded that light was an electromagnetic wave.

Maxwell mathematically determined that the velocity of the electromagnetic wave was equal to the ratio of electrostatic units to electromagnetic units. This ratio has the physical dimensions of length over time, or velocity. Kohlrausch and Weber had previously in 1856 determined this ratio by measuring the same quantity of electricity first in electrostatic and then in electromagnetic measurement. The quantity of electricity measured was the charge of a Leyden jar. The quantity of electricity in electrostatic measure was found as the product of the capacity of the jar and the difference of potential of its coatings. To determine the value of this charge in the electromagnetic measure, the jar was discharged through the coil of a galvanometer. By observing the extreme deviation of the galvanometer, the amount in magnetic units could be determined. The ratio of electrostatic units to electromagnetic units was then calculated.

The value of the ratio obtained by Kohlrausch and Weber was 310,740,000 meters per second. Previously the velocity of light had been measured as 314,000,000 meters per second by H.L. Fizeau in 1848 and as 298,360,000 meters per second by J.L. Foucalt in 1850. Maxwell was well aware of the equivalency of the ratio of electrostatic over electromagnetic units to the measured velocity of light. Later Maxwell himself measured this ratio and obtained a similar value.

Faraday had in his *Thoughts on Ray Vibrations* published in 1846 proposed an electromagnetic theory of light. This was without a firm mathematical framework and with no data to calculate the ratio of electrostatic to electromagnetic units. In addition, the velocity of light had not yet been measured. However, Faraday had a marvelous insight for his time. This was recognized by Maxwell.

Maxwell's equations were based on the work of Faraday, Ampere and others. Maxwell added a new concept which made his equations unique, the displacement current. He noted that in a capacitor, two conducting plates separated by an insulating dielectric, such as glass or air, the electrical charge on one plate will cause an opposite charge on the other plate. Maxwell proposed two electric currents, one in a conductor and one he called a displacement current in a dielectric like free space. He also postulated that a varying displacement current produced a varying magnetic field just like the current in a conductor.

Maxwell's idea of a displacement current and his complicated mathematics were not universally accepted until his electromagnetic theory of light was proven by experiments by Heinrich Rudolf Hertz (1857-94) in 1886-8, nine years after Maxwell died of abdominal cancer on 5 November 1879.

Hertz was a German Jewish professor at the Technische Hochschule at Karlsruhe, Germany. He performed a series of experiments around 1887 that systematically proved Maxwell's assumptions. One of these assumptions was that a displacement current in a dielectric produced a magnetic field. He published his findings in November 1887 in a paper, " On Electromagnetic Effects Produced by Electrical Disturbances in Insulators."

He then performed a series of experiments to prove Maxwell's theory that electromagnetic waves traveled through space with the finite velocity of light. He produced electromagnetic waves from an electric discharge, traveling through space at the speed of light. Hertz made a transmitter consisting of an induction coil, a spark gap and two capacitor plates. In the induction coil two separate windings are wound on an iron core. The primary coil has a few turns of coarse wire connected to a battery through a vibrator. The secondary coil consists of many turns of fine wire. The vibrator alternately makes and breaks the primary circuit. A large voltage is obtained by the rapid make and break of the vibrator and many turns in the secondary coil. The spark at the gap produces an electromagnetic wave which can be detected by a device shown in Figure 2-1. Hertz transmitted a distance of 60 feet.

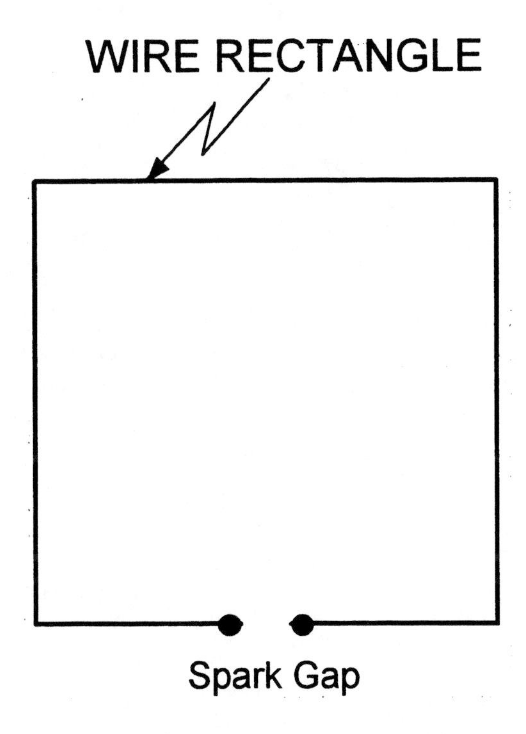

Figure 2-1 A Hertzian Spark Detector

Hertz moved his detector and found the wavelength by observing when the spark of the detector went off. He found in his first experiments that the wavelength was 2.8 meters. This falls in the radio frequency range now known as VHF or very high frequencies. He calculated the velocity of this new wave as the speed of light confirming Maxwell's equations.

Hertz described this experiment in a paper, "On Electromagnetic Waves in Air and Their Reflections," published in 1888. He paid tribute to Maxwell for his theoretical work. Later he designed equipment to produce a much shorter wavelength of 25 centimeters which would now be called microwaves. He tested the new equipment and showed that the electromagnetic wave exhibited characteristics of light like reflection, refraction, and polarization.

Hertz and Oliver Heaviside (1850-1925) converted Maxwell's equations into the modern form taught in textbooks. The equations of Maxwell were consolidated into four basic equations. In 1890 Hertz published a paper, "The Fundamental Equations of Electromagnetics." This became the last chapter of his book, *Electric Waves.* An English translation was published in 1893 and reprinted in 1962 by Dover Publications. Hertz died in 1894 at the age of 37 of bone cancer.

The spark detector used by Hertz was sufficient for proving Maxwell's equation but it was not practical for communications. A number of investigators had discovered that finely powdered metal was normally highly resistant to an electric current. However, when the metal powder was subjected to a nearby electrostatic charge, the powder became conductive. When the metal powder was shaken it became resistant again. Sir Oliver Lodge, a professor at the University of Liverpool, named this device in 1894 a coherer. He added an electric bell which when energized by the coherer action tapped the coherer tube to restore the metal powder to its original condition. Lodge did not use this device for telegraphic communication.

A young Italian, Guglielmo Marconi (1874-1937), read both Hertz's work and Lodge's description of the coherer. He combined Hertzian waves with a coherer detector in a number of experiments in 1895. He connected a modified Hertz transmitter to an antenna and earth as shown in Figure 2-2. He used the same type of induction coil and spark gap that Hertz had used. He made a receiver with a coherer that used silver-nickel

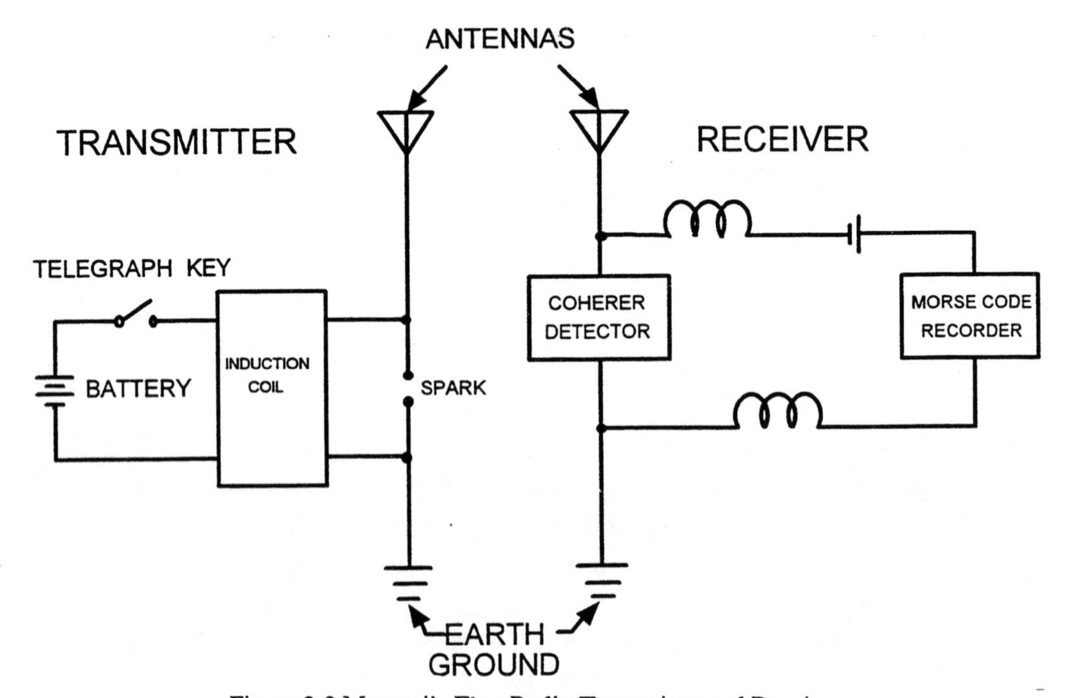

Figure 2-2 Marconi's First Radio Transmitter and Receiver

filings. Marconi also used an antenna and a ground connection for the receiver. He sent Morse code signals a mile and a half on his father's estate. He had demonstrated that radio communication was possible.

Marconi was not only an inventor but he was a shrewd businessman. His Anglo-Irish mother, Annie Jameson, had taught him English and he moved to London to take advantage of that center of finance. Marconi received his first radiotelegraphy patent in 1897 and the British financiers formed an organization that became the Marconi Wireless Telegraph Company, Ltd. Marconi linked England and France by his equipment in 1899. He then envisioned radio communications across the Atlantic. He built equipment on the coast of Cornwall in Wales and across the Atlantic in Newfoundland. Marconi transmitted the letter "S" in Morse code (three dots) from Poldhu in Wales to St. Johns, Newfound-land, on December 12, 1901, Figure 2-3. The photograph shows Marconi, accompanied by his assistant, George Kemp, reading the Morse code ink recorder shortly after the transmission. The riding of the electromagnetic wave had begun.

Figure 2-3 Marconi reading a Morse recording, 1901
IEEE Center for History of Electrical Engineering

2.2 THE BASIC RADIO PATENT WARS

Tesla saw the possibilities of wireless communications in 1893 and he filed his basic U.S. radio patent applications, No. 645,576 and No. 649,621, in 1897 and they were granted in 1900.

Marconi won the Nobel Prize in 1909 for his work in radio and he became recognized in the west as the man who made radio transmission possible. Tesla sued Marconi in August 1915 for patent infringement. The patent war went on for decades and was typical of numerous advancements in the field of radio and other technologies. On June 23, 1943 the U.S. Supreme Court ruled that Tesla's fundamental radio patents took priority over all other contenders. This has not been generally recognized by the scientific community or the general public in the western world.

In the east and particularly in Russia the man who is credited with the invention of radio is Aleksandr Stepanovich Popov (1859-1906). Popov became interested in electromagnetic waves following their discovery in 1887. In 1894 Oliver Lodge constructed an indicator of electromagnetic waves, which he called a coherer. Under the action of electromagnetic waves, the grains of powder stuck together and the sensitivity of the apparatus declined sharply. Popov, as well as Lodge, improved this indicator, equipping it with an electric bell-like apparatus that automatically tapped the powder tube when the impulse of current was produced and thereby restored its sensitivity for receiving the signal.

Popov in 1895 equipped a radio receiver with a wire antenna and sent radio signals up to eighty meters. By the beginning of 1896 Popov had substantially improved his receiver and had obtained important results in transmitting and receiving radio signals. Popov published a description of his work in January 1896. Marconi published a notice of his wireless invention in the fall of 1896. Marconi was issued a patent in 1897. The diagram of his apparatus appeared to coincide with the description by Popov in January 1896. A commission of competence, established in 1908 by the Physical Section of the Russian Physico-Chemical Society concluded that Popov "was justified as being recognized as the inventor of the wireless telegraph." Ever since Eastern Europe and Russia have considered Popov as the inventor of radio.

2.3 THE FIRST RADIO BROADCAST

The radio waves generated by Hertz, Marconi, and other pioneers were damped as shown in Figure 2-4a. While this type of wave could be used to transmit the dots and dashes of the Morse code, it was too noisy for voice. A continuous wave (CW), shown in Figure 2-4b, was necessary for both voice and music radio transmission.

The first successful CW transmitter was the invention of the Danish engineer, Vlademar Poulsen in 1902. His transmitter was called the Poulsen arc. He took a carbon-arc lamp, modified the electrodes, placed the arc in an atmosphere of hydrocarbon

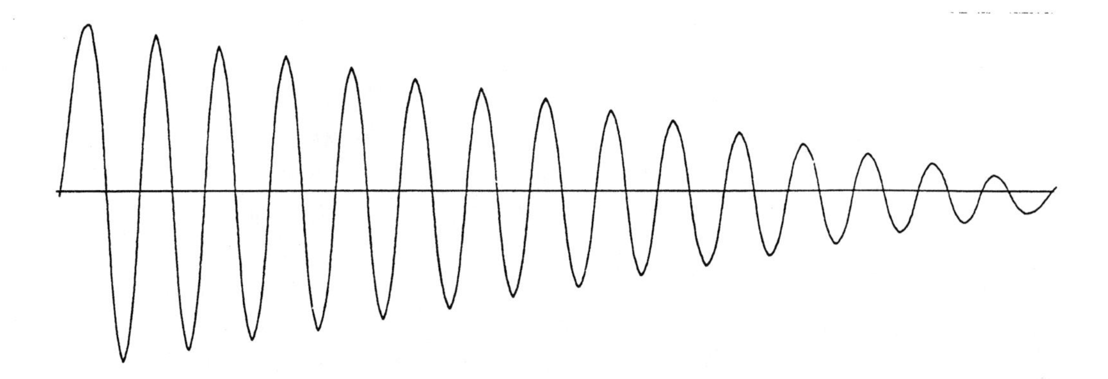

a - Damped Wave

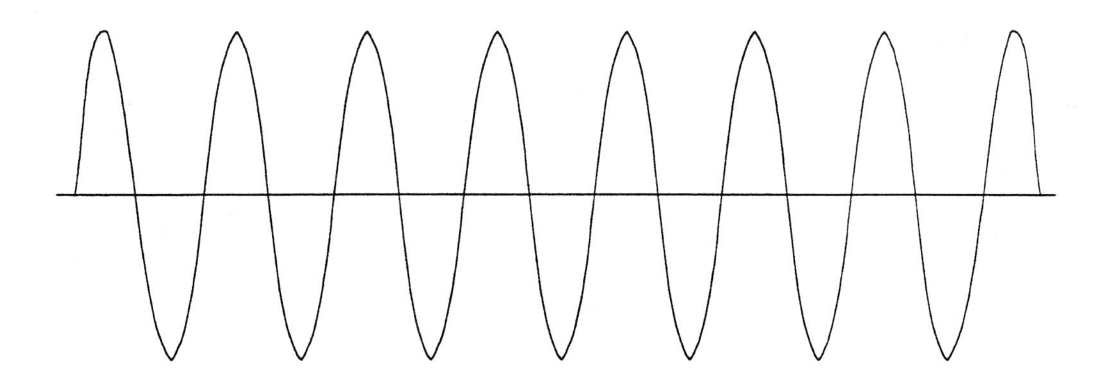

b - Continuous Wave

Figure 2-4 Damped and Continuous Waves

vapor and added a transverse magnetic field. The result was a generator of continuous radio waves that was used until displaced by other methods. The Poulsen arc had a disagreeable hissing noise and was not suitable for voice transmission.

The second CW transmitter was developed by Reginald Aubrey Fessenden, 1886-1932, who used high frequency alternators to make the first radio broadcast in 1906. Previously, from 1890-1895, Nikola Tesla had built high frequency (up to 20,000 Hertz) multi-polar alternators. Fessenden contracted with General Electric in 1904 to build a high frequency alternator in the 50-100 kHz range. General Electric hired a Swedish engineer, Ernst Alexanderson, to build the generator. In November 1906 General Electric delivered the alternator. After modification and rebuilding by Fessenden a generator was ready for radio transmission by November 1906.

On December 24, 1906 Fessenden presented the world's first radio broadcast. He used a modified Alexanderson generator operating at 100,000 Hertz and a modified Edison carbon microphone. He first sent a general CQ call in Morse code to all ships at sea. Then Fessenden stepped up to the microphone to give a short speech. This was followed by a broadcast of Handel's *Largo* from an Edison phonograph. Fessenden then took out his violin, stepped up to the microphone and played *O' Holy Night.* He sang the last verse at the same time. Fessenden wished his listeners, "Merry Christmas." He repeated the performance on New Year's Eve but this time there were vocals from his associates. Ships in the North and South Atlantic received the programs and radio operators mailed letters back to Fessenden praising the first radio broadcasts.

Radio broadcasting did not flourish until the vacuum tube was used to generate a continuous wave. Frank Conrad, a Westinghouse engineer, in 1918 started to play records over his vacuum tube amateur radio station. In 1920 Harry P. Davis, a Westinghouse Vice President, decided that Conrad's broadcasts were a good way to generate the sale of radio equipment. On October 16, 1920 Westinghouse applied to the Department of Commerce for a special license to begin regular broadcasting from Pittsburgh under the call sign KDKA. The first transmission on November 2, 1920 reported returns of the Presidential election between Warren G. Harding and James M. Cox. Within two years there were almost five hundred stations in the United States.

The achievements of Fessenden faded as the vacuum tube and other devices made radio broadcasting practical. Fessenden became a forgotten man for many years until in the 1990's researchers wrote about the first radio broadcaster.

2.4 THE VACUUM TUBE AND THE DEVELOPMENT OF RADIO

Thomas Edison discovered in 1883 that electric current in a carbon incandescent lamp flowed from the filament to a positively charged plate inside the vacuum of the lamp. He observed that the amount of current that flowed from the filament to the plate was in proportion to the amount of brightness of the filament. He described this in a patent obtained in 1883 for a modified incandescent lamp to measure the flow of electrical current. This phenomenon was later called the "Edison Effect" but Edison himself did not do any further work on it.

An employee of the Edison Company in England, John Ambrose Fleming (1849-1945), knew about Edison's patent. Fleming was appointed scientific advisor to the Marconi Company. He was given the task of finding a new detector of wireless waves. Fleming decided to do research on a lamp using a filament and a plate like Edison's. He applied an alternating current of a wireless to the filament and found that the lamp allowed only direct current to pass through to the plate. He named the lamp a valve that allowed only the negatively charged electrons to pass. In 1904 Fleming patented a vacuum tube with a metal cylinder (the plate) surrounding the filament.

Fleming published a description of his invention in a publication for the Royal Society in 1905. It was a crude device that needed much more work to make it practical. The Marconi Company for which he worked was not interested because Marconi wanted to develop the galena crystal as a detector of radio waves.

Lee de Forest, an American inventor, read Fleming's article on his device. He had an assistant take a Fleming tube to an automobile lamp manufacturer for duplication. On December 9, 1905 de Forest applied for a patent on a detector for wireless telegraph systems, essentially using the Fleming tube. Lee de Forest then applied a negative battery voltage to the filament and a positive battery voltage to the plate. He called this an "Audion" which he proposed as a new detector of wireless waves. In January 1906 he filed for a patent on his "Audion." So far, it still was basically the same as Fleming's device. However, Lee de Forest soon made a basic discovery that opened up the radio age in the 20th century.

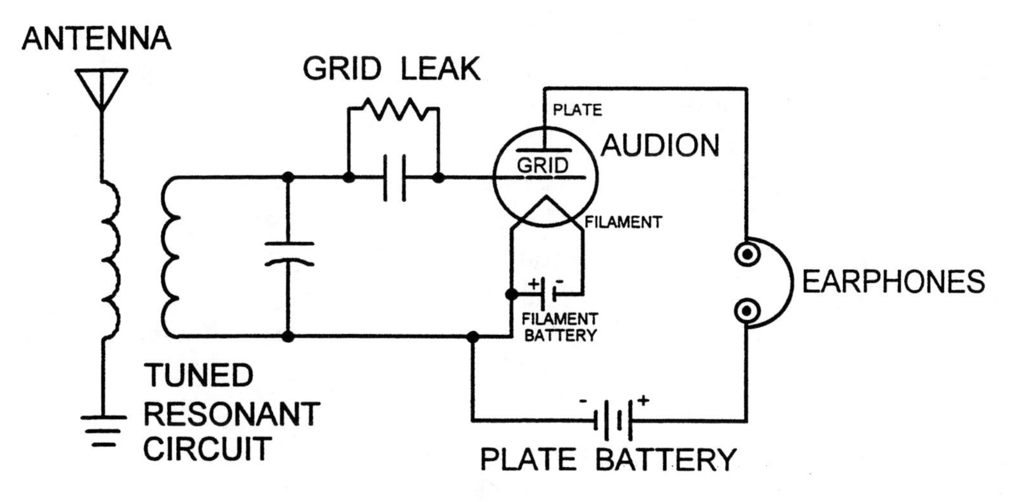

Figure 2-5 Lee deForest's Audion Used as a Detector and Amplifier

Lee de Forest in the fall of 1906 added a third element to the two elements of the "Audion." Lee de Forest called the new element, a grid. It was a nickel wire bent in zigzag fashion placed between the filament and plate but as close to the plate as possible.

When the grid was positively charged it would attract the electrons coming out of the filament and accelerate them towards the plate. The more positive the grid the greater the plate current. A small radio signal between the filament and grid would be amplified in the plate current. This new device detected the radio frequency and also amplified it. Later the grid was made into a cylindrical mesh around the filament. The openings in the mesh were wide to allow the electrons to flow through from filament to the plate. An early circuit using de Forest's Audion for a detector and amplifier is shown in Figure 2-5. His tube and its further developments became the foundation of the radio industry in the first half of the 20th century.

Another engineer who used the vacuum tube to advance the science of radio was Edwin Howard Armstrong (1890-1954) one of the greatest inventors of the 20th century. He was born in New York City on December 18, 1890. His family moved to Yonkers in 1900. While going to Yonkers High School, he experimented with induction coils, coherers, galena crystals and built wireless transmitters and receivers using the technology of his time. He communicated with other young radio amateurs, transmitting and receiving over greater and greater distances.

In 1909 he entered Columbia College where he studied electrical engineering. He joined a group of young radio amateurs in 1912 when they changed their name to the Radio Club of America. In that same year while still a student at Columbia, Armstrong made his first great invention, regeneration. Armstrong experimented with de Forest's three element Audion. Lee de Forest himself did not understand how his invention worked. He mistakenly thought that the flow of current in his tube depended on the ionization of gases in the Audion. Armstrong discovered the true action of the Audion which actually needed no gases but could operate in a vacuum. He found that the current coming out of the plate oscillated in a steady rhythm. He fed the current back from the plate to the grid which increased the signal coming from the antenna.

Armstrong found that this feedback or regeneration as he called it, produced much greater sensitivity and amplification in a receiver. He discovered that the regeneration principle could be used in a vacuum tube transmitter. This was also a great advance. As had been pointed out before, spark transmitters did not propagate a continuous wave which was necessary for music or speech. Fessenden had General Electric make a high-speed alternator for him which produced a continuous radio frequency. This transmitted music and speech in 1906. However, the cost to manufacture and maintain this special high-speed alternator was as great as its size. Armstrong in 1913 built a small inexpensive tube transmitter using the feedback system.

Armstrong had now developed an extremely sensitive receiver and a tube transmitter, both using the feedback principle. He filed a patent application on October 29, 1913 for a wireless receiving system using the feedback principle. He left out the design of a tube transmitter in this patent application. On December 18, 1913 Armstrong made another patent application but this time for a vacuum tube transmitter.

In September 1915 Lee de Forest filed a patent application on an audion that oscillated. This could be used as either a receiver or transmitter and was based on the regenerative principle. In the application he claimed to have discovered this principle in 1912, a year before Armstrong. Thus began another patent war that lasted for years. The U.S. Supreme Court in 1934 ruled in favor of Lee de Forest. However, most radio engineers still consider Armstrong as the true inventor of regeneration.

The regenerative receiver was used through the early 1930's. It was extremely sensitive but it required the operator to make adjustments to avoid the receiver breaking out into audible oscillation. The regenerative receiver was replaced by a new type of radio receiver, the superheterodyne, which came into production in 1924. It was invented by Armstrong and is still the basic circuit for all radio, television, and communication receivers.

Fessenden in 1901 proposed a receiver which he named the heterodyne. He derived the name from the Greek *Heteros* (external) and *dynamis* (force). He used a steady signal generated in the receiver. This signal was mixed with the received signal and the resulting difference frequency was in the audible range. The heterodyne receiver was not used widely because the local oscillators were not stable enough. Armstrong invented the superheterodyne which was a great improvement over the heterodyne receiver.

In 1918 Edwin Armstrong was a Captain in the U.S. Army attached to the U.S. Signal Corps' laboratories in Paris. Here he developed the superheterodyne circuit. He used a local oscillator mixing with the incoming signal. However, the resulting difference was an intermediate frequency (IF) which was not audible. The intermediate frequency was amplified greatly by the IF amplifier circuit. The local oscillator had only to be stable enough to place the IF signal in the frequency band passed by the IF amplifiers. In this way Armstrong overcame the limitation of the heterodyne receiver. After much amplification in the IF amplifiers, detection followed to make the signal audible. A very small radio signal can be detected and amplified by the superheterodyne receivers which are universally used today.

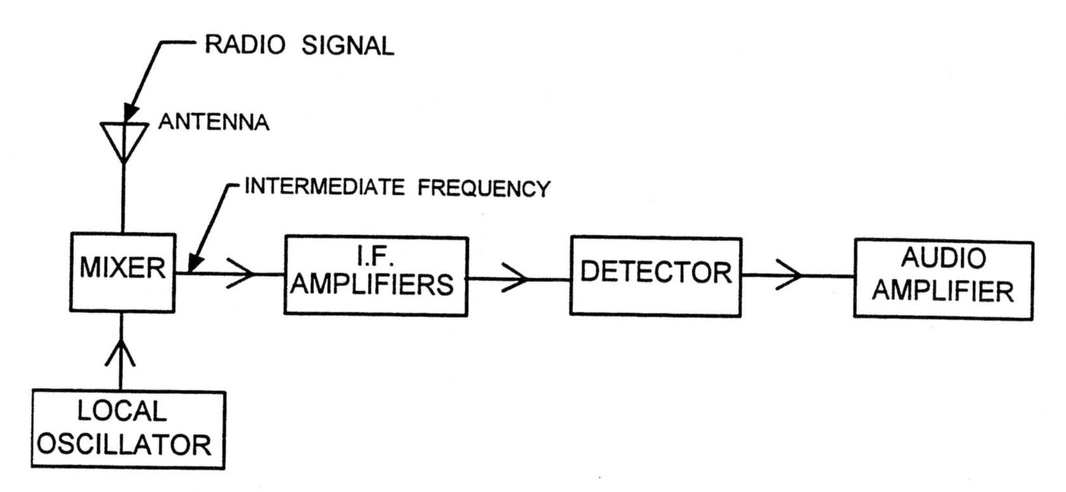

Figure 2-6 Basic Superheterodyne Receiver

A block diagram of a superheterodyne receiver is shown in Figure 2-6. This type of very sensitive radio became very popular. By 1933 there were almost 20 million receivers in the United States, mostly superheterodynes. Automobile radios were introduced in 1930. By 1933 there were 500,000 automobile radios. Armstrong's superheterodyne circuit spread the use of radios.

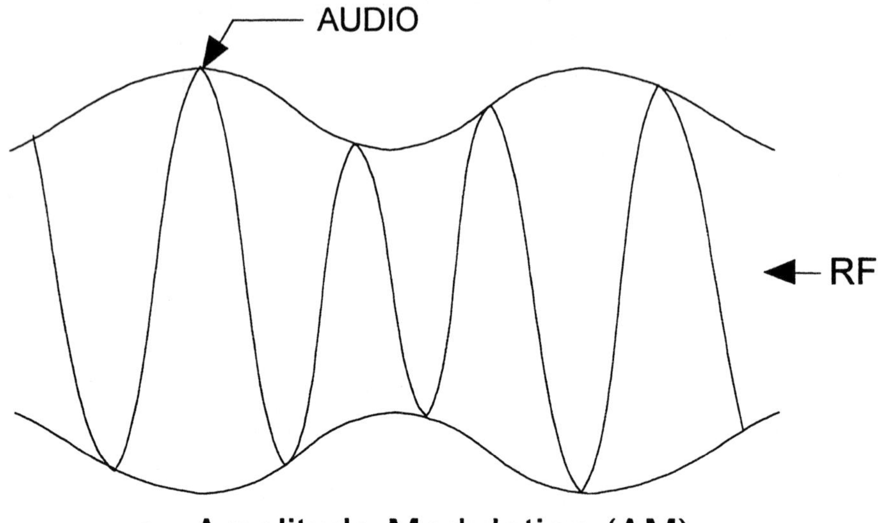

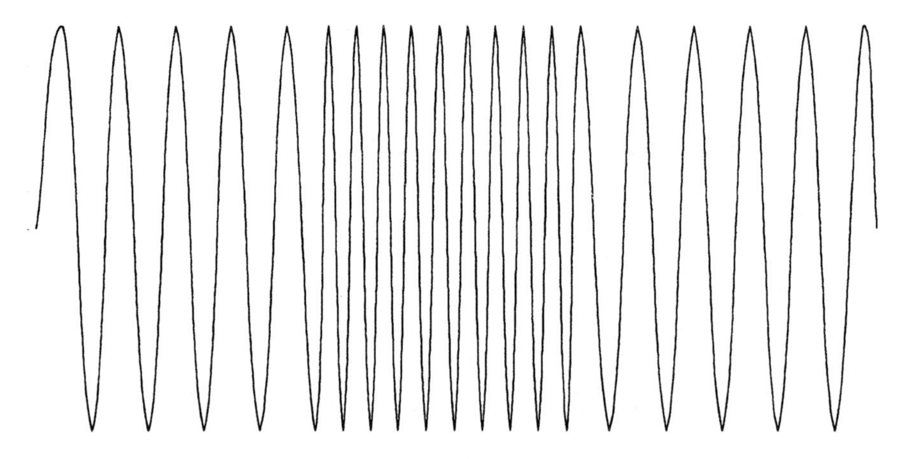

Figure 2-7 Amplitude and Frequency Modulation

Armstrong's third great invention was wide band frequency modulation presented by him in 1935 before the Institute of Radio Engineers. At that time broadcasting stations used amplitude modulation. In this method the music or speech varies the amplitude of the radio frequency carrier. This is shown in Figure 2-7a. Frequency modulation is shown in Figure 2-7b where the music or speech varies the frequency of the radio frequency carrier. Armstrong designed a circuit that varied the frequency of the transmitter in accordance with the amplitude of the audio signal. He also designed a receiver with a limiter to strip off all amplitude noise and a detector to convert frequency variation into amplitude variation. The amplitude variation was then amplified and fed into a loudspeaker.

The great advantage of the FM system was the elimination of atmospheric static and man-made electrical noise. This interference varied greatly in amplitude. Armstrong eliminated this amplitude variation by using frequency modulation and a radio frequency carrier of constant amplitude. Besides eliminating the crackle of atmospheric static and man-made electrical noise, Armstrong wanted a high fidelity signal. He used a wide band of 20,000 Hertz to broadcast speech and music on a radio frequency of 41 megahertz.

Armstrong's own FM transmitter in Alpine, New Jersey, went on the air in 1938. Soon a network of New England FM stations developed called the Yankee network. Then police and other safety communication systems adopted a narrow band type of FM. During World War II the Signal Corps adopted FM for the Signal Corps.

At the end of World War II there were about 47 FM stations in the United States and over 500,000 receivers.

After the war the FCC made a decision to move the FM band from 41 megahertz to a new band of 88 to 108 megahertz. Armstrong and the manufacturers of FM equipment bitterly opposed this move but they lost. Armstrong maintained a deep distrust and hatred of the FCC for the rest of his life. Sometime in 1950 at Alpine, New Jersey, he threatened to turn the dogs on an FCC engineer inspector. The author was that engineer.

RCA tried to buy out Armstrong's FM patents but he would not sell. RCA's engineers developed their own FM patents which Armstrong felt infringed on his own patents. On July 22, 1948 he sued RCA in Federal District Court. The legal struggle went on for a number of years and seemed to be one that would drag on for many years. Armstrong became very discouraged and on January 31, 1954 he committed suicide by jumping out the window of his apartment in New York City.

His widow continued the legal fight against RCA and one by one she won every case over a period of thirteen years after her husband's death. In the meanwhile FM broadcasting stations increased slowly until by 1990 there were almost 5,000 in the United States. Frequency modulation is now used in most mobile communications, FM broadcasting, and the sound of all TV stations.

2.5 Using Electromagnetic Waves to Detect Aircraft and Ships

Just before World War II in 1935 the British government was investigating the possibility of making a death ray that would stop enemy bombers from destroying London. A Committee for the Scientific Survey of Air Defense met on January 28, 1935. Professor Watson Watt was head of the Radio Research Laboratory of the Government Department of Scientific and Industrial Research. He had been asked by Dr. H.E. Wimperis, Director of Scientific Research, Air Ministry, of the possibility of a death ray. Watson Watt turned the question over to his assistant Arnold Wilkins who made calculations that showed radio beams could not be used to directly destroy enemy bombers. Watson Watt then asked Wilkins, "If the death ray is not possible how can we help the Air Ministry?" Wilkins replied that British Post Office engineers had noticed in 1931 that disturbances to VHF (very high frequency radio transmissions) had occurred when aircraft flew overhead. Wilkins thought this phenomenon might be useful for detecting enemy aircraft.

Watson Watt then wrote a memorandum on February 12, 1935 to the Committee for the Scientific Survey of Air Defense. The memorandum stated that although there was no possibility of a death ray, the detection of aircraft might be feasible and proposed that it should be tried. This was the birth of radar in England and Watson Watt has long been considered the father of radar in that country.

However, the possibilities of using radio wave reflection to detect ships and aircraft had been explored in several other countries. In 1922 two American engineers A.H. Taylor and Leo C. Young were experimenting on high-frequency radio communications for the Aircraft Radio Laboratory of the Anacostia Naval Air Station. The laboratory was located in a wooden shack on the eastern shore of the Anacostia River in Washington, DC. A radio transmitter in a truck across the river sent signals to the receivers in the laboratory shack. Taylor and Young found that a passing ship reflected the radio signal. They wrote a report dated September 27, 1922 to the Commanding Officer, Naval Air Station, Anacostia, outlining a proposal to detect ships by radio reflection.

All over the world scientists and engineers had observed that radio signals were reflected from ships and planes. They saw the possibilities in the 1920's and 30's. This happened in Germany, Japan, Holland, Italy, France, and Russia. However, it was in England only that at the beginning of World War II, on September 3, 1939 there was an organized large-scale operating radar system. Although at first it did not work properly, the radar screen did a decisive job in the summer of 1940, during the battle of Britain.

The wavelength that Wilkins picked for radar operation was 25 meters. This was based on the wingspan of bombers which at that time was 25 meters. Electromagnetic waves can be defined by either wavelength or frequency. One can be found from the other by the simple relationship, wavelength in meters times frequency in Hertz equals 300,000,000 meters per second, the speed of light. The unit Hertz is defined as one cycle per second. For example, the wavelength of 25 meters corresponds to a frequency of 300,000,000 divided by 25 or 12 megahertz where megahertz stands for a million.

Robert Watson Watt in 1935 went to London and asked Hugh Dowding, Air Officer for Research and Development, for 10,000 pounds to develop the idea of radar. Dowding refused and he wanted to first see a demonstration. Watson Watt told this to his assistant Arnold Wilkins who came up with a demonstration scheme that would not need money. Wilkins knew that station GSA of the BBC in Daventry was broadcasting on 49 meters with a strong signal. Wilkins thought that 49 meters was close enough to 25 meters for an experiment. Wilkins loaded up receivers and a cathode ray tube in a van. He and Watt had used this equipment to study thunderstorms. He set up the equipment in a field near Daventry and an RAF bomber flew back and forth on a prearranged path while its radio interference was observed. On the morning of February 26, 1935 Watson Watt, Wilkins and the official Air Ministry observer, A.P. Rowe, watched the cathode ray tube. They were able to detect the bomber as it flew back and forth. Rowe wrote a report which was relayed to Air Marshal Dowding who immediately approved 10,000 pounds for the development of radar.

Watson Watt now proposed a group of Chain Home stations along the southeast coast of England. He had a series of twenty 250-foot towers built on the southeast coast of England. The antennas were mounted on top of the towers. Operations were to be conducted on a single wavelength in the frequency band from twenty-two to twenty-seven megahertz. His idea was to use cathode ray tubes to detect planes 80 or so miles away and to feed the information through human filters to fighter pilots.

A cathode ray tube is evacuated and the heated up cathode emits a stream of electrons which goes between two vertical and two horizontal plates. A specific voltage is applied to the vertical plates which causes the electron beam to sweep from left to right at a uniform speed and then return. A special coating on the inside face of the cathode ray tube fluoresces when electrons strike the coating. This results in a straight horizontal line across the center of the cathode ray tube face. The time in microseconds it takes for the electron beam to go from left to right is readily calculated from the values of the components of the electronic sweep circuit.

A signal from a receiver applied to the horizontal plates of the cathode ray will cause the electron beam to be displaced vertically on the face of the cathode ray tube.

The basic simplified use of the cathode ray in radar detection is shown in Figure 2-8. A narrow high-powered transmitted pulse is reflected from an aircraft back to a receiver where it is amplified and fed to the horizontal plates of the cathode ray tube. This results in a small blip or signal. The distance between the two blips can be calibrated in miles or meters. This is so because the time in microseconds of the sweep circuit can readily be converted to distance units. Since an electromagnetic wave travels 186,000 miles in one second then 100 microseconds (or .0001 second) would equal 18.6 miles. The face of the cathode ray tube can then be calibrated in units of distance.

Watson Watt's great contribution was to convert the well-known phenomenon of radio reflection into an integrated system. This included the chain home stations, radar units and operators, telephone lines to a central filtering system where enemy planes are plotted on a map, controllers who communicate to fighter squadrons and direct the fighters to enemy planes.

The system was ready by September 1940 for the battle of Britain. Radar played a decisive role when German bombers first tried to destroy the British fighter force and then attempted to destroy London. The Germans came very close to attaining both

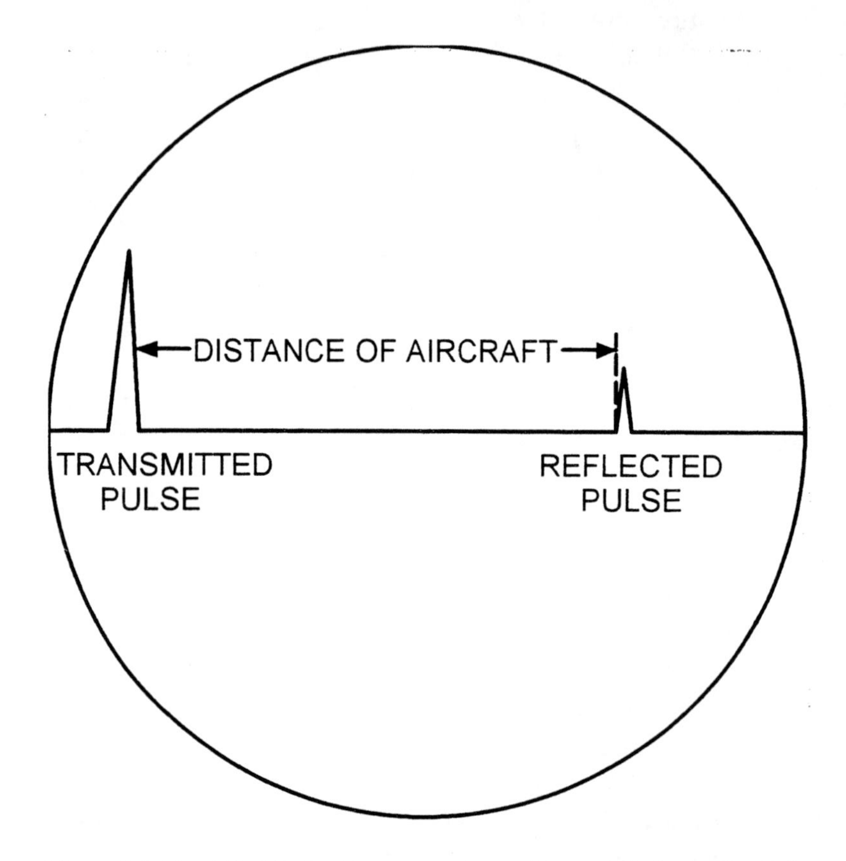

Figure 2-8 Simplified Radar Representation on Face of Cathode Ray Tube

objectives which would have opened up England to an invasion. Radar made the difference as the limited British fighter force could be where and when the German planes were. Without radar there is no question that the Germans would have wiped out both fighter command and London. Radar saved Europe and perhaps the world from a new dark age.

It was realized very early that higher frequencies in radar would mean a narrower beam and much smaller antennas. A narrower beam meant more definition and smaller antennas could be rotated in azimuth and also moved in elevation. This allowed much better direction finding with readings in azimuth and elevation.

By 1941 radar frequencies were using 200 megahertz (200 MHz) as operating frequencies. Examples are the SCR 268 and the larger SCR 270 in the United States. At the time of Pearl Harbor radar in Hawaii picked up the Japanese planes but the infrastructure that Watson Watt had originated was not yet in place in Hawaii. As a result the radar information was not used at Pearl Harbor.

The British looked for a new way to generate very small wavelengths (microwave) and be light enough to carry in planes for night fighters. While radar using the Chain Home towers was very effective during the day it was not accurate enough for night fighting. The British reasoned that the ground radar would put the night fighters in the general vicinity of enemy bombers. Then airborne radar would enable fighter pilots to see enemy fighters in their scopes.

The University of Birmingham was assigned the job of developing a new way of generating very short radio waves by the Inter-Services Committee for the Co-Ordination of Valve Development in 1939. Professor John Randall and his graduate student Henry Boot developed a magnetron which used the principle of resonant cavities and a magnetic field. Randall and Boot constructed a small solid block of copper which was the heart of their device. They scooped out a number of precisely measured evacuated cavities in the copper block. An electric current was made to flow through the copper and a magnetic field was applied. Electrons were caught in the cavities and were made to circulate around there by the magnetic field. As they bounced back and forth, resonating, they emitted electromagnetic waves just a few centimeters long.

On February 21, 1940 they turned their magnetron on and measured its power by hooking up a set of floodlights to the output of the new device. The power was 400 watts and the wavelength was 9.8 centimeters (3,061 MHz). The British government decided to send the magnetron to America for further development. A mission left England aboard the Duchess of Richmond in August 1940 taking the magnetron to the United States. The magnetron was turned over to the Radiation Laboratory at MIT in Cambridge which developed airborne radar using the magnetron. The equipment employed a parabolic reflector which focused the microwaves into a narrow beam.

In April 1942 the airborne magnetron radar equipment became operational in RAF night fighters. In the United States this equipment was incorporated into an American night fighter, the P-61 Black Widow. This plane did not see action until 1944. The magnetron was used in all types of aircraft and ships during the war all over the world.

After the war radar found civilian use in detecting storms and weather prediction. The magnetron was used in a chain of microwave tower stations every thirty miles to relay information. Radar was also used to control aircraft at civilian airports. A modified radar system is also used to catch auto speeders. Radar has become an indispensable part of all modern military forces. The magnetron is also used in microwave cooking. The microwaves heat up food very quickly.

2.6 THE BACKGROUND OF TELEVISION DEVELOPMENT

Many scientists, engineers, and technicians contributed to the concept of television over a long period of time. A telegraph operator named May discovered in 1873 that light varied the resistance of selenium to an electric current. The more light shining on the selenium wire the more electric current flowed through the wire. An English experimenter, Willoughby Smith, wrote about the phenomenon in engineering journals.

This stirred the imagination of scientists and engineers. In 1875 George Cary proposed a television transmitter consisting of a mosaic of selenium cells and a receiver using a similar mosaic. This system was to use wires to transmit the signal between the receiver and transmitter. It would be many years before the photoconductivity of selenium was used in a television camera.

Another phenomenon connecting light with electricity was the photoelectric effect where light falling on a specific material will produce an electric current. In 1887 Hertz accidentally discovered this when he noticed that a spark could be made to jump over a gap more readily if one of the electrodes was illuminated. A German, Hallwachs, studied the photoelectric effect and proposed that light set free electrical particles from the electrode surface. Later, Sir J.J. Thompson identified the electrical particles as electrons. In 1890 Elster and Geitel built the first practical photoelectric cell that converted light to electricity. Thus one method was defined by which a television camera converts a picture into an electric current.

It was realized early that it was impractical to convert a picture all at once into an electric current. In 1880 LeBlanc developed a principle of scanning wherein a picture is divided into lines and each line into segments. In 1884 Paul Nipkow proposed a scanning disk shown in Figure 2-9. He realized that an image could be progressively scanned point by point and line by line if it were projected through a large rotating disk with a number of small holes in a spiral arrangement. The 36 apertures in the disk scan the image in a succession of lines. The Nipkow scanning disk was used in early experimental television work even up to 1940.

The mechanical system of scanning was gradually replaced by electron scanning. Electron motion is much superior to the movement of mechanical parts, because the electron has an extremely small mass and can be moved at very high speeds with only moderate amounts of energy. Electron scanning was first developed in the cathode ray tube. The electrostatic cathode ray tube has been described in the section on radar in this chapter.

Around 1897 Professor F. Braun built a cathode ray tube and showed that magnetic coils could control the electron beams in tracing their paths on a fluorescent screen. This was to be the means of scanning control for the future television receivers.

The earliest experiments with the use of cathode ray tube as a television receiver was those of Professor Max Dieckmann in Germany described in a German patent application September 12, 1906. He used a Nipkow type scanner transmitter adjacent to a cathode ray tube as a receiver. Professor Boris Rosing, a lecturer at the Technological Institute of St. Petersburg, had two British patents of a television system, 1907 and 1911. His cathode ray tube had a pair of magnetic deflecting coils and a pair of modulating plates.

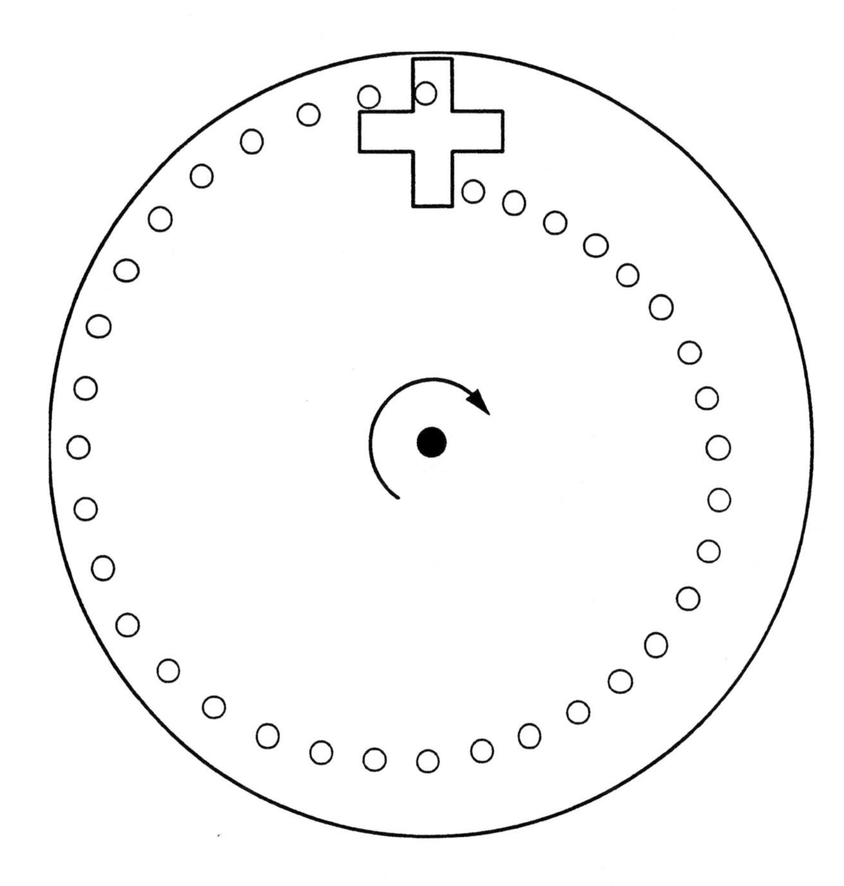

Figure 2-9 The Nipkow Scanning Disk

Vladimir K. Zworykin, a student of Professor Rosing, worked with him on Braun cathode ray tubes as television receivers around 1910 in St. Petersburg. Zworykin emigrated to the United States in 1918 and did his early work with the Westinghouse Research Laboratories. He applied for a patent application in 1923 on a completely electronic television system. In 1924 he made a demonstration using a cathode ray tube in the receiver and an early form of camera tube at the transmitter. However, the electronic camera did not give satisfactory results. Zworykin reverted to the use of mechanical scanning for transmission and concentrated on the design of better cathode ray tubes for the receivers.

2.7 THE CATHODE RAY TUBE FOR BLACK AND WHITE TELEVISION RECEIVERS

The cathode ray tube for television receivers was developed by Zworykin and others into an efficient device shown in Figure 2-10. The electron gun consists of a cathode and a filament, and a beam-controlling aperture, G_1. The cathode is heated by the filament to a temperature high enough to cause it to emit a stream of electrons through the controlling aperture. By varying the voltage difference between the cathode and the G_1 aperture the density of the electron beam can be modulated with a video signal. An accelerating electrode, G_2, with a small diameter defines the diameter of the electron beam.

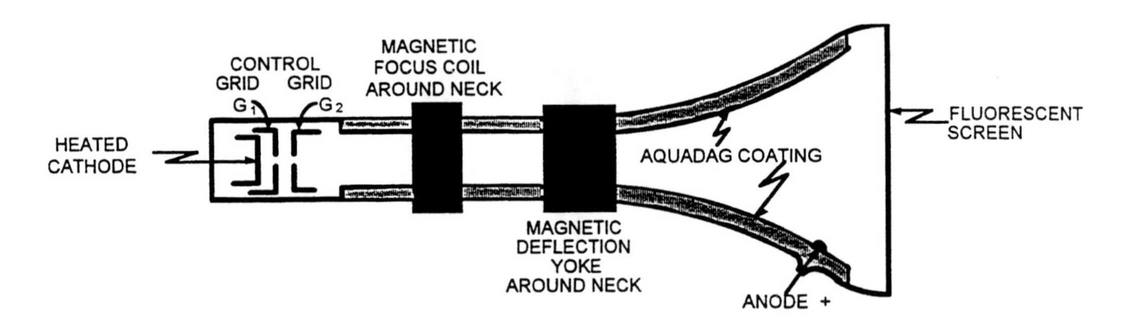

Figure 2-10 Simplified CRT in Black and White TV Receivers

The focus coil is wound inside a soft iron ring concentrating the magnetic field and making it possible to produce the required field strength with a small amount of current. The magnetic deflection consists of two pairs of coils. The two coils of one set are placed above and below the beam axis and are connected in series to deflect the beam horizontally. The coils of the other pair are mounted left and right of the beam to deflect the beam vertically.

The magnetic deflection coils cause the electron beam to produce the necessary scanning. The electron beam is accelerated by the positive voltage on the anode. When the electron beam strikes the fluorescent screen the CRT will glow at that spot. The cathode ray tube shown in Figure 2-10 was later modified for color.

2.8 TELEVISION CAMERAS

Zworykin and others developed a number of different types of television cameras. In a television camera tube an electron beam in a cathode ray tube traces a pattern of horizontal and vertical scanning lines over the light sensitive surface of the photocathode. The photosensitive surface is charged to different potentials corresponding to differences in brightness of the scene focused on the surface by the camera lens and optical system.

There are a number of different types of television cameras. In 1923 Zworykin developed an electronic pick up device. He called it the iconoscope (from the Greek for "image" and "to see"). The iconoscope was not satisfactory because of a spurious flare. The iconoscope was replaced by the image iconoscope around 1939. This in turn was replaced by the orthicon and then by the image orthicon in 1946. The image orthicon became the standard studio camera for a number of years. All of the above cameras used a mosaic of separate small areas of photoemissive material. Light falling on this material causes electrons to be emitted.

Another completely different type of camera was the vidicon which first came out in 1950. This uses a photoconductive material like antimony trisulfide. Light falling on this material increases its conductivity. This light is converted into an electrical signal. Later a solid state light detector for television cameras, a charge coupled device (CCD) was used. This made for a very rugged small camera which works at low levels of light.

2.9 POST-WAR II BLACK AND WHITE TELEVISION SYSTEMS

The number of scanning lines in television in the United States increased from 30 lines in 1925 to 525 lines in 1945 increasing the resolution considerably. Standards were established to produce pictures with good definition and without flicker.

Television images consist of picture elements (dots arranged along lines). The lines are assembled to cover the field (the picture area). A succession of fields is transmitted to create the illusion of continuity and motion in the image. The television image is scanned with two sets of lines, one interlaced with the other. One set of these lines, having blank spaces between lines is known as a field. The other half of the area (the space between the lines) is filled in by the lines of the next successively scanned field. Hence all points in the picture have been scanned when two successive fields have been scanned. Two successive fields, comprising all the lines in the image are known as a frame.

To insure continuity in the image, it is necessary that the fields succeed one another at a rapid rate. If the fields are presented at a rate slower than the 15 per second, the apparent motion will be disjointed. Above that rate the retina of the eye will retain successive images giving the illusion of continuity.

In practice the rate of successive fields must be much higher than the minimum valve of 15 per second because of flicker. Flicker appears because the light on the screen is cut off between successive fields. If the rate of scanning successful fields is too low, the light on the screen will appear to blink on and off. If the successive fields are scanned at a sufficiently rapid rate, the sensation from one picture persists throughout the dark interval between fields and the screen appears as if it were continuously illuminated.

Experience has shown that rates between 48 and 60 fields per second are required to produce flicker-free pictures of adequate brightness. In the U.S. 60 fields per second are used because the electric power is 60 cycles and this makes filtering easier. In England the electric power is 50 cycles per second and 50 fields per second are used.

Synchronization

Not only must the scanning rates of the TV transmitter and receiver be the same but the scanning at the receiver must be exactly synchronized with the scanning at the transmitter. In order to keep the transmitter and receiver scanning in step with each other, special synchronous signals are transmitted for the receiver scanning system. These are rectangular pulses used to control both transmitting and receiver scanning.

The synchronizing pulses are transmitted as a part of the complete picture signal for the receiver, but they occur during the time when no picture information is transmitted. The picture is blanked out for this period while the electron-scanning beam is retracing. A horizontal synchronizing pulse at the end of each horizontal line begins the horizontal retrace, and a vertical synchronizing pulse is provided at the end of each field to begin the vertical retrace, thus keeping the receiver and transmitter scanning synchronized.

The Sound Associated with TV

The sound is transmitted simultaneously with the picture signal in a common channel to permit a complete visual and sound reproduction of the televised program. The sound is transmitted as a frequency modulated signal while the picture signal is amplitude-modulated.

2.10 The Color Television War

Color television takes advantage of the reaction of the eye to color. The retina of the eyes contains three sets of cones that are separately sensitive to red, green, and blue light. The cones are uniformly dispersed in the color-sensitive area of the retina, so that light falling on them produces a sensation of color that depends on the relative amounts of red, green, and blue light present. When the ratio of red, green and blue light is approximately 30:60:10, the sensation produces white light. Other ratios produce all the other color sensations.

The television camera produces three signals proportional respectively to the relative intensities of red, green, and blue light in each portion of the scene. In the picture tube of the receiver the relative intensities of the red, green, and blue are reproduced. The many ways of accomplishing this led to the color wars beginning around 1949 and ending in the United States in 1953.

In 1950 three companies presented three separate and distinct color TV schemes before the United States Senate Advisory Committee. The companies were Color Television, Inc. (CTI), Columbia Broadcasting System (CBS), and Radio Corporation of America (RCA). The Committee stated that only one system would be chosen.

In the CTI system, a line system, the color was assigned to successive lines of the image. The CBS system was a field system, the color values being assigned to successive fields of the image. The RCA system was a dot system with the color assigned to successive picture elements or dots of the image.

The fight came down to a struggle between the two giants of television, CBS and RCA. RCA's system was compatible with black and white television systems. This means that a black and white television receiver would still operate with a color transmission. Only the received picture would appear in black and white. The CBS color system was not compatible with existing black and white systems. CBS developed their system first but the government decision was delayed until there were too many black and white sets in use to make them immediately obsolete by adopting the CBS system. RCA's dot sequential system won out and was adopted as the sole color system in the United States.

2.11 CABLE TELEVISION

The original purpose of cable television was the improvement of the reception of television broadcast stations. After World War II it was found that reception of the new television stations was not satisfactory in all areas of the United States. Antennas were erected on mountain tops and other high points. Cables connected the antenna to television receivers in the community. Early cable systems were known as community-antenna-TV (CATV) systems. When other services, like Satellite Reception, Public Access, Local Origination Studios, and Microwave Pick Up of Remote Programs were added the CATV systems became known as cable television. Usually off-air reception or terrestrial broadcasting accounts for less than half of the modern cable system channel capacity.

SATELLITES FOR CABLE TELEVISION

A significant part of cable television programs in the United States comes from satellites. In 1990 there were four satellites available with each satellite containing 24 transponders (receive-transmit) channels thus providing a total of 96 channels. The programs are sent from an earth transmitter station in the 6GHz* and up to a satellite where it is transmitted down on a channel in the 4GHz band and beamed in a pattern covering the United States, Central America, and South America.

THE CABLE TELEVISION HEAD-END

All of the cable services, satellite, television stations, public access, local origination studios, and microwave pickups are processed in the head-end and after suitable frequency conversions are combined in a combiner. They are then fed into a television modulator which is an extremely low-power television transmitter in the head-end. The television modulator takes audio and video signals and places them on video and audio television carrier frequencies which drive the cable.

THE COAXIAL CABLE SYSTEM

The coaxial cable system consists of three types. They are the main trunk cables, the feeder or distribution cable, and the drop cable entering the subscriber's home. The trunk and distribution cables consist of a solid aluminum sheath with polyethylene foam surrounding a copper-clad aluminum center conductor. The trunk cables are the largest with an outer diameter of 0.5 inches or more. The feeder cables which come off from the trunk cables at special clusters of subscribers are about 0.412 inches. The trunk and distribution cables are not flexible and cannot be used to enter subscribers' homes. The drop cables which come off the feeder cables have a very small outer diameter and contain a solid dielectric. The losses in the drop cables are high per foot and must be kept short. However, they are flexible which is necessary in the subscriber's home. The drop cable connects to the set top converter. In some newer TV cable systems the coaxial cables are being replaced by fiber optics which will be described later.

THE SET TOP CONVERTER

The set top converter makes the necessary frequency conversions and its channel selector converts the desired channel to an unused frequency in the television receiver (Channel 3 in some areas). In newer cable ready television receivers the functions of the set top converter are incorporated in the receiver.

---

*GHz stands for 1000 million Hertz

2.12 THE HDTV WARS

At the normal viewing distance of direct-view screens in general use, the maximum resolution of 525 and 625 line systems has been adequate. Any increase in resolution would not be discernible and consequently not justified.

However, the trend to larger direct-view screens and improvements in project systems results in a greater perception of fine detail by viewers and a demand for an increase in resolution.

In 1968 the Japanese Broadcasting Company (NHK) initiated a research program for high definition television (HDTV) requiring a wider transmission bandwidth than that available for 525-line terrestrial broadcasting. In 1974 NHK gave its first public demonstration of HDTV. By 1980 the United States and other countries became interested in this new technology. The 525 line countries like the United States opted for a 1125 line 60 fields per second system. The 625 line countries (PAL AND SECAM) like Great Britain opted for a 1250/50 standard. Japan with its headstart began in June 1989 a regular HDTV program for one hour a day transmitting via satellite. This was increased in 1990 to eight hours a day.

The Japanese used the NHK system known as MUSE (Multiple-Sub-Nyquist-Sampling-Encoding) compression system which reduces the 20 MHz HDTV bandwidth to 8 MHz bandwidth of the 12GHz satellite service. MUSE is only one of several proposed HDTV systems. MUSE uses DBS (direct broadcast from satellites). It uses 1125 lines and 60 frames per second.

HDTV terrestrial broadcasting has been undertaken by the European Economic Community under the name Eureka. The MAC (multiplexed analogue component) time compression system will be used to deliver an HDTV series to terrestrial stations. HDTV direct broadcasting from satellites to home receivers for the PAL and SECAM audiences started in 1990 using channels in the 12 GHz and 22 MHz bands. The MAC system of compression was used to get the bandwidth down to the 8 MHz bandwidth in the satellite service.

In the United States the FCC in 1987 announced it would issue the standards for the HDTV service after testing the many alternative proposals then before it. The FCC announced that it would accept the recommendations of the Advisory Committee on Advanced Television Systems. This Committee was organized in 1985 and set up an industry-supported laboratory, the Advanced Television Test Center (ATTC). Field tests of proposed terrestrial HDTV broadcast systems transmitted over two UHF stations near Washington, DC began in December 1988.

After a great deal of testing, it was decided to use a digital system. A grand alliance of TV equipment manufacturers and broadcasters was formed to recommend one standard for a new generation of television transmitters and receivers. The computer industry objected and there was a delay.

On November 25, 1996, the U.S. TV and computer industries agreed on a technical standard for digital television. This was submitted to the FCC for final approval by the end of 1996.

Early in 1997 the FCC allocated a second digital channel for each TV station in the nation. Stations would then apply to the FCC for licenses to broadcast digital television.

In 1998 some TV networks will broadcast digital HDTV programming. New TV receivers capable of receiving these programs will probably cost about $1,500 more than conventional analog receivers.

For at least seven years, broadcasters will continue to send analog transmission on their original channels.

It should be noted that digital television even without high definition eliminates snow and ghosts which can appear in analog TV. Digital TV does not gradually get worse as propagation deteriorates in the way that analog TV does. Digital TV gives either a good picture or no picture.

Flat Display for TV

In 1963 George Heilmeier at RCA David Sarnoff Research Center in Princeton, New Jersey, found that liquid crystal conducts light when an electric charge is used. This was applied in the RCA laboratory on an experimental basis to electronic calculators, digital watches, and even a TV display unit.

RCA turned this discovery over to its semiconductor production division which let this discovery lapse. Sharp, a Japanese electronic company, bought the license to manufacture and apply liquid crystals. Over the years Sony developed the liquid crystal concept for commercial electronic calculators and digital watches. Sony used these developments to build up a manufacturing capability to produce flat displays for TV receivers in 1990. These are 2 inches thick and can displace the cathode tube for some television receivers. The flat display is also used in personal computers.

In 1994 Textronic, Inc. of Beaverton, Oregon, developed a new technology using a plasma-addressed flat display which can be made in large sizes for wall-mounted flat-panel television. This may be used in future HDTV receivers.

2.13 Safety Levels of RF Electromagnetic Fields

It is well known that microwave frequencies can heat up animal and other tissues. These frequencies are used in microwave ovens to cook food. After World War II tests were made on animals such as rabbits. It was found that eyes and testes were particularly affected by microwave radiation at a minimum level of one milliwatts per square centimeter. It was decided to set a safety level of 10 milliwatts per square centimeter.

This allowed a safety factor of ten. The levels applied only to the microwave frequencies (above 1,000 megahertz).

A more general set of limits was issued in ANSI (American National Standards Institute) C95.1, 1982. This covered radio frequencies from 0.3 megahertz through 100,000 megahertz. The limits varied with the frequency band.

This was superseded by ANSI/IEEE C95.1-1992 which was produced after a number of years of study by prominent engineers and scientists. This standard can be purchased from Global Engineering Documents, 285 McGaw Avenue, Irvine, California 92714.

There have been a number of lawsuits from workers on microwave systems and users of radio cellular portable phones.* There may be a case for workers who have been exposed right in front of a high powered microwave beam from a high gain parabolic antenna. After all we use microwaves to cook meat. However, at any distance the power is rapidly dissipated so that the general public is not in danger from microwave transmitters.

Sometimes the fear of microwave transmitters has led to a delay in solving problems in other technologies. For example, special Doppler radar can help detect local atmospheric turbulence at airports. This would alleviate the problem of airplanes crashing due to unexpected thrusts of air on take off or landing. Objections by local people have delayed the installation of the special radar equipment on the ground even though there is no proven evidence of a danger from the microwave transmitters at any distance.

As far as low powered cellular portable phones, they come within the limits of C95.1-1992. There has been no proof that these portable phones cause brain cancer or other harm.

2.14 THE REPLACEMENT OF THE VACUUM TUBE

The vacuum tube became the workhorse of radio, radar and television until about 1960. Then new inventions, the transistor and the integrated circuit rapidly replaced the vacuum tube and made it obsolete. This is described in the next chapter.

* Radio cellular systems are described in Chapter 8.

Bibliography

Abramsen, Albert — *Zworykin: Pioneer of Television,* Champion, Illinois, University of Illinois Press, 1955

Belrose, John S. — "Fessenden and the Early History of Radio Science," Proceedings of the Radio Club of America, Inc., November 1993, Pages 6-21

Benson, K.B. and Donald G. Fink — *HDTV, Advanced Television for the 1990's,* New York, McGraw-Hill, 1991

Elliot, George — "Who Was Fessenden?" Proceedings of the Radio Club of America, November 1992, Pages 25-37

Fisher, David E. — *A Race on the Edge of Time: Radar-The Decisive Weapon of World War II,* New York, McGraw-Hill, 1988

Garratt, GRM — *The Early History of Radio From Faraday to Marconi,* London, The Institution of Electrical Engineers, 1994

Geddes, L.A. — "Fessenden and the Birth of Radiotelephony," Proceedings of the Radio Club of America, November 1992, Pages 20-24

Lewis, Tom — *Empire of the Air, The Men Who Made Radio,* New York, Harper Collins, 1991

Morrisey, John W., Editor — *The Legacies of Edwin Howard Armstrong,* New York, Published by the Radio Club of America, 1990

Ritchie, Michael — *Please Stand By: A Prehistory of Television,* Woodstock, NY, Overlook Press, 1994

CHAPTER 3

HOW MANY TRANSISTORS CAN DANCE ON THE HEAD OF A PIN?

3.1 INTRODUCTION

In medieval times scholars argued about, "How many angels can dance on the head of a pin?" In the 20th century engineers and scientists (modern versions of medieval scholars) increased the density of electronic devices, called transistors, to several million on a tiny sliver of silicon, a chip, the size of a thumbnail.

3.2 THE ROOTS OF THE TRANSISTOR

In the early 20th century a new branch of physics called quantum mechanics opened up an Alice in Wonderland picture of the internal structure of the atom. This led to the invention of the transistor and devices such as the laser, the MRI, and even the structure of DNA. Quantum mechanics started with the proposal by Max Planck late in 1900 that energy levels at a fundamental level were not continuous but came in discrete amounts called quanta. Albert Einstein built upon this concept when in 1905 he proposed that light is also a quantum of energy. This was later given the name of photon. That concept led to the idea that light and all electromagnetic radiation can be considered both as a wave and a particle.

In 1923 the French physicist Louis-Victor de Broglie applied the duality of particle and wave to a description of matter. He derived an equation which implied that every bit of matter undulates as waves. The wave nature of matter led to a picture of a complex atomic structure.

There are electrons (negative particles) in a probability cloud of orbits around a positively charged nucleus. This is the picture in every atom. Each of the electrons occupies a specific orbit with a definite energy. The farther the electronic orbit from the atomic nucleus, the higher the energy of the electron. Electrons ascend and descend among the orbital energy levels as if they were rungs on a ladder. If an electron in a low-lying inner orbit gains energy it jumps to a higher empty orbit, leaving behind an empty track. There is a forbidden energy gap between adjacent electron orbit bands.

W. Heisenberg, a young German physicist developed the Principle of Uncertainty at the atomic level. One example is the measurement of the position and momentum (mass times velocity) of an electron. If the position is known precisely, then the momentum cannot be measured. The more precise one measurement is made the less precise the other measurement.

The concepts of the duality of wave and particle and the Uncertainty Principle together with some probability aspects of quantum mechanics raised doubts in some scientists such as Albert Einstein about the completeness of the theory. On the other hand many scientists such as E. Schrodinger, Niels Bohr, P. Dirac, A. H. Compton, L. de Broglie, Max Born, and Heisenberg, not only contributed significantly to quantum mechanics but strongly defended it.

In October 1927 the Fifth Solvay Congress in Brussels became the focus of discussion and debates about quantum mechanics. Some of the greatest physicists of the time attended as shown in the photograph, Figure 3-1. They listened to a spirited discussion about quantum mechanics between Albert Einstein and Niels Bohr who defended the new ideas. Einstein was reported to have said, "God does not play dice!" Bohr retorted, "Stop telling God what to do."

However, the theory explained many phenomena in science such as the spectrum or frequency of light emission from hydrogen and other chemical elements. In a short time most physicists came to believe in quantum mechanics even though many of its concepts were counter intuitive. Twenty years after the Fifth Solvay Congress, the transistor was invented. Its development depended heavily on quantum mechanics.

Quantum mechanics gave a detailed picture of the inner workings of the atom. In the 1940's scientists and engineers took this knowledge and developed semiconductors called transistors. They started with pure crystals of germanium and later silicon. Then controlled amounts of two types of impurities were added to the structure of these crystals. This was the start of the vast technology of the semiconductor industry.

The quantum idea of high and low energy levels separated by a forbidden energy band applies to semiconductors. The energy levels in the semiconductor include all the vast number of electrons that are in a solid of macroscopic size. There is a band gap of energy previously described in the picture of an individual atom. Electrons above the gap are in high-energy states called the conduction band and those electrons are relatively mobile. The electrons below the gap occupy states of relatively low energy known as the valence band. Those electrons are tightly bound to the atoms and cannot contribute to the flow of electric current. Energetic electrons in a semiconductor can leap from valence to conduction bands.

The idea of valence band, band gap, and conduction band is applied to semiconductor devices such as the transistor. An electron advancing from valence band to conduction band leaves behind a space called a hole. The hole makes room in which the remaining valence electrons can move. The space left is what makes motion possible and the hole enables the electrons that remain in that band to join in the flow of current. Missing negative charge is the same as added positive charge so the holes act like

Figure 3-1 Quantum Mechanics, Discussions and Debates, 1927
International Institute of Physics and Chemistry
Courtesy of AIP Emilio Segre Visual Archives

mobile positive charges.

Pure crystals of germanium or silicon can be doped with impurities chosen to add either extra electrons in the conduction band (N type) or holes in the valence band (P type). When the two types are put together to make an N-P junction the result can be used to make an electronic switch which can be turned on and off. This device called a diode can also be used to change alternating current to direct current. While the diode has two elements the real technological revolution was the invention of the transistor with three elements.

On December 23, 1947 a semiconductor speech amplifier was demonstrated at Bell Laboratories using a germanium point contact transistor. Figure 3-2 shows the three inventors of the transistor with the apparatus at Bell Telephone Laboratories which led to

Figure 3-2 The Inventors of the Transistor, 1947
International Institute of Physics and Chemistry
Courtesy AIP Emilio Segre Visual Archives

the discovery of the transistor. Seated is Dr. William Shockley (1910-1989) who initiated and directed the Bell Laboratories Transistor Research Program. Standing are Dr. John

Bardeen (1908-1991), left, and Dr. Walter K. Brattain (1902-1987), key scientists in bringing the invention to reality. The transistor they developed produced amplification as high as 100 to 1 but was very noisy.

Silicon rapidly replaced germanium because silicon is much more plentiful and also silicon operates at a much greater temperature range. Originally silicon was produced as a large monocrystal which was hit with a hammer producing chips. The term stuck long after the hammer was retired. Soon the point contact transistor was replaced by the junction transistor shown in Figure 3-3. The N and P type of crystals can be combined into an NPN or a PNP transistor as illustrated in that drawing.

3.2 Bipolar Transistors

These types of transistors are called bipolar because their operation depends on the transport of both electrons and holes. The three regions of the bipolar transistor are called the emitter, base, and collector. The base is very thin and is very lightly doped. In the NPN, as shown in Figure 3-4, an electric field applied between the emitter and collector will cause the electrons to flow from the emitter through the base to the collector. Very few of the electrons will be bound to the holes in the base. These are pumped out and returned to the collector. A very negative current flow into the base will cut off current from the emitter to the collector. Thus the transistor can be used as a switch. A small varying current flowing into the base will vary the current from the emitter to the collector. In this way the transistor acts as an amplifier.

The three scientists who invented the transistor at the Bell Laboratories in 1947 were awarded the Nobel Prize in Physics in 1956 for the invention. Bardeen later in 1972 was the first person to win a second Nobel Prize in the same field. He shared this second Nobel Prize in Physics for his work on the theory of superconductivity.

Walter Brattain was born on February 10, 1902, in Amoy, China. He received his Ph.D. from the University of Minnesota in 1929. He joined the staff of Bell Telephone Laboratories in that year as a research physicist. He retired in 1967 and began teaching at Whitman College. Here he and a number of colleagues did research on phospho-lipid bilayers as a model for the surface of living cells.

The third inventor of the transistor, William Shockley (1910-1989), left Bell Laboratories in 1954 and founded a semiconductor laboratory and factory near Stanford University in California. He sent out a call to a dozen handpicked young Ph.D.'s in Physics and Chemistry, to join him in a "warehouse" in Mountain View, at Shockley Semiconductor Laboratory. Bob Noyce and Gordon Moore were two of the people who answered Shockley's call.

Noyce, Moore, and six others got financial backing from Fairchild Camera and Instrument Company to form Fairchild Semiconductor in 1957 to develop a better transistor. Two dozen companies have spun off from it. Three of the leading companies

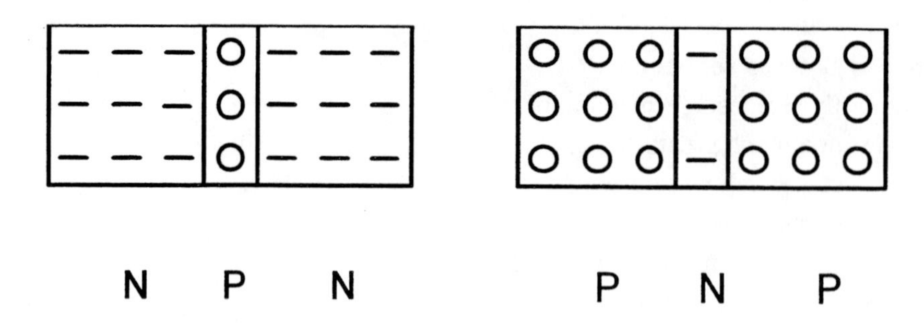

NOTE: — REPRESENTS NEGATIVELY DOPED SILICO

O REPRESENTS POSITIVELY DOPED SILICON

Figure 3-3 NPN and PNP Transistors

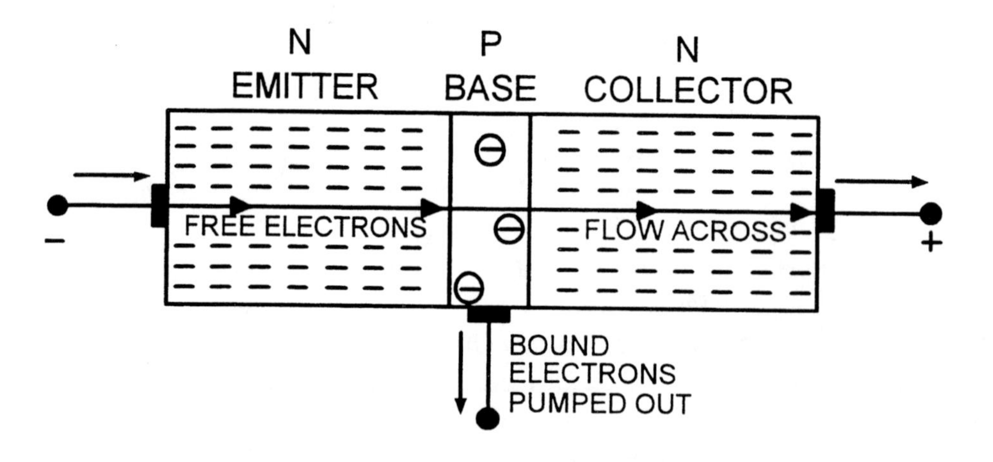

Figure 3-4 Flow of Electrons in a NPN Transistor

were Intel, Advanced Micro Devices, and National Semiconductor. Noyce and Moore became the leaders at Intel.

The industry that Shockley started spread some 20 miles from Palo Alto to the City of San Jose. This became known as Silicon Valley and is located about an hour's drive south of San Francisco.

Shockley himself later in life preached a philosophy of retrogressive evolution. He deemed Blacks genetically inferior to Whites and unable to achieve their intellectual level. He suggested that Blacks were producing faster than Whites and were reducing the intelligence of the general population. Many of Shockley's scientific colleagues denounced his racial theories. Shockley considered that his racial theories were more important than the invention of the transistor. At the age of 68 Shockley contributed "more than once" to a controversial sperm bank for a project offering to pass along the "genes of geniuses."

3.3 Bipolar Transistor Manufacturing Technology

The transistor invented by Bardeen, Brattan, and Shockley quickly replaced the tube in radios, television, and electronics in general. This happened because the transistor was more reliable and required much less power and space. The tiny transistor was soldered into printed circuit boards and an industry was born.

A new manufacturing technology was created to produce the great number of transistors required. The technology is important to understand because it was the forerunner of the integrated circuit manufacturing which in turn led to the chip with its concentration of millions of transistors in a tiny area.

The process of making transistors starts with the manufacture of monocrystalline silicon. The manufacture of one large single silicon crystal usually begins with a liquid chemical compound, trichlorosilane, which contains a great deal of silicon extracted from sand. Trichlorosilane is chemically transformed into pure polycrystalline silicon with many crystals all jumbled up.

The polycrystalline silicon is melted in a crucible. A small seed of monocrystalline is lowered into the crucible just far enough to touch the surface of the melted silicon. The seed is cooler so the molten silicon begins to crystallize on the seed, reproducing the crystalline structure of the seed. As it grows, it remains monocrystalline. The seed is slowly but continually lifted from the melt, and more and more of the molten material accumulates to build up a large crystal. The crystal and the crucible are rotated in opposite directions as the crystal is pulled upward in order to keep growth uniform.

The finished product is a 2"-6" diameter cylinder which is cut into thin disks or wafers. Each wafer can be cut up into tiny chips. In each of the chips all of the atoms are lined up in exactly the right pattern, or crystal lattice, to produce a semiconductor. This pure silicon must be processed to have P-type and N-type material in just the right places. There are a number of ways to do this but the diffused junction process was widely used to make transistors and is now used in most integrated circuits. This is illustrated in Figure 3-5.

The process starts with a slice of monocrystalline silicon. The heated slice is exposed to gaseous phosphorus which is an N-type dopant. The phosphorous diffuses or soaks in throughout the slice. The phosphorous takes the places of a small proportion of the silicon

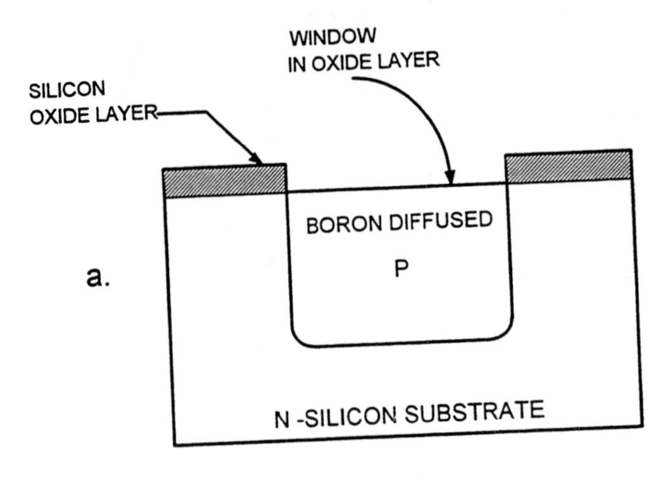

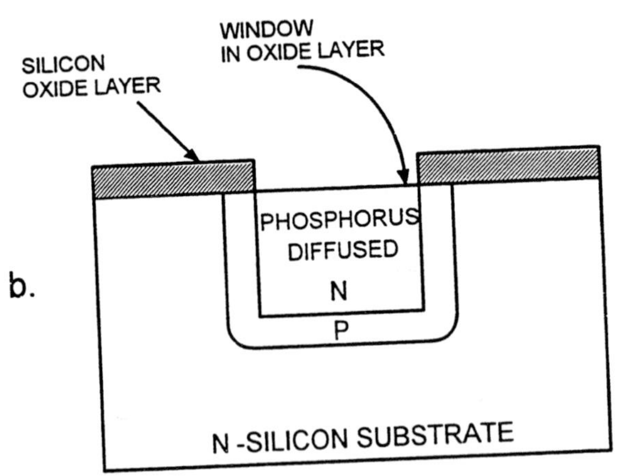

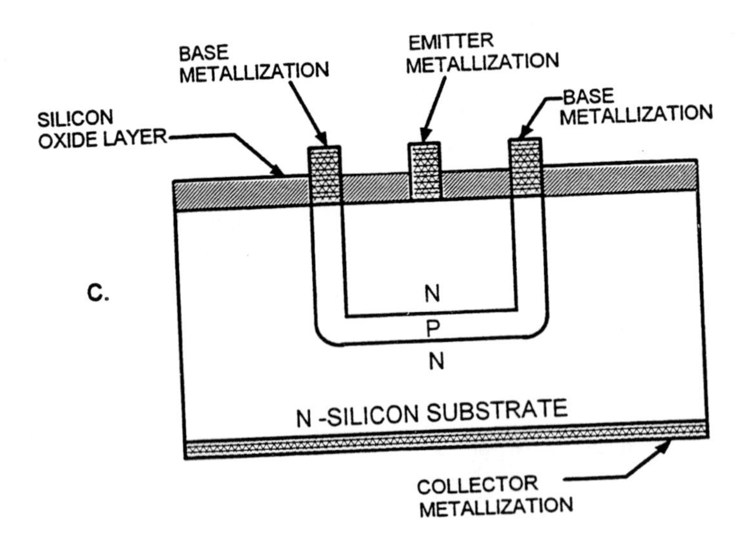

Figure 3-5 Construction of NPN Transistor

atoms in the silicon crystal. This N silicon is called a substrate (underlayer). The top of this N silicon substrate is oxidized so that a sealing layer of protective substance, silicon oxide (glass) completely covers the N silicon substrate.

The oxide layer is covered with a material called "photoresist." The photoresist is sensitive to light like a photographic film. A "photomask," a kind of master photographic negative, is placed over the photoresist and light is shown on it. The light shows through selected areas of the mask, and causes chemical changes in those selected areas of the mask. The light hardens the photoresist which prevents the action of etchants. The slice is now washed away, exposing the oxide layer in selected areas. The photoresist resists the attack of etchants which are used to cut windows in the silicon oxide layer.

A slice has thousands of windows each about 20 thousandths of an inch across for each transistor. The slice, with the etched silicon oxide layer acting as a stencil, is placed in a furnace and heated to a temperature near the melting point of silicon. Boron gas, a P-type dopant is pumped into the furnace and it strikes the exposed windows of the silicon slice. The boron diffuses into the silicon crystal in the area of each window. The boron atoms take the places of a small proportion of the silicon in the crystal. The slice is removed from the furnace and a P-type layer has now been formed as shown in Figure 3-5a.

A new layer of silicon oxide is then produced over the entire surface of the slice including the windows. New smaller windows are created within the areas of each old window. The slice is put back into the furnace and exposed to gaseous phosphorous which is an N-type dopant. The phosphorous atoms diffuse into the window areas and overpower the effects of the boron in a small region within the P-region, converting the silicon in this small area to N-type. The structure is now an NPN transistor shown in Figure 3-5b.

Connections now have to be made to the outside. The surface of the emitter and the base are plated with a thin layer of aluminum as shown in Figure 3-5c. This metallization process employs photomasking similar to that used in making the different layers. The bottom of the splice is plated with a layer of gold which provides a good `electrical connection for the collector. The layer of gold allows for the attachment to the metal of the transistor package. This type of transistor is called a planar transistor invented by J. A. Hoerni, of the Fairchild Corporation in 1959.

Thousands of transistors are formed on the same slice which is then sawed apart after the processing is complete. The transistor element is attached to the gold plated platform by a heat process which alloys the collector to the bottom of the transistor case. The platform itself serves as the collector lead and the platform is connected to one of the external leads. The emitter and base leads are insulated from the platform by glass.

Thin gold wires are fused to the metallized areas that contact the emitter and base regions. The other ends of these wires are attached to the external leads which go to the outside world. A protective cap is welded to the header completing the transistor.

The transistor external leads were soldered on circuit boards, plastic cards embossed with flat snakelike wires. The transistor often broke off with any physical stress on the electronic equipment. With the use of thousands of circuits in computers designed for space vehicles, this became a problem.

3.4 INTEGRATED CIRCUITS USING BIPOLAR TRANSISTOR

In 1958, Jack St. Clair Kilby, shown in Figure 3-6, came up with a method of solving the transistor connection reliability problem. His solution was to make the crystal serve as its own circuit board. On August 28, 1958 he produced the first working model of a silicon integrated circuit, using pre-etched grown junction transistors. This was improved by Robert Noyce of Fairchild Semiconductors in 1959. He produced the first silicon planar integrated circuit which made the integrated concept commercially feasible.

In March 1961 Fairchild Semiconductors produced the first commercial integrated circuit. This allowed the simultaneous fabrication of transistors, diodes, resistors and capacitors on a single silicon chip. The chip did not have soldered wires, reducing failure points and making it very reliable. Since the chip is extremely small, electrical signals take short paths from circuit to circuit, saving time.

Mass production of chips with thousands of transistor circuits reduced the price drastically. In the 1950's transistors cost over $100 each. Thousands of transistors on a chip brought the cost down to pennies each. A chip with 5,000 transistors operates a digital watch. A chip with 20,000 transistors is the heart of a pocket calculator and a chip with above hundreds of thousands or more transistors operate personal computers.

All of the above are the realm of very large scale integration or VLSI. VLSI is defined as an integrated circuit containing 1,000 or more gates. A gate is a logic circuit that has 2 inputs that control one output. A typical VLSI chip can have 450,000 transistors laced together with 60 feet of vapor-deposited conducting "wire." The circuits etched on the tiny chip are the equivalent of mapping every street on a small town. Gordon Moore postulated what became known as Moore's Law regarding the increase of transistors on a chip. One version of Moore's Law is that the number of transistors on a chip per unit area doubles every 18 months.

3.5 THE MANUFACTURE OF VLSI USING BIPOLAR TRANSISTORS

Computers are used to generate a map which contain all of the circuits including as many as 12 different layers or more within the chip. The software disc is sent to a foundry where a glass photomask of the chip is made. This is a kind of master photographic negative.

The actual manufacturing begins with pure melted silicon grown into long crystals as described before in the manufacture of transistors. The crystals are sliced into thin wafers with diameters as large as six inches.

Figure 3-6 Jack Kilby, The Inventor of the Integrated Circuit.
Photograph taken in 1980
International Institute of Physics and Chemistry
Courtesy of AIP Emilio Segre Visual Archives

The wafers are insulated with a film of silicon oxide and coated with a soft light-sensitive plastic called photoresist. This material also has the property of resisting the action of the etchants used in the chip manufacturing process.

Light is passed through the photomask on to the photoresist coating. The exposed photoresist hardens into the proper outline. Acids and solvents strip away the unexposed photoresist and silicon oxide. The patterned silicon is prepared to be etched by super hot gases.

The etchants are used to cut windows in the silicon as described previously in the manufacture of transistors. Each slice has many thousands of tiny windows. The slices are then treated with the proper impurities to make transistors similar to the manufacture of transistors themselves.

More silicon is laid down, masked, and stripped. The steps are repeated building gaps or "windows." Vaporized metal is condensed into the wafer, filling these gaps and forming conducting pathways for the circuits.

Circuit elements like resistors and capacitors are also represented in the photomask. One method of producing resistors is to deposit a thin metal film of metal on an insulating layer of silicon oxide. The metal may be tantalum, tin oxide or nichrome.

A thin-film capacitor can be produced by laying down thin metal films above and below a thin layer of insulating silicon oxide. This forms a capacitor that is part of the integrated circuit in the chip.

Each chip is diced from the wafers by a diamond saw, then bonded wired to gold frames and sealed in small ceramic cases. Wires of gold or aluminum connect from the chip to the outside prongs.

The original design of the chip was drawn up on drafting paper where a four by eight-foot sheet represents a strip approximately 0.0585 inches wide. This is just a tiny part of the chip. CAD (Computer Aided Design) is used to store diagrams, rules on how to link them, and data on the intended function of the chip. This enables the computer to design a chip circuit, display it on a screen and report on its performance.

3.6 The Integrated Circuit using MOS Transistors

Up to now we have been talking about the bipolar transistor. There is another family of transistors, the metal oxide semiconductor field effect transistor or MOSFET that was introduced in the early 1970's. These enable many more transistors to dance on the head of a pin than the bipolar since the MOSFET is much smaller than the bipolar transistor. The MOS, an abbreviation of MOSFET, uses a single carrier, either an electron or a hole, but not both at once. If the carrier is the electron only, then the transistor is called NMOS. If the carrier is a hole, then the transistor is called PMOS.

3.7 THE PMOS TRANSISTOR INTEGRATED CIRCUITS

Figure 3-7a shows the basic structure of a PMOS transistor. There are three main

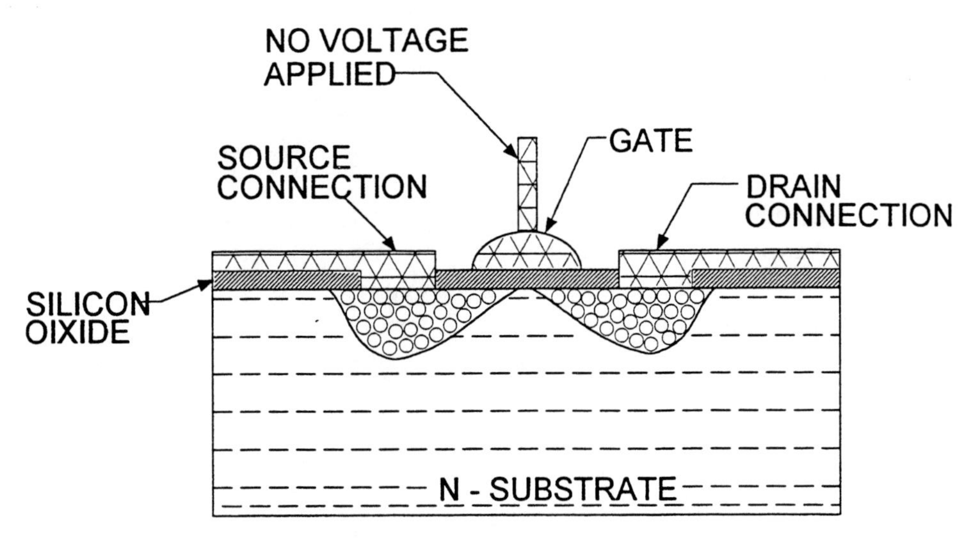

a - NO VOLTAGE APPLIED TO GATE

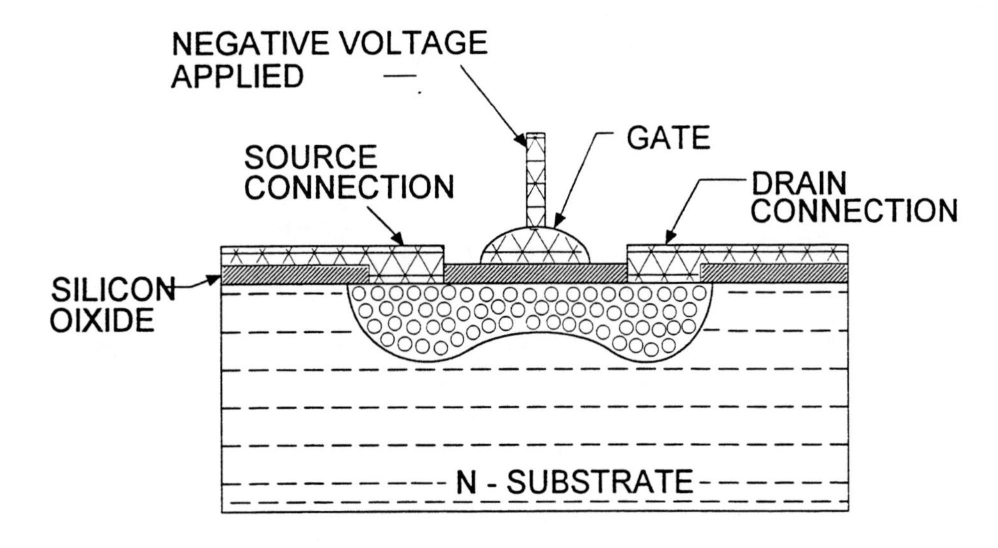

b - NEGATIVE VOLTAGE APPLIED TO GATE

Figure 3-7 Operation of a PMOS

elements, the source, the gate, and the drain. A layer of insulating silicon oxide is put down over an N-type substrate. Then two P-regions called the source and the drain are diffused side by side in the substrate. A metal strip, the gate, lies on top of the silicon oxide, over the gap between the P-regions. Two strips of metal are made to penetrate through windows in the oxide to contact the source and drain.

The PMOS acts in a general way similar to a PNP bipolar transistor. The source can be compared to the emitter, the gate to the base, and the drain to the collector. However, the bipolar transistor is controlled by current while the MOS is controlled by voltage.

When no voltage is applied to the gate as in Figure 3-7a, holes cannot flow from the source through the N substrate to the drain. The N-region between the source and the drain blocks this flow. When a negative voltage is applied, as in Figure 3-7b, to the gate terminal, the field around the gate repels the electrons away from the gate. At the same time the holes are attracted to the gate. The insulating oxide prevents the hole from reaching the gate, so they distribute themselves right under the oxide, forming a bridge from source to drain through which working current can pass. This bridge is in effect a strip of P-region called the channel. The more negative the gate voltage, the thicker the channel becomes and the more holes flow.

3.8 THE NMOS TRANSISTOR

This acts in a similar way to a PMOS transistor and in a general way it is comparable to an NPN transistor. The source and drain are made of N-material on a substrate of P-material. A positive voltage on the gate repels the holes under the gate forming a channel through which the electrons flow. The more positive the gate, the thicker the bridge between the source and the drain. This is similar to the action in the NMOS except that electrons are used to carry the current instead of holes in the PMOS.

3.9 THE CMOS

The PMOS and NMOS can be connected back to back as in Figure 3-8 to form a CMOS or complementary metal oxide semiconductor. The gates are connected together. A negative voltage will turn on the PMOS section and completely cut off the NMOS section. The CMOS draws much less current than the NMOS or PMOS. The CMOS is used a great deal in integrated circuits.

Figure 3-9 is a photograph of an ATM (Asynchronous Transfer Mode) concentrator chip used in telecommunications. It has about 81,000 transistors with a die (chip) size of 6.6 x 6.6 millimeters (0.26 x 0.26 inches). The chip is implemented with 0.8-micron CMOS technology. It is packaged in a pin grid array carrier with 145 pins. The chip is an example of CMOS technology which is used extensively in the telecommunication industry.

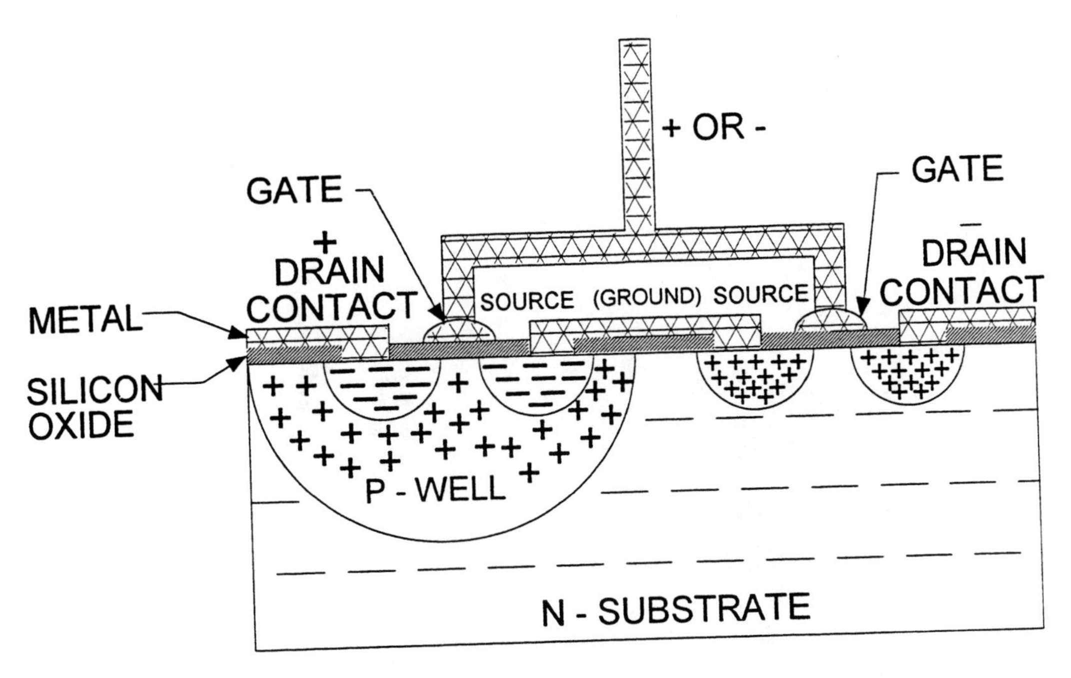

Note: There are two power supplies, one positive and one negative with respect to ground. Drawing is shown with no voltage applied to gates.

Figure 3-8 CMOS Configuration

3.10 Comparison of Bipolar and MOS Transistors

MOS transistors take up much less room on a chip than a bipolar transistor. The manufacturing process is simpler since the bipolar requires more diffusions than the MOS transistor. Circuits using MOS transistors are simpler than those using bipolar transistors.

A MOS transistor can act as a resistor so that additional resistors do not have to be added to the integrated circuits. A MOS transistor can be used as a resistor by making its channel region longer and narrower than usual, and connecting its gate to a constant low voltage supply so that it stays turned on. MOS transistors used in this way are hardly larger than normal MOS transistors. In bipolar integrated circuits, resistors must be extremely long and occupy much more room than a bipolar transistor.

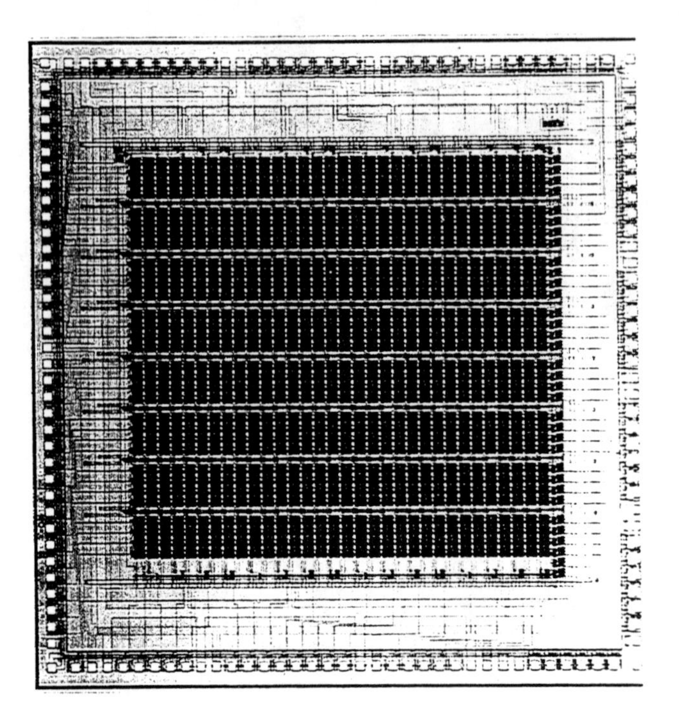

Figure 3-9 Photograph of Pattern of CMOS Chip
Courtesy of Professor Necdet Uzun, Polytechnic University

3.11 A Faster Transistor for Integrated Circuits

The bipolar transistor is faster than the MOS transistor. The MOS transistors discussed here are one division of a class called FETs or Field Effect Transistors. Another type of FET is the Gallium Arsenide Field Effect Transistor or GaAs FET. The gate in this type is composed of metal and gallium arsenide. A voltage is applied to the gate. The height of the conducting channel is decreased as the gate voltage is increased. As the conducting channel is decreased, the velocity of charge carriers under the gate increases (according to the laws of fluid mechanics). The velocity continues to increase with gate voltage until a speed of 300,000 feet per second is reached. The GaAs FETs can operate at extremely high frequencies (1000 billion Hertz) because of the high speed of the carrier.

3.12 INCREASING THE NUMBER OF TRANSISTORS ON A CHIP

Sophisticated new methods of design and manufacture are used to obtain chips with many millions of transistors. The more transistors on a chip, the more it can accomplish.

DESIGN

The design of chips has become increasingly dependent on CAE (computer-aided-engineering). CAE provides specific data management to support the design of very complex, high performance products.

ASIC (application-specific-integrated circuits) use short design times and rapid production cycles to produce chips used for one purpose such as the ATM concentrator chip for telecommunications in Figure 3-9. ASIC designs are usually array-based or utilize standard cells stored in a computer.

The array-based design involves rows of predesigned logic gates fabricated with routing channels between each row to allow interconnection of the gates to attain a higher-level function. The CAE system in a computer prescribes the interconnection of the gates. An extension of the row array-based design is the sea-of gates array. This design technology provides denser combinations. Homogeneous planes of logic gates can be interconnected by wiring over intervening gates.

Standard cell design is incorporated in a library of predesigned cells that have been utilized for a specific manufacturing technology. The standard cells are stored in computers. These computers automatically layout and interconnect cells and modules to achieve a complete design including functions, pin placement, etc.

Large volume products such as microprocessors and dynamic random access memories (DRAM) are custom designed to achieve the highest possible density and performance. These custom designs can involve many designers and a couple of years of effort. Special unconventional design techniques are used to achieve very high density. An example of this is the location of a transistor on the walls of the trench capacitor that is used in a memory chip to store data.

MANUFACTURING

Manufacturing technologies for integrated circuits have been changing every three to five years. One trend over the years has been a gradual increase in wafer size and also chip size. This was basically for economic reasons. The size of the wafer of silicon went up from a diameter of 2 inches in 1970 to 8 inches in the nineties. The wafer thickness remained a fraction of an inch. The memory chip sizes have been increasing to accommodate more and more circuits. In 1974 the memory chip size was 0.02 square inches. In 1984 the memory chip was 0.069 inches square and by 1989 the size was 0.12 inches square.

There are general fabrication stages for silicon integrated circuits but within each are different techniques. The manufacturing process starts with purified monocrystalline. This is sliced into wafers. Each wafer is divided into small areas or chips. Lithography is applied to define the small areas used in integrated circuits. Lithography is the use of electromagnetic radiation to expose the pattern of an integrated circuit onto a wafer of semiconductor material.

There are three general methods of lithography: optical, X-ray, and electron. Most chips are produced by optical projection using ultraviolet light. The surface of the slice is coated uniformly with a thin film of photosensitive material called photoresist. The "negative" of the required pattern is made in metal by an electron beam on a glass mask that is transparent to ultraviolet light. A collimated beam consists of parallel rays.

The image of the circuit on the mask is projected onto the wafer using a projection camera that works like a photographic enlarger but in reverse. Some cameras reduce the image while others project it at full size. The most common type of optical projection camera operates by a step-and-repeat method. The exposed area is a square or a rectangle the size of a single chip typically 10 to 20 square millimeters. A millimeter is 0.0397 inches. The wafer is moved in steps between exposure sites. Extremely complex lenses are used to project the finest lines in the same detail over the whole area of the chip.

The ultraviolet light hardens the photoresist in specific areas and makes those surfaces impervious to etchants or etching agents. Etchants can then be used to eat through unexposed areas of the chip making windows into the silicon dioxide. These windows can then be the areas for impurity doping.

Another lithograph technology is the use of a collimated beam of X-rays. This can produce integrated circuits with smaller features because the wavelengths of X-rays are smaller than ultraviolet light. One way to produce collimated X-ray beams is the synchrotron. A synchrotron is a device in which magnetic fields are used to accelerate electrons to close to the velocity of light. The synchrotron emits a collimated beam of X-rays. Synchrotrons are very expensive and research is underway in the United States, Japan, and Russia to produce collimated X-rays much more cheaply. Figure 3-10 shows a synchrotron used to produce collimated X-rays for chip research at the IBM facility in East Fishkill, NY, 1991.

X-ray lithography does not use lenses but shadow prints an image onto the wafer. The image is obtained by shining low energy, soft X-rays through a mask made from an extremely thin silicon or silicon carbide membrane. A metal pattern on the mask made by electron beams absorbs the X-rays. X-ray lithography not only produces sharper images but the exposure is not affected by small particles such as flakes of human skin. The flakes of human skin sometimes spoil the chips made by optical lithography.

A third lithography technology is electron-beam direct patterning without a mask by

Figure 3-10 IBM Synchrotron for Making Chips courtesy of IBM Corporation

using a controllable electron beam and an electron-sensitive resist. A computer controls the electron beam which can produce circuits with very small features. Electron beams write slowly and the necessary exposure is longer than optical or X-ray lithography. This disadvantage at present rules out scanning electron lithography for mass production. However, it can be used for custom chips with a very high density of transistors on a chip.

Etchants: integrated-circuit fabrication has traditionally employed wet chemical processes to etch lines and features. There has been research in other types of etching such as dry plasma etching, reactive ion etching and other methods to produce extremely fine lines.

Impurity Doping: impurities containing excess electrons are used to replace silicon atoms in a crystal lattice to produce N-material. Similarly impurities containing the absence of bonding electrons are used to replace silicon atoms to produce P-material. The absence of bonding electrons is known as holes. There are two general methods of impurity doping, thermal diffusion in high-temperature furnaces and ion implantation.

In the diffusion process the silicon is exposed to a concentration of the dopant while maintaining a high temperature. Boron gas, a P-type dopant, is pumped into the furnace and it strikes the exposed windows of the silicon slice. At high temperatures the silicon lattice contains missing silicon atoms. The boron atoms migrate to vacant lattice site producing a P area with holes. Similarly phosphorous gas is used to produce an N area.

Ion implantation is used when greater precision of dopant concentration is required or when a reduced temperature is advantageous. Ion implantation makes use of intense uniform beams of high-energy positive or negative ions of the desired dopant. The beam is made to impinge on the silicon substrate which has appropriate masking so that the dopant beam impinges on the proper area of the silicon surface.

3.13 Increasing Memory Storage

The technology of increasing circuits on a chip has been used a great deal in storing more and more information. Information is stored in the form of "zeroes" and "ones" or 0 and 1. Combinations of bits can be used for storing information. There has been a demand for random access memory or RAM chips that can store millions of bits of information. One bit of information is stored in a memory cell. There are two kinds of RAM chips. One kind can store information indefinitely while the other, dynamic RAM (DRAMs), require a charge replacement technique called "refresh" to make the storage permanent.

There has been a history of rapidly increasing bits on a RAM chip as the chip design and manufacturing technology have become more sophisticated. In 1970 the first RAM MOS IK-BIT chip was introduced with 1024 bits. This was followed by a 4K-Bit RAM chip in 1973, the 16K-bit chip in 1975, the 64K-bit in 1979, the 250K-bit in 1982 and the 1M-bit chip in 1985 and the 4M-bit in 1988 and the 16M-bit in 1991. The 1M bit has 1,048,576 bits. Most 1M bit chips and above are dynamic RAMs.

Higher densities of memory cells have led to lower costs per bit and have driven the chip technology revolution to place more and more transistors on a DRAM chip. In 1992 IBM and Siemens using advanced technology announced a DRAM chip with a 64 million-bit memory. Nearly a million of the chip's memory cells fit into an area the size of the head of a pin. A memory cell contains a minimum of one transistor per cell.

IBM, Siemens and Toshiba jointly produced a 256M-bit DRAM chip in 1995 and plans were made for a gigabit (billion) RAM. This answers the question, "How many transistors can dance on the head of a pin?" By the end of the 20th century millions of transistors can fit into the area of the head of an ordinary pin.

The increase of memory cells on a chip does not only decrease the cost per bit but allows the use of much more complicated programs which can be loaded into very small computers. This allows programs that recognize handwriting and eventually voice inputs. The high-density DRAM chips will change the application of computers in very important ways.

Smallest Feature Size

In 1980 the smallest feature size in a chip was 5 microns. A micron is a millionth of a meter; 180 microns is the diameter size of a human hair. By 1990 the smallest feature size was 1 micron. In 1994 the smallest feature was 0.5 micron and by 1997 it was 0.25 microns. As the size becomes smaller, second-order effects like leakage current, threshold voltage variations, hot charge carriers and channel punch through become a threat. New concepts by the end of the 20th century will probably result in still smaller feature size.

3.14 Some Applications of Chip

Since the mid-1960's the chip has become increasingly an internal part of the 20th century civilization. The increase in transistor density from less than a dozen to well over a million on a chip has transformed civilization to a degree that people do not realize. The chip has made possible reliable computers, personal computers, lap computers, and calculators. It also made possible digital watches, increased efficiency in automobiles, control of robots, and control of communications. The chip has made possible cellular telephones, satellite communications, digitized telephones, electronic mail and newspapers, home banking and many other new technologies.

THE CALCULATOR

One of the first uses of the chip was the hand-held calculator which varied from very simple ones that performed arithmetic to sophisticated units that handled scientific and engineering calculations.

The simple calculator made anyone capable of doing arithmetic with a speed that previously only people with special gifts could do. For example, any child can divide two numbers, each with eight digits, quickly and accurately with the calculator. In pre-calculator times only specially gifted people could perform this feat quickly in their heads.

The scientific calculator rapidly eliminated the slide rule that had been the mainstay of scientists and engineers for many years. In the early 1970's the slide rule quickly became a relic stored away in drawers and closets.

THE CHIP IN PERSONAL, LAP, AND NOTEBOOK COMPUTERS

The chip made possible computers for everyone. Before the chip, computers were very large and used primarily by large scientific and business organizations. The technology of chip manufacturing made possible first the personal computer and then the smaller lap computer and eventually the notebook computer that could recognize handwriting and fits into a pocket. As more and more circuits were packed into a chip, the small computers became more and more powerful.

A variation on the personal computer was the workstation which was tied in to a main computer. A chip in the workstation connected the operator into the capabilities of the main computer which could be located far away. Workstations have many uses such as in the stock market, airline reservations, computer aided design and manufacture, and many other uses.

THE CHIP AND THE DIGITAL WATCH

One use for the chip is in the digital watch using a quartz crystal. The crystal oscillates at a frequency of 32,768 pulses per second. Circuits in the chip halve the pulse rate 15 times to arrive at intervals of one second. An externally switched fast/slow capacitor in the crystal oscillator enables the user to set the readout in accordance with a standard time signal. Power may be supplied by a mercury or silver oxide battery which is replaced annually.

There are in general two kinds of readouts in the digital watch. One is the light-emitting diode or LED. An assembly of these on a monolithic chip illuminates appropriate bars of a seven-segment display for each digit of the readout. The LED is self-illuminating which requires a great deal of power. The second type of display, a liquid crystal display, LCD, depends on ambient illumination. Glass plates confine a thin layer of liquid crystal. On the inside surface of the front plate a transparent metal coating in a seven-segment pattern receives a signal from the readout counter. A highly reflective metal coating on the inside surface of the back plate operates at ground potential. When a high-frequency pulse energizes a segment, the electric field established through the liquid causes that region to become turbulent and thereby scatter incident light so that the segment appears diffusely illuminated against an illuminated background.

THE CHIP IN AUTOMOBILES

Sensors, actuators and a specialized computer on a chip make up an electronic engine control system used on almost all-automobile engines. This system controls engine and transmission functions. An example is the control of the air-fuel ratio in order to meet emission standards. An oxygen sensor in the exhaust sends information to the chip. Circuits in the chip decide how much change if any is needed to signal the actuators to take the appropriate action.

A memory chip in the car can store information about malfunctions that have occurred and perhaps momentarily disappeared. The information may be recalled by a repairman from the memory chip. A chip can display driver information in a variety of electronic displays instead of indicator lights or analog gages. A chip can also tune the radio by push buttons.

THE CHIP IN CAMERAS

The camera chip automatically sets the focus. An infrared beam is projected from a small window on the camera to the object whose picture is being taken. The infrared beam is reflected from this object back to the camera. The reflected beam is located by a rotating mirror behind a second window. The angle of the mirror and the distance between the mirror and the infrared emitter are fed into a chip which calculates the distance from the camera to the object by simple trigonometry. The chip then sends instructions to actuators which move specific elements within a complex lens to obtain the correct focal length. Alternatively the actuator may change the distance between the lens and the film to obtain the correct focus.

The chip also sets the exposure automatically. A sensor in the camera such as a cadmium sulfide or silicon photodiode determines the brightness or amount of light available and sends the information to the chip. Another input to the chip is the ISO film speed which is automatically read from the film by the camera. The circuits in the chip calculate the correct aperture and shutter speed for the proper exposure. The chip sends this information to actuators which control the lens aperture or opening and the shutter speed. If the amount of light available is not sufficient, the chip will automatically activate the flash circuit.

The chip also activates light signals on the camera to let the operator know what is going on in the automatic system. The chip has made the modern automatic camera of the late 20th century do all the work in focusing, setting the aperture and speed and automatically rewinding the film. This has simplified the taking of a picture to, "point and shoot."

Chip Control of Household Devices

In the home chips control the sequence of operations in washing machines making them automatic once the clothes and cleaning powder and softeners are loaded into the machine.

Chips are used in VCRs to set the time on and time off for recording TV programs.

Chips are used for remote control of TV receivers including station selection, volume adjustment and scanning what stations are available in the area.

Another use of the chip in the home is controlling the microwave oven including cooking time.

In some homes chips are used to control protective alarm systems.

Smart Cards

The smart card is a piece of plastic with a computer chip on its face. There are many uses from security and health care to retailing and transportation. In Europe they are used to unscramble satellite TV signals. As of 1992 there were 114 million smart cards in Europe but only 1 million in the United States. However, the use in the United States is rapidly increasing.

The present day credit card has a magnetic strip. This contains one line of type. The smart card is equipped with a silicon chip sometimes displayed at left center but sometimes embedded in the plastic. Smart cards may have a photo of the owner of the card. The smart card can hold three pages of typewritten data. That means several accounts could be loaded onto one smart card.

An example is that the same card could be used to check out library books, make purchases on credit and give emergency-room personnel a patient's blood type, insurance data and doctor's name. Each account would have a separate ID number so that only the appropriate information would be available.

The smart card provides a considerable amount of information to businesses about the customer who uses the card. A number of supermarkets are testing out the smart card.

Ohio is using the smart card to replace food stamps in order to reduce fraud. It also reduces the stigma of pulling out food stamps at the checkout counter.

In England in 1995 electronic money was introduced using a smart card called MONDEX. A customer buys a smart card and the store clerk inserts it in a machine when a purchase is made. The machine enters the purchase amount in the smart card where it is subtracted from the previous money amount.

The Chip that Eliminates TV Ghosts

These chips can remove annoying multiple images from the screen. It was the reduction in the chip costs that made it possible. The chips compare a ghost-canceling reference signal, transmitted with the video signal, with a stored unimpaired reference, and automatically reconfigure filters to compensate for distortion introduced in the transmission path.

A ghost-canceling reference (GCR) signal will be embedded in the vertical blanking interval of the transmitted National Television System Committee (NTSC) video signal and will undergo the same distortion as the video signal.

Television receivers equipped with special digital signal processor chips compares the received GCR signal with an unimpaired stored GCR waveform. The circuits in the receiver will then automatically configure compensatory filters to correct for transmission path distortion.

This technology was introduced in Japan in 1989. In the United States ghost-canceling circuits are scheduled sometime at the end of the 20th century. However, the advent of digital television may in the future make this chip unnecessary.

The V Chip

In 1996 the U.S. Government passed a law requiring a V chip to be installed in all new TV receivers. The V chip will allow parents to eliminate certain objectionable programs for children. This will be based on a rating system developed by the television industry. The law was passed before the V chip itself was designed. Congress and the President were confident that chip technology could be developed very quickly to do the job.

3.15 Controlling the Wild Stallion

Chip manufacturing and associated circuit boards produce a number of health problems including miscarriages among pregnant workers, ozone layer depletion, and the storage of hazardous chemical wastes. It is necessary to minimize these problems while still producing very inexpensive chips that fuel many of the new technologies of the late 20th century. A history of the different health hazards will now be described together with solutions tried in the past and also new technologies which may alleviate some of the problems.

Miscarriages

Digital Equipment Corp. (DEC) in Hudson, Massachusetts, received complaints in 1983 from employees that there were an unusual number of miscarriages. DEC hired scientists at the University of Massachusetts to investigate. They reported in 1985 that the women in the fabrication department suffered more miscarriages than women in other departments of the company. However, the number of individuals was small.

IBM in 1988 followed this up using Johns Hopkins University investigators to look into the problem. A report was issued in October 1992 which indicated that women working on mixing two ethylene glycol ethers, used in thinning photoresists, suffered twice the normal rate of miscarriages. IBM decided to purchase the chemicals already mixed instead of mixing them on site.

The Semiconductor Industry Association (SIA) hired scientists at the University of California in Davis to investigate the problem on a larger scale using thousands of women employees in many companies. The report in December 1992 also implicated ethylene glycol ether as the culprit.

The semiconductor industry is financing research into replacements for ethylene glycol ethers. This is a difficult problem because of economic and other considerations.

Ozone Layer Depletion

The largest use of CFCs (chloro-fluoride carbons) in the electronic industry was removing resin flux residue from printed-circuit boards after components like chips were soldered in place. Flux is applied after components are inserted in a board, but before their connections are soldered. Flux acts as a deoxidizer, ensuring better adhesion between the solder and the components. However, resin fluxes may eat away at the board and the components, causing metal changes resulting in short circuits. The flux must be removed and before the 1970s a chemical trichloroethylene (TCE) was used for that purpose. Around 1975 TCE was determined to be a potential human carcinogen. The electronic industry needed a substitute flux cleaner.

Around 1975 chemical companies sold Freon or CFCs to Motorola and other electronics companies as a safe replacement for TCE. The CFCs were economical and at the time were considered non-harmful. However, in 1985 a hole in the ozone layer was discovered over Antarctica. The ozone layer protected humans and other forms of life from harmful ultraviolet radiation. In 1987 representatives of 23 countries in Montreal agreed to cut CFC usage in half by 1999 and to require companies starting in 1987 to report the amount of chemicals being discharged into the atmosphere. In that year IBM Corporation's San Jose facility had the second largest CFC release in the United States with nearly 70,000 kg. At that time the electronic industry consumed only 20 percent of all CFCs used. Refrigeration, air conditioning, auto painting, and printing were among the users at the time. Those chemicals had previously been banned for use in aerosols.

As more data was collected about the depletion of the ozone layer, it was decided to move the CFC deadline from 1999 back to 1995. Also a U.S. law passed in 1990 required that all products manufactured with ozone-depleting chemicals be conspicuously labeled with a warning which stated that a substance was used which harms public health and the environment by destroying ozone in the upper atmosphere. This gave a very strong impetus to electronic companies to find substitutes for CFCs.

Two approaches have developed in eliminating the use of CFCs in cleaning residue from personal computer printed circuit boards. Both of these approaches eliminated the resin flux and employed different types of flux. One approach used an organic flux that could be cleaned with water. Ray Turner of Hughes Aircraft Company in 1990 discovered lemon juice as a flux that could be washed off with water. It was found that this flux had a number of advantages over resin flux. It was cheaper, more environmentally friendly and more active than the resin flux it replaced. Soldering became faster and exposure of components to heat is less so that system reliability is increased.

The other general method of handling the problem is to use a no-clean process. One example is a mixture of 98 percent alcohol and 2 percent synthetic solids to form a flux. This is blown into circuit boards with a modified sprayer. The resulting flux does not require cleaning. This is only one of a number of no-clean processes which are being used by different companies.

CHEMICAL HAZARDOUS WASTE

Heavy metals and toxic gases are part of the process of chip and circuit board manufacturing. Washing the chips and boards produces poisonous liquid waste. If discharged into sewers, it can end up in rivers and bays. The chemicals may endanger fish, people who eat fish, birds and mammals. If the hazardous waste is stored underground in tanks, there is a danger of the material leaking into wells and water supplies.

A San Jose, California newspaper in January 1982 in an article stated that an underground waste storage tank in the Silicon Valley had leaked into a nearby well. An ensuing investigation revealed that most of the underground waste storage tanks in the Silicon Valley were leaking. The companies involved had all followed the waste storage guidelines of the time.

Various techniques are being used to prevent hazardous materials from leaving the fabrication plant. These include recycling and/or the conversion of toxic materials to inert compounds.

The subject of hazardous material disposal is of course not limited to semiconductor manufacturing. At the end of the 20th century, the chemistry industry in the United States and many other industrialized nations are employing new technologies to prevent hazardous materials from poisoning the environment.

The storage of hazardous chemicals in tanks is only a temporary solution. Over a period of time, tanks can leak and always remain a time bomb in case of fires or earthquakes.

New techniques are being developed to handle hazardous wastes by treating them in a closed system which reduces the waste to its original elements.

Molten Metal Technology of Waltham, Massachusetts, has a patented resource recovery system called Catalytic Extraction Processing (CEP). CEP enables a wide variety of materials and chemical intermediates to be transformed into usable commercial products with minimum residual generation. The process involves retrograding chemical substances to synthesize new products. It can handle both organics and inorganics, in a solid, liquid or gaseous form.

In CEP, waste materials are injected into a multiphase reactor containing a liquid catalytic metal solvent. Upon injection, they revert to their elemental form (e.g., carbon, hydrogen, sulfur, nickel, etc.). Highly efficient reaction allows selective combinations to form new products. The introduction of special reactants into the system and the catalytic properties of the solvent assist the rearrangement of these atoms and allow them to be engineered into new commercial products. When there are no commercial markets for the products engineered from feed material, vitrification of the resulting residue ensures an environmentally benign byproduct.

A diagram showing a CEP unit is shown in Figure 3-11. The heart of the CEP is a molten metal bath operated near 3000° F. At these temperatures, CEP achieves unprecedented feed destruction. Compounds fed into the molten metal bath are disassociated into their constituent elements. CEP then accomplishes elemental recycling by rearranging these elements into usable products.

3.16 THE ADVENT OF THE CHIP SUPERFACTORY

Towards the end of the 20th century, huge factories were built to produce very sophisticated chips.

An example is the factory or “FAB” built by Intel in 1995 in Chandler, Arizona, near Phoenix. It cost 1.3 billion dollars and covers 1.5 million square feet. Hundreds of machines are used that were purchased for 1 million to 8 million dollars each. A wafer may go through 400 separate steps that take months to complete. Each wafer travels miles inside the factory, moving from one machine to another. A machine can be outdated in three years by advancements in technology.

Factories are producing chips with more and more layers. The new Intel factory manufactures chips with 20 layers to meet the demands of increasing sophistication.

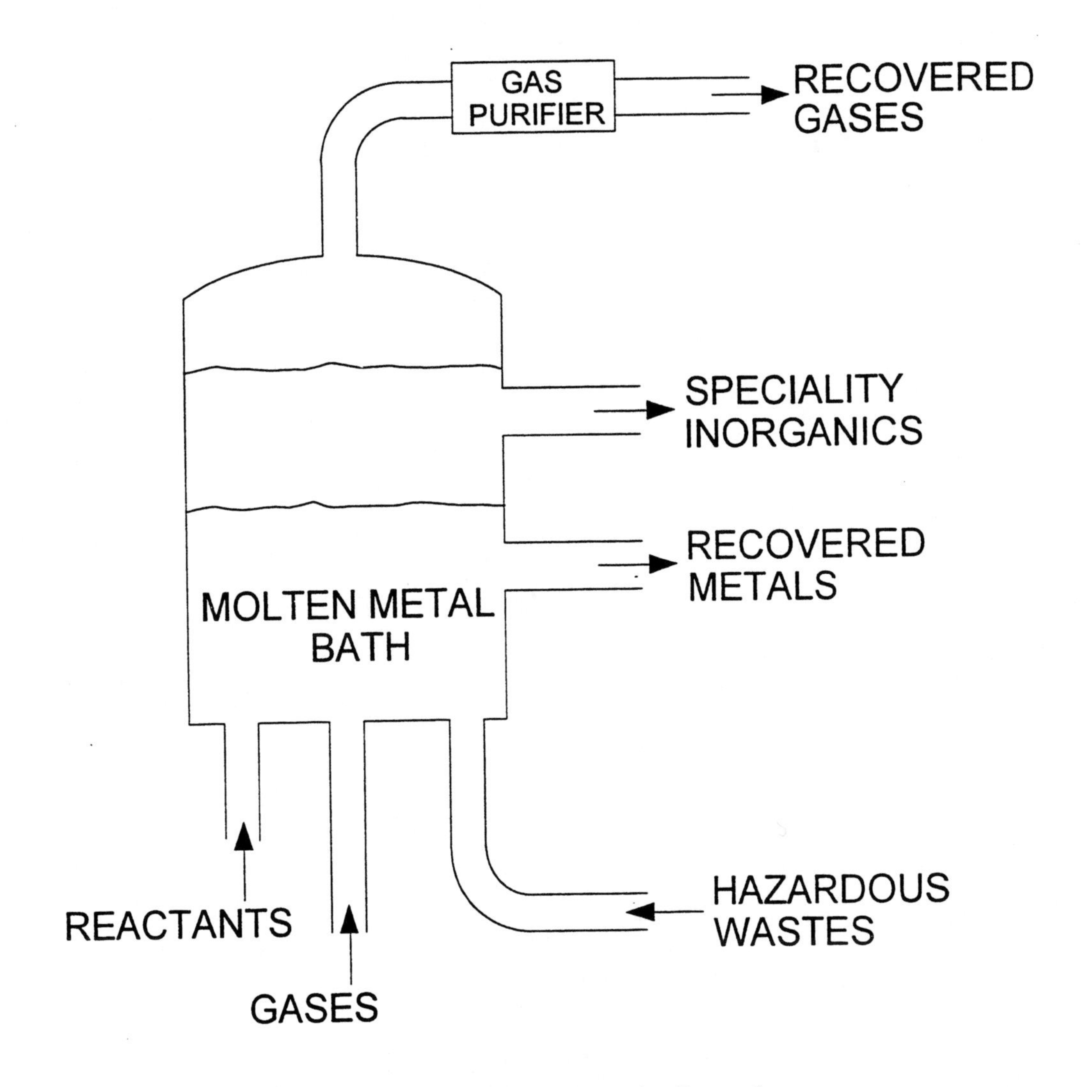

Figure 3-11 Catalytic Extraction Processing
(Courtesy of Molten Metal Technology)

Intel is not the only company building chip superfactories. Hyundai in 1995 announced that it was constructing a plant in Eugene, Oregon, that would employ 1,000 people. The cost would be about 1.3 billion dollars. Six other corporations announced that they also are considering setting up huge factories in the Portland, Oregon area where there is ample land and water. The large amounts of water needed emphasize the necessity of using different technologies of hazardous waste disposal to avoid the environmental problems of Silicon Valley. The best way is to ensure that toxic materials do not leave the plant in any significant way. The discharge of water into rivers or bays must be continuously monitored to avoid future disasters.

These factories are the most expensive industrial projects in the history of technology. The basis of this huge industry started in the early 20th century with an abstract form of physics, quantum mechanics. It reached a tangible form with the invention of the transistor in 1947. The integrated circuit in 1958 led rapidly to the chip which in turn made the personal computer possible.

For the 21st century a balance must be found so that we can keep the benefits of the chip technology and control the harmful effects on people and the environment. Only in this way can the transistors continue their dance on the head of a pin.

BIBLIOGRAPHY

Adler, R.B. Smith, A.C. and Longi, R.I. — *Introduction to Semiconductor Physics,* Vol. I, J. Wiley & Sons, 1964

Cadler, Nigel — *Einstein's Universe,* Viking, 1979

Hey, T. and Walters, P. — *The Quantum Universe,* Cambridge University Press, 1987

Hunter, Lloyd P. — *Handbook of Semiconductor Electronics,* 3rd Edition, McGraw Hill Book Co., 1970

Moore, Ruth — *Niels Bohr, The Man, His Science, and the World They Changed,* Alfred A. Knopf, 1966

Morris, Peter Robbin — *A History of the World Semiconductor Industry,* Peter Peregrinnus Ltd., London, 1990 (This is a part of the Institution for Electrical Engineers History of Technology)

Perkowitz, Sidney — *Strange Devices,* The Sciences, January, February 1995, New York Academy of Science, 2 East 63rd Street, New York, New York

Perry, Tekla S.	"Cleaning Up," *IEEE Spectrum*, February 1933
Tiwary, Gyanendra	"Below the Half-Micron Mark," *IEEE Spectrum*, 1994

CHAPTER 4

ADDING A BRAIN

4.1 THE ROOTS OF THE COMPUTER REVOLUTION

Charles Babbage (1792-1871), Figure 4-1, is considered by many to be one of the founding fathers of the computer age. He was educated in mathematics at Cambridge University in England. His interests were varied and covered many branches of science. These included cryptanalysis, probability, geophysics, astronomy, altimetry, ophthalmoscopy, statistical linguistics, metrology, actuarial science, lighthouse technology, and the use of tree rings as historic climatic records.

Babbage was an expert on calculation tables used by mathematicians, scientists, astronomers, navigators, engineers, and actuaries. In the late 18th and 19th centuries these published tables were mainly calculated by hand, copied, and then printed. The resulting tables were full of errors compounded by these separate steps.

Charles Babbage sought to build a steam-powered machine that would calculate and print out tables of numbers eliminating the human error problem. From 1821 to 1831 he worked on the design of a huge machine, "Difference Engine Number One," to compute and print out mathematical tables to twenty decimal numbers. The machine would have weighed several tons and contained thousands of parts. The entire machine was never built but in 1832 Babbage constructed a small section and successfully demonstrated it. The machine is now in the Science Museum in London and can still operate. It is regarded as the first automatic calculator.

During Babbage's lifetime a Swedish engineer named George Scheutz working from a magazine account of Babbage's project built a smaller version that was used for many years in the Dudley Observatory in Albany, New York. It printed out mathematical tables to eight decimal numbers.

Babbage around 1840 conceived the idea of the Analytical Engine which was designed as a programmable general purpose machine for finding the value of practically any algebraic equation. This computer is considered the forerunner of the modern computer although it was a mechanical and not an electronic device. Babbage's

Analytical Engine was programmable using punched cards that contained the instructions for the machine and made it a general-purpose machine. The Analytical Engine contained the equivalent of memory, central processor, and software of the modern computer.

Figure 4-1 Charles Babbage 1871
Charles Babbage Institute, University of Minnesota

The punched card technique employed was originally used to control the patterns woven with thread in looms.

Augusta Ada Byron, Lady Lovelace, Figure 4-2, an amateur mathematician, helped Babbage with his computer. She was born on December 10, 1815, the daughter of the famous British poet, Lord Byron. Shortly after the birth, Lady Byron accused her husband of incest with his half sister, Augusta. Lord Byron left England quietly.

Ada, as the daughter became known, studied mathematics, and married a man who became Lord Lovelace. Many years later the U.S. Department of Defense named a computer language Ada after the first computer programmer. The design of the Analytical Engine called for a huge machine the size of a locomotive and it was not

Figure 4-2 Augusta Ada Lovelace
Charles Babbage Institute, University of Minnesota
Crown Copyright

built in its entirety. Babbage's son, Henry P. Babbage, built a large hand-operated calculator based on his father's design. It was completed in 1910 and was used to calculate and print the value of Pi to 23 places. This machine exists to this day. Babbage's machines were not used but the idea of punched cards was taken up in the late 1800's.

Punched cards were used by Herman Hollerith in 1890 to solve a problem that was drowning the U.S. Census Bureau. The process of taking the required 10 year census had been handled manually with scores of clerks needed to count and classify information from millions of record sheets. In the late 1800's there was a huge influx of immigrants into the United States and this amount of information was swamping the clerks at the U.S. Bureau of Census.

Hollerith, an engineer from Columbia University, invented an electric tabulating system which revolutionized data processing. Data from the census was coded on punched cards using a specially designed cardpunch. Each of the categories of information was represented by specific holes on a small area of the card. Hollerith invented an electrical device to transform the holes into data information. A hand-operated press forced a large cluster of spring pins on the card. The pins would pass through holes that had been punched and dip into a mercury cup to close an electrical circuit. Counters connected to each of these circuits accumulated total number of individuals in the various categories. With this tabulator a census clerk could process more than 7,000 cards a day.

Hollerith on the basis of his contract with the U.S. Bureau of Census founded the Tabulating Machine Company in 1896. Punched cards became a basic part of data processing in a number of large industries. The Tabulating Machine Company combined later with three other companies to form IBM. In Britain the British Tabulating Machine Company was founded in 1907 and performed the same function as Hollerith's company. The British Tabulating Machine Company later became the ICL.

In 1936 Alan Turing wrote a paper proposing that a machine could in principle be built to compute all mathematical statements. He visualized a machine whose hardware would follow instructions. The instructions themselves and the numbers to be manipulated were to be supplied on a tape that would feed the machine. The tape could be changed for each problem that the machine was to solve. Turing showed that it was theoretically possible to build a computing machine which could automatically carry out any set of instructions which could be computed in finite form. The machine could do anything that a human computer could do. He also realized that numbers were only symbols similar to letters or musical notes. The chess pieces also followed rules of logic which could be manipulated by the machine. Turing gave a definition to a modern computer before it was built.

In 1945 Turing wrote a report, subtitled "Proposed Electronic Calculator" on the prospect of building a full scale electronic model of a Universal Turing Machine. This was essentially the electronic digital computer which would change the world in the last quarter of the 20th century.

Turing designed a new British digital computer, the Automatic Computing Engine (ACE). He believed that a computer someday could mirror the workings of the human mind. He thought that the computer would arrive when it passed a certain test. He envisioned a human being in a room communicating with a computer in an enclosed adjacent room. The computer had arrived when the human being could not tell whether there was a computer or a human in the adjacent room.

Turing never saw his machine completed but a smaller version was built during his lifetime. A full scale ACE was completed after his death and became operational in 1957. Turing himself was a homosexual who was arrested and sent to prison. He committed suicide in 1954 but his idea of a universal Turing machine that could solve not only numerical but logic problems of a general type lived on.

Other engineers and scientists in different countries took up the idea of a computer. Konrad Zuse (1910-1995) was working in 1936 on an automatic computing machine. He was a German civil engineering graduate who hated solving tedious and difficult equations. Zuse was the first one to use the binary system in computers. Up to that time the decimal system had been used in calculators. Humans naturally count in tens since we have ten fingers. However, Zuse found that electrical computers could more readily use the binary system of counting. An electrical switch has two states, on and off. The "on" state can represent "1" and the "off" state "0" thus enabling the switch to operate in a binary system. Zuse used electromechanical relays which were switches operated by an electric current.

Zuse combined the electromechanical switches into logic gates which modified and combined binary digits (O's and 1's). A combination of logic gates can carry out any task in arithmetic. Two binary numbers passing through one particular set of logic gates will be added. Other combinations of logic gates performed different arithmetic functions.

Zuse's Z-3 machine, demonstrated in 1941, used electromechanical relays and was the first fully operational program-controlled calculator. It had 2,600 relays and used discarded movie film with punched holes to program the machine. The machine was slow compared to modern computers. It took about a third of a second for an addition, and three to five seconds to multiply two numbers.

A friend of Zuse's, Helmut Schreyer, proposed using electronic valves, or vacuum tubes to replace electromechanical switches. Zuse calculated that this would enable his computer to operate 1,000 times faster. Zuse and Schreyer decided that it would take two years to develop. They asked the wartime German government to give them the support they needed. However, Hitler turned them down because he thought Germany would win the war before the two years were up. In 1942, Zuse began work on the Z-4, an improved version of the Z-3. Figure 4-3 shows Zuse and the Z-4 computer tape reader in 1944.

Specific World War II problems in both England and the United States forced the rapid development of electronic computers using vacuum tubes. In England the problem was related to breaking the codes of German military communications. The German messages were encoded by machines such as the Enigma and the Lorenz. Decoding keys

Figure 4-3 Konrad Zuse with Z4 Computer Tape Reader, 1944-1945
Charles Babbage Institute, University of Minnesota

relating the German plain language letters and the encoded transmitted letters were required to break the code. The keys were periodically changed. British crypto experts needed a method to find the key quickly before events made the encoded messages valueless. It was decided to build an electronic machine using 1,000 vacuum tubes to find the encoding key. The machine was completed in December 1943. It could read 5,000 characters on a tape in one second. The name of the computer was the Colossus and it handled letters rather than numbers

In the United States an electronic computer, the ENIAC (Electronic Numerical Integrator Analyzer and Computer) was developed to solve a numerical ballistic problem. Artillery pieces are aimed by setting the angle of the barrel so that the shell travels the desired distance to the target. The trajectory of the shell depends on the weight of the shell, the propellant charge, and air resistance. This last varies with temperature, humidity, altitude and the shape of the projectile. The trajectory equations were too complex to be solved by gunners in the field.

The Ballistic Research Laboratory in Maryland was responsible for tabulating fire tables that gave the correct settings for various conditions. They tried using the available analog mechanical computers which took three months to complete one firing table. The Ballistic Research Laboratory fell hopelessly behind in its calculations. In order to catch up, the laboratory assigned Lieutenant Goldstine to establish a computing substation at the Moore School of Electrical Engineering at the University of Pennsylvania in Philadelphia. In 1942 Professor John Mauchly of that school had written about the use of vacuum tubes for high-speed calculations. Lieutenant Goldstine met Professor Mauchly and they discussed the possibility of a vacuum tube computer to calculate firing tables quickly and accurately.

On June 5, 1943 the Ballistic Research Laboratory commissioned the development of a vacuum tube computer which became the ENIAC. Mauchly became the chief consultant and J. Presper Eckert, Figure 4-4, a colleague of Mauchly, was the chief engineer. Lieutenant Goldstine was the supervisory consultant with the Ballistic Research Laboratory.

ENIAC was completed in November 1945, after the war was over. It was 100 feet long, 8 feet high, 3 feet deep and weighed thirty tons. It contained around 18,000 vacuum tubes and cost over $500,000. It could perform 5,000 additions per second. ENIAC operated up to twenty hours reliably without any failures. Although it was completed too late for World War II, it performed calculations for the H-bomb and other projects.

The major problem with ENIAC was programming which had to be set up for each problem by plugging in cables. In effect each new problem required a rewiring of ENIAC. Problems requiring two minutes to solve took days to program. The bottleneck was no longer the time of computing but the time it took to program by setting switches and wiring plug boards.

Eckert and Mauchly conceived the idea of storing programs internally. The stored-program computer was called EDVAC (Electronic Discrete Variable Computer) and was put out on contract by the Ordinance Department in the autumn of 1944. John von Neumann, Figure 4-5, joined the team and in June 1945 wrote a paper called "First Draft of a Report on the EDVAC." This was the first formal description of the design of a general-purpose digital electronic computer.

The computer architecture that von Neumann proposed consisted of several basic components: A Memory Unit for storing both data and program instructions, an Arithmetic Unit for performing numerical calculations, a Control or Processing Unit that directs the movement of information and the sequence of actions. In addition he called for input and output sections for feeding information into the system and a way of recording results. He

Figure 4-4 J. Mauchly (Left) and J.P. Eckert
Charles Babbage Institute, University of Minnesota

believed that the system should operate serially to simplify the construction of the computer. The most important problem in making this type of computer practical was the memory. There was no practical existing method at the time of storing vast amounts of programming instructions. Maurice Wilkes of the University Mathematical Laboratory at Cambridge in England decided to investigate the possibility of using a mercury delay line memory. This technology using sound waves circulating in mercury delay lines was used

Figure 4-5 John von Neumann
Charles Babbage Institute, University of Minnesota

in radar in World War II. Electrical pulses were converted to sound waves and sent down a column of mercury. Sound travels much slower than electrical signals and this provides the delay. The sound pulses are stored and kept circulating by detecting them at the receiving end and feeding them back to the input.

The Cambridge University Electronic Delay Storage Automatic Calculator (EDSAC) was developed by Wilkes and is considered to be the first useful computer using the stored-program. This computer was used mostly for numerical computations at Cambridge University from 1950 to 1958.

4.2 THE COMMERCIAL COMPUTER MAIN FRAME

J. Presper Eckert and John Mauchly, the inventors of ENIAC, started the first electronic computer business in the world, the Electronic Control Company in Philadelphia in 1946. Later it was renamed the Eckert-Mauchly Computer Corporation. They had left the University of Pennsylvania in March 1946 over a patent rights dispute. Around that time the Census Bureau of the United States was having difficulty analyzing the results of the census. The punched card system developed by Hollerith had reached its limit by 1946. Congress had mandated that the Bureau of Census had to finish the census in ten years. The Census Bureau needed a new machine and they contracted with Eckert and Mauchly to build the Universal Automatic Computer (UNIVAC). Eckert and Mauchly pioneered magnetic tape for data storage in the UNIVAC. The two scientists ran into financial problems and on February 1, 1950 the Eckert-Mauchly Computer Company became a subsidiary of Remington Rand. Remington Rand was at that time a manufacturer of business machines, typewriters, and punched card tabulating machines. One year later the first UNIVAC was delivered to the Census Bureau.

IBM which had been selling Hollerith Tabulating Machines to the Census Bureau started to lose a great deal of business. Thomas Watson, Sr. had been the leader of that Corporation since 1914 and he was against IBM making business computers. However, his son Tom Watson, Jr. realized that the future of IBM lay in exactly that direction. His opportunity came with the Korean War when he was placed in charge of IBM's defense activities. The Defense Department told Tom Watson, Jr. that they needed computers. Watson, Sr. reluctantly allowed the construction of the Defense Calculator which started operations in April 1953. IBM quickly made a commercial version of that computer and IBM was in the computer business guided by Tom Watson, Jr.

In 1954, IBM built the first moderately priced business computer, the 650. It used existing punch card installations that IBM customers were used to. IBM rapidly became the world's largest computer manufacturer. Thomas Watson, Sr. died in 1956 and his son led the company into the computer age. He adopted innovations like magnetic tape for storing data which had been introduced by the UNIVAC. The manufacturer of UNIVAC, Remington Rand merged with Sperry Company to form another computer giant in the United States.

In the meanwhile, England had gone into the commercial computer business. It started with the needs of a wholesale food and catering business which operated 150 teashops and a number of hotels and restaurants. This was the J. Lyons & Company, Ltd., a forward looking business outfit that sought to improve its operations. They hired two Cambridge mathematics graduates, Raymond Thompson and John Simmons, to make the office operations more efficient. Thompson and Simmons began work on LEO, the Lyons Electronic Office, in 1949. In 1953 the LEO's came off the production line. Lyons used the LEO to compute the optimum mix for their various brands of tea, to process the payroll and do tax rolls. Thus England came into the commercial computer age.

THE PROGRAMMING PROBLEM

One of the basic problems in the commercial use of computers was writing programs (or software) to tell the computers how to carry out particular tasks. Computers operate with a binary system with only two states like on-off, "1" or "0," high or low. This simple code can be used to express decimal numbers, letters of the alphabet, musical notes, logical and arithmetic commands. The 0's and 1's of binary code can be handled by large numbers of electronic switches. The problem comes in translating English commands into the binary code or machine language of the computer. Programs consist of thousands of instructions which must be translated into as many as millions of lines of the binary codes.

In the early 1950's programmers developed Mnemonic or Assembly Language Program. A mnemonic is an abbreviation of what the instruction does. For example, AD for add, SU for subtract, LD for load operations, ST for the store operation, MOV for the move operation. These mnemonics or assembly language is fed into an assembler in the computer, which converts the mnemonics into machine language. An example of assembly language is as follows:

INSTRUCTION IN ENGLISH	ASSEMBLY LANGUAGE	MACHINE LANGUAGE
Move contents of Register C to Register D	MOV C, D	0100 0111

A computer has an instruction set in which the instructions are stated in assembly language format, which the assembler will convert into machine language.

The assembler made life easier for programmers but there was still the problem of making errors in thousands of lines of code. Programmers spent hours trying to debug the program. A partial solution was the development of computer languages closer to human languages. One of the most popular of these was *BASIC* (Beginner's All-Purpose Symbolic Instruction Code) invented by John Kemeny and Thomas Kurtz at Dartmouth in 1964. Many other languages were developed like COBOL (Common Business-

Oriented Language) for business, and FORTRAN (Formula Translating System) for scientific work. Later C language was developed for use with the AT&T UNIX computer.

High-level languages like COBOL and FORTRAN needed a mechanism to convert them to machine language. A piece of software called a compiler took the higher language program and compiled it into the machine language which the computer could execute. Each higher language had its own compiler. Once the particular compiler program was written, from then on a programmer writing in that higher language did not have to know how the computer worked.

IBM decided to make large main frame computers an integral part of every large business. They invested five billion dollars to develop a family of six computers called the System/360 which could all use the same software and peripheral equipment. The development was announced by IBM on April 7, 1964 and by the end of the 1970's IBM 360's were found in libraries, banks, universities and hospitals all over the world. Soon, however, IBM was challenged by a new development, the microcomputer.

4.3 THE MICROCOMPUTER

The integrated circuit described in Chapter 3 was invented in 1958 by Jack Kilby at Texas Instruments and Robert Noyce of Intel. Both companies used special purpose integrated circuits for calculators. Other special purpose integrated circuits were used for purposes like missile guidance systems. Ted Hoff, a former Stanford physics professor, and then an employee at Intel came up with the idea of a general microcomputer using an integrated circuit as a microprocessor and other integrated circuits as memory and input and output circuits. This microcomputer could function as a general-purpose computer as powerful as some main frame computer but much smaller and cheaper.

Intel came out with the first commercial microprocessor in 1971, the INTEL 4004. It handled four bits in parallel at a time. A bit (binary digit) is either a "1" or a "0." In 1974 Motorola produced the Motorola 6800, an eight bit microprocessor. Over the next few years new microprocessors were developed using 16 bits, then 32, and in the 1990's 64 bits. As the number of bits in parallel was increased, the microprocessor became more powerful. The microprocessors made possible small microcomputers such as the IBM Personal Computer which came out in 1981. By the 1990s, the components were as shown in Figure 4-6. Soon the (B) drive was eliminated.

The basic operation of the microcomputer is demonstrated in Figure 4-7. The heart of the system, the microprocessor, has two main functions. It controls the timing and sequence of operations in the computer and it executes the arithmetic and logic operations of the data being processed. It is also known as the Central Processing Unit (CPU).

The control and timing logic part of the microprocessor is required because all of the different functions of the microcomputer must be timed or synchronized to work together. To assure accurate transfer of data and proper operations, signals must change at special points in time. This is controlled by a clock signal. Operations throughout the components of the computer change only when a clock signal appears. In this manner the timing of all parts of the system is synchronized by the clock signal. Many of the microprocessors have an on-chip oscillator, which needs an external quartz crystal for timing of the clock signal. Also there are different cycles that the computer goes through, such as instruction fetch cycle and an execution cycle. These cycles are controlled from the control and timing logic. By the end of the 1990's many personal computers had a frequency well above 100 MHz.

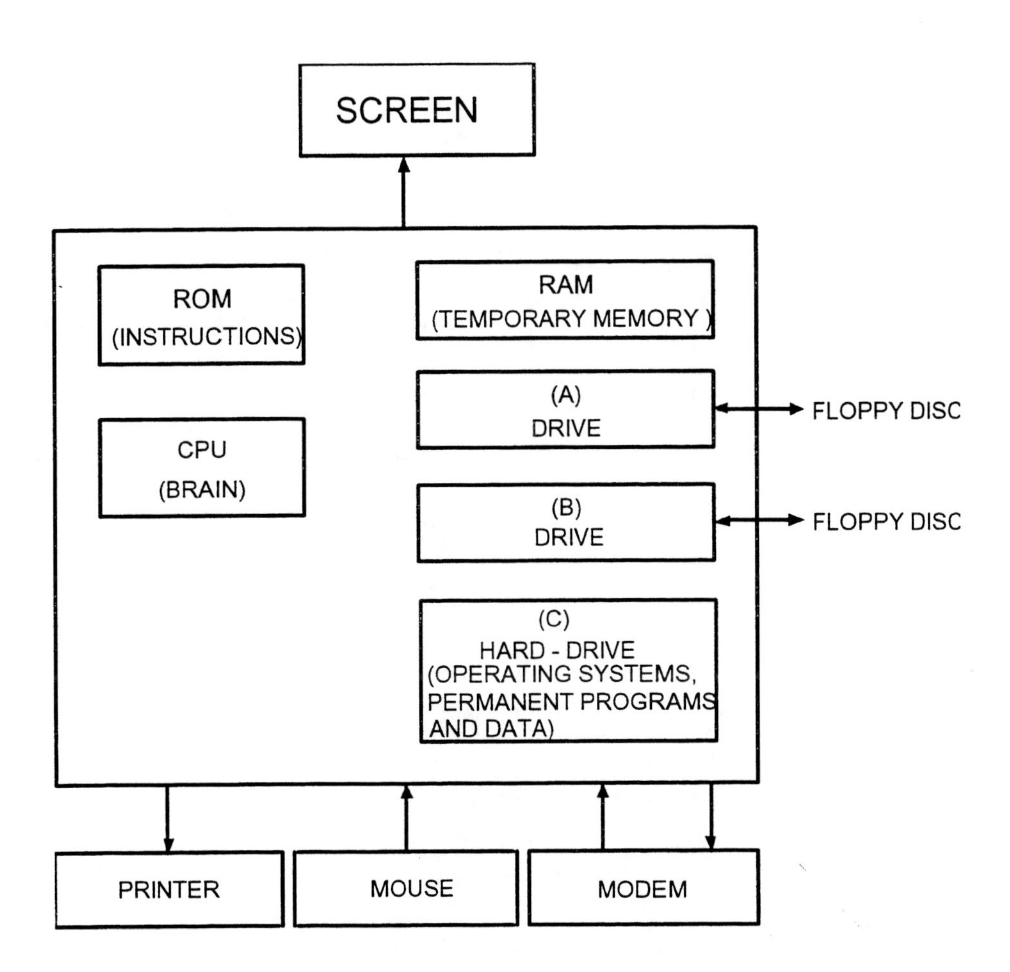

Figure 4-6 Components of a Microcomputer

Another part of the control unit of the microprocessor in Figure 4-7 is the program counter. The purpose of the program counter is to give the memory address of the next instruction. At the start of a program, the program counter will be loaded with the memory

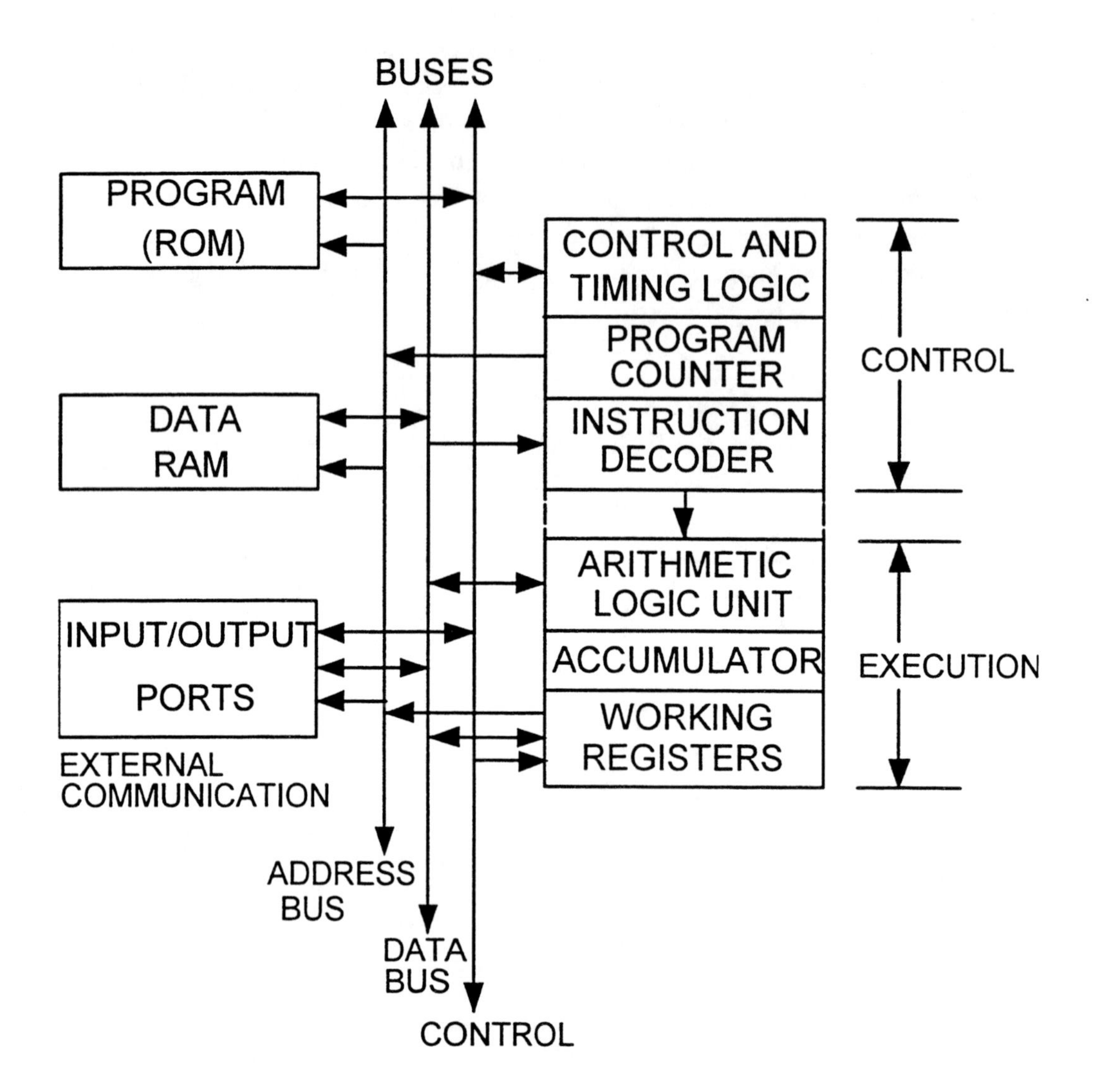

Figure 4-7 Architecture of a Microcomputer

address of the first instruction. When the system starts running, the address will be transferred to the address bus and the instruction will be read from memory into the instruction decoder. Part of the instruction is called opcode (operation code) and specifies the operation to be performed. Typical operations would be subtract, divide, add, and so on. Each of these will have a different opcode. The instruction decoder then sets up all of the logic linkages required to perform the desired operation. The actual performance of the operations is accomplished in the ALU (arithmetic logic unit). A typical instruction execution consists of one or more instruction cycles. The first cycle, the instruction fetch cycle, calls in the opcode from memory to the instruction decoder. In simple instructions this is followed by an execution cycle when the required operation is carried out. If data are required, extra cycles are needed to read in the data words following the opcode.

Microprocessor execution involves the ALU, the accumulator, and working registers. The ALU, the arithmetic logic unit, carries out the desired arithmetic or logic functions. Typical functions provided by the ALU are add, subtract, and, or, exclusive or, complement, and clear. The accumulator is a special register which stores the results of the ALU operation. The working registers are used to store data or intermediate results.

The microcomputer memory is basically divided into ROM, a read-only memory containing instructions permanently stored when the unit was made and RAM, a random-access memory where words may be "written" (stored) or "read" (recovered) in any order at random. Figure 4-8 shows the basic operation of a ROM. The ROM stores a bit by the presence or absence of a connecting link between a row line and a column line in the memory array. The two address bits, A_0 and A_1, can by means of the decoder select one of four rows to be read out. If the A_0 and A_1 address lines both have logical 0, the decoder will produce row 1 as an output. In this example, row 1 is 1000, which will appear at the output of the ROM. Similarly, an address input of 0 for A_0 and 1 for A_1 will produce an output corresponding to row 2, or 1100 in this particular ROM. The open or closed link pattern in the ROM is produced by a mask during the manufacture of the chip.

There are two basic types of random access memories, static and dynamic. A static memory stores the information in flip-flop or latch type structures and maintains the information without requiring refreshing or restoration of data. A latched register is a type of temporary storage for digital information where the information will be released upon receipt of an electronic or digital signal. Most RAM chips lose their data when power is removed. However, in the 1990's flash chips were developed which hold their data when power is removed. A dynamic memory uses a storage medium, such as a capacitor, which loses its information over a period of time and must be refreshed at intervals to ensure the retention of data.

The transfer of data, control signals, and addresses among the various units of the microcomputer takes place over sets of parallel wires known as buses. Most microcomputers have at least three buses: one for control signals, one for data, and one for addresses. The control bus controls timing and transfer of data on the other bus lines. A read-write line in the control bus determines the direction of the data bus.

The data bus performs the function of transferring data to and from the microprocessor and the memory or input/output lines. The data bus is bi-directional and is controlled by the microprocessor. A read-write line in the control bus determines the direction of data in the data bus. If the control line is set to write by the microprocessor, this allows data to be sent by the microprocessor via the data bus to other parts of the system. When the control line is set to read, the microprocessor receives data from the data bus.

The third general type of bus is the address bus which provides a signal for the memory to select one particular location of address within the memory for connection to the data bus.

Digital data in and out of the microcomputer are transferred by way of data ports. The data port consists of latched registers connected to the data bus and selectable by the microprocessor for data transfers. Each input or output port has a pair of handshake or control lines. One handshake line goes from the computer to the external device which indicates that it is ready to transfer data via the port. The second line is an input to the computer system from the external device and may be used to indicate either that the external device is ready to accept data or that it has placed data on the port lines for the computer to read.

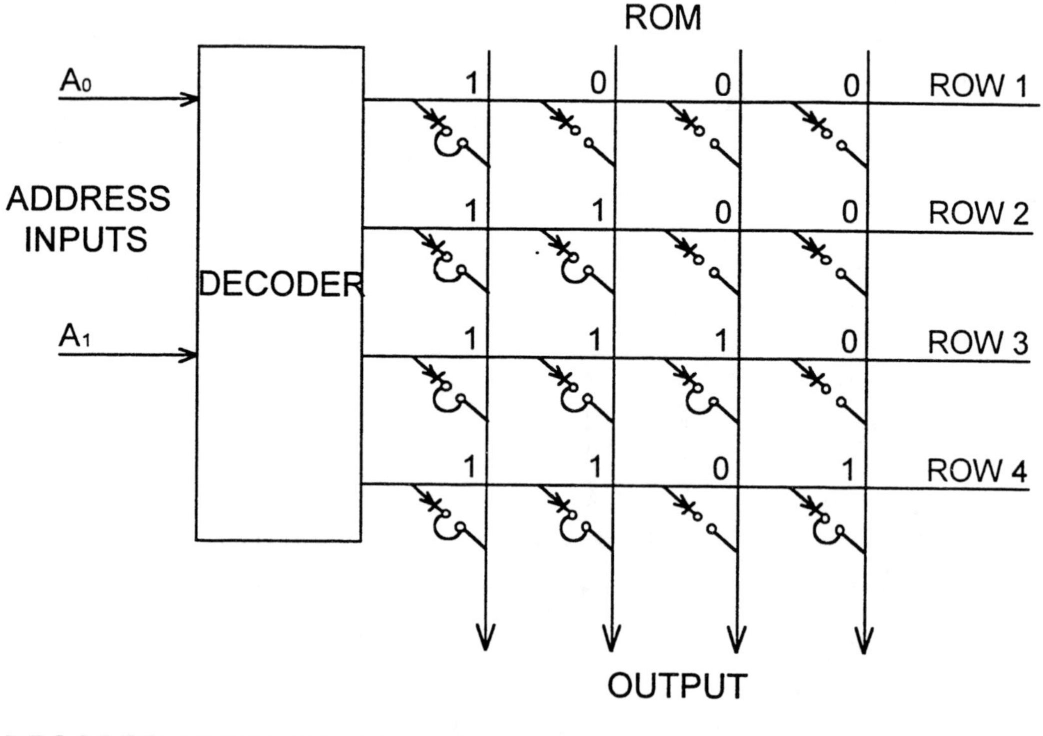

DECODER ADDRESS A1	A0	
0	0	ENABLES ROW 1 FOR OUTPUT
0	1	ENABLES ROW 2 FOR OUTPUT
1	0	ENABLES ROW 3 FOR OUTPUT
1	1	ENABLES ROW 4 FOR OUTPUT

Figure 4-8 Basic Operation of a ROM

The computer peripherals such as modems, operate with digits in serial form while the computer itself operate with digits in parallel. The parallel output of the computer must be converted by an interface card to obtain a series form. Similarly the serial input from peripherals is converted to a parallel form by an interface card in the computer.

MODEMS

Modems (Modulator-Demodulator) are used with the input output lines to transfer the input and output through telephone lines or radio to other computers or databases. Modems are used for changing the digital signals of the output lines to audio tones for transmission on radio or telephone circuits. Modems also change the audio tones back to digital signals for the input lines. The parallel digital signal output of the microcomputer is changed to serial form before transmission through the modem. By the end of the 1990's many modems used 33.6 kilobits per second or higher.

PRINTERS

Another peripheral component of a computer system is the printer which in rare cases requires a series input from the computer. Most printers use a parallel input.

The first type of printer for personal computers was the dot-matrix printer. This type of printer creates characters on paper with a printhead consisting of minute wires that strike a ribbon. Different dot patterns of the wires form different letters. The quality of the printing is less than some other printers but they are inexpensive.

A laser printer uses a laser beam to create characters on paper. Laser printers produce very high quality output that looks like results of a typesetting system. However, they are expensive in the range of $1,000 or more.

Another kind of computer printer is the ink-jet type. Ink-jet printers work by spraying ink on the paper to form characters and images. The printing quality is almost as sharp as laser printers but the price is considerably less.

4.4 THE SELLING OF THE PERSONAL COMPUTER

At first people did not understand the importance and use of Ted Hoff's invention of the microprocessor in a general-purpose inexpensive small computer. A tiny calculator company, MITS in Albuquerque, New Mexico, ran into a debilitating price war. The owner, Ed Roberts, in desperation in 1974 decided to build a small inexpensive computer using the new 8 bit Intel 8080 microprocessors. He designed a small computer called the Altair which he sold as a set of parts. It had no keyboard and no screen. Programming was accomplished by turning on small switches on its front panel. The Altair sold for $500 and less and soon made Ed Roberts a wealthy man.

The Altair incited a group of computer hobbyists in San Francisco in March 1975 to form the Homebrew Computer Club. Many of these young people started to build their own computers. One of those was Stephen Wozniak, one of the future founders of Apple Co. The members of the Homebrew Computer Club saw that the personal computer was one of the great technological revolutions in the history of humankind.

Inspired by the Homebrew Computer Club, Wozniak built a computer on a circuit board and called it Apple I. A friend of his, Steve Jobs, joined him with suggestions on how this computer could be improved and then marketed. Wozniak and Jobs worked out of the garage of Jobs' parents. They contacted a former engineer from Intel, Mike Markkula to help them. Markkula had retired from Intel, a millionaire at 32. Markkula went to Jobs' garage and looked at the latest version of the efforts of the two computer enthusiasts, Apple II. Markkula decided to help Jobs and Wozniak form a company to manufacture and market a computer. He put up $90,000 and had the two computer enthusiasts move their manufacturing operation from the garage to a warehouse in Silicon Valley. This was the start of the Apple Co., a model for all computer entrepreneurs.

Wozniak and Jobs worked very hard to make the Apple II prototype into a computer that everybody could use, not just hobbyists. They realized that they needed a great deal of advertising to get their message across. Markkula went to see Regis McKenna, owner of a large advertising company, and agreed to pay all of the bills for the first three months. Markkula also hired engineers and managers from Silicon Valley to convert the company from hobbyists to a modern corporation.

A number of other companies such as Radio Shack and Commodore and others also began to manufacture personal computers. Wozniak and Jobs realized that they had to come out with something new that would set them apart from their competitors. At the time, programs were stored on ordinary audiocassette tapes which, while cheap, were extremely slow. Disc drives existed which were a hundred times faster than cassette tape but were very costly and unreliable.

Wozniak, at the urging of Markkula, designed a much cheaper and more reliable disc drive. The new disc drive required one tenth of the chips that the conventional design used. Apple now had a good competitive position in the burgeoning computer market. Apple also had a larger internal memory than other computers. This meant that it could run longer programs more quickly than its competitors. However, there was a lack of available programs to run and this was a limiting factor in the computer business.

In 1979 two Harvard Business School students wrote a financial spreadsheet called VisiCalc which was soon in great demand. VisiCalc met a great need in businesses and Apple sales soared. Other software programmers soon expanded the use of the Apple Computer into video games, educational programs, word processors, income tax formats, music, art, and many other fields. The availability of different types of software made Apple and other personal computers very versatile.

IBM soon joined the parade in 1981 when they introduced the IBM PC. Clones or cheaper copies of the IBM PC soon flooded the market. By 1983 a number of personal computer companies went broke and IBM and Apple were engaged in a fierce struggle. Apple was saved by a new development in Palo Alto, California. A new type of computer had been developed there by a research and development laboratory financed by Xerox.

Xerox was a company that had made a fortune photocopying paper. They were afraid of a paperless future where computers stored and transferred information electronically. Xerox decided to enter the computer age even though they had no experience in that field. Xerox founded the Palo Alto Research Center or PARC and hired a group of computer scientists. This became known as Xerox PARC and a great deal of computer research was accomplished there. In 1973 Xerox PARC developed a new type of computer, the Alto, which had a hand-held mouse. Icons such as a garbage can could be placed on the screen and deletion was then accomplished by moving the mouse until the pointer reached the garbage can. Other computer functions were accomplished by using different icons. This system replaced special commands which had to be typed in on conventional computers. The computer scientists at Xerox PARC also developed laser printing and other modern computer technology. However, Xerox did not market the Alto or other innovations developed at Xerox PARC. Xerox decided to concentrate on their copying business.

Xerox asked Apple to take over the manufacture of the Alto computer and they invited Steve Jobs to see that computer. Steve Jobs was impressed by the Alto but he decided to design a new type of computer using the principles developed at Xerox PARC. His idea was to develop a computer that would prevent IBM from dominating the personal computer market. On January 23, 1984 the Macintosh was unveiled with a great deal of fanfare. IBM now had a very serious competitor for the most popular personal computer.

4.5 Disc Drives and Discs

Disc drives are devices that read and transmit information from and into a disc. The drives include both chips with their electronic circuits and mechanical devices to read from and write on discs.

Floppy Discs

Floppy discs are square discs that are inserted into personal computers. Data and programs are stored magnetically. Older personal computers used 5.25-inch square floppy discs which stored between 350 K and 0.7 MB. (B stands for byte, eight bits, one character. MB stands for megabytes or million bytes.) In the 1990's all newer personal computers used 3.5-inch square floppy discs with a storage density of 1.44 MB.

HARD DRIVE DISCS

Hard discs have much larger magnetic storage capacity than floppy discs. Some internal hard drive discs have the capacity of 1,000 MB or more. Operating systems like Windows 95 use about 60 MB so that it is necessary to use a hard drive disc with a large capacity for additional storage. Hard discs can store a large number of files.

CD-ROMS

CD-ROM players were installed internally in personal computers in the 1990s. CD-ROM's have very large capacities and can store text, images, audio and video information. CD-ROM's include encyclopedias, atlases, travel guides, art and photo collections. Unlike the floppy and hard discs which store digits magnetically, the CD-ROM depends on lasers to store and read the digits. The information is etched into the CD-ROM surface as a series of pits. In the computer the disc spins while a laser reflects off it, reading the pits as O^{s} and the intervening planes as 1^{s}.

4.6 SOFTWARE

Software is a set of computer programs, operating systems, and procedures. Software is distinct from hardware which is the physical parts of a computer system including circuit components, disc drives, etc.

Bill Gates was one of the pioneers in the development of software. He created much of the software for the original MITS computer, the Altair. He went on to develop a whole series of programs especially for business operations. Bill Gates founded Microsoft Corp., the world's largest personal computer software company. Microsoft makes money from each machine that uses its software including personal computer operating systems. Bill Gates in 1992 became the richest American according to Forbes. He was then only 36 years old.

OPERATING SYSTEMS

An operating system of a computer provides the environment in which programs perform. It is an internal integral part of the microcomputer. A program properly developed for a standard operating system can be used on any computer using that operating system with a negligible amount of effort.

The CP/M (Control Program for Microcomputers) was the first standard operating system for microcomputers. Its principal advantage was that different types of microcomputers could use it. The CP/M was in wide use in the early days of the microcomputer.

One of the most popular operating systems was the MS-DOS (Microsoft Disc Operating System). In the 1980's most of IBM's personal computers and clones used this system.

UNIX is AT&T's standard operating system which is a much more flexible system than either CP/M or MS-DOS. It has advanced features such as two or more users sharing the same hardware simultaneously, and multitasking, where one user runs two or more programs at the same time. Versions of UNIX became available for a number of personal computers.

Other operating systems used in computers are Windows and several Macintosh systems.

Programs on floppy discs are another form of software. The programs are inserted into a computer and convert it for many completely different uses. For example, Lotus 1-2-3 is used for business spreadsheets.

Another form of software is the database management system. This is a suite of programs that allows the user to store, retrieve, select, and associate information of various kinds in an efficient manner. It allows users to access information stored in a computer's discs without knowing exactly where each piece of information is stored, and it makes it possible to see the relationships between different types of information without manual comparisons. An example of one database management system is the dbase series. There are other systems and there is a standard syntax for retrieving information from databases kept in different types of computers. This syntax, developed by IBM, is called SQL (Structured Query Language).

4.7 REDUCED INSTRUCTION SET COMPUTER (RISC)

The instruction set is the group of operations that the procedure can perform, such as moving data, adding numbers, and testing for a zero value. RISC supports only the most frequently used operations.

RISC's use large numbers of transistors on a chip. RISC microprocessors have a clock speed of at least 150-megahertz. They execute as many as 300 million instructions per second. One distinctive trait is a large number of registers on the processor chip. Registers are fast-access on-chip storage areas that can be used to temporarily hold intermediate values, memory addresses or copies of instructions.

RISC combines six strategies to attain high-performance microcomputers. The first is the simple instruction set which incorporates only frequently used operations. Other less frequently used operations are obtained from a subroutine. In one non-RISC about 10 percent of the instructions account for about 90 percent of the execution time. The remaining 270 instructions are rarely needed. RISC also limits the number of addressing modes. The only instructions that can access memory are load and store. All other instructions access registers. A RISC is in general characterized typically by 50 instructions, fewer than four addressing modes and fewer than four data types.

The second RISC strategy is pipelining. Without pipelining the processor completes all the four stages of the cycle, fetch, decode, execute, and store before starting the next instruction. With pipelining the micro steps of one instruction are executed in parallel with the micro steps of subsequent instructions.

The third feature of a RISC is a large number of registers on the processor chip. The register file reduces the frequency of off-chip memory access, which takes longer than register access.

The fourth RISC strategy is high-level language support. The fifth RISC strategy is fast operations on numbers represented in floating-point operation. The final strategy is the efficient use of a compiler.

4.8 VIRTUAL REALITY

Ivan Sutherland in 1965 proposed reaching through the screen, immersing the operator into a simulated computer generated world.

The idea was to use a stereo head-up display (two small television monitors in a helmet), and a method of tracking head movements. This was to give the user the illusion of being inside a three-dimensional world. In the late 1960's the technology was not available to make the virtual reality practical.

By the early 1990's such technology became available and Fred Brooks and Henry Fuchs of the University of North Carolina were able to demonstrate the idea of virtual reality. An operator puts on a helmet with goggles, which gives him a projected view of a computer model of a building. The operator walks on a treadmill which tells the computer where he is in space, and the image is adjusted so that he has the illusion of walking through a virtual building. A system in the helmet tracks the movements of his head. If he looks up he sees what should be above him. If he looks behind him he sees what the computer pictures behind him. The user is immersed in a simulated world of thought. The operator can turn around 360 degrees and look all around him as he moves.

An architect can design a building on a computer and then employ virtual reality to enter the building, and proceed to examine each floor. The architect inspects the features of the kitchen, living room, bedrooms, and playroom of the building. The operator is able to go up and down the stairs and even explore the backyard without stepping off the treadmill. Virtual reality is a way of expressing the output of a computer in 3 dimensions instead of 2 dimensions as on an ordinary computer screen.

4.9 Mobile Computers

This is defined as a complete computer that can easily be carried by one person who operates it. There are a number of mobile computer types. Flash type memory chips are used which do not lose data when the power is turned off.

Laptops

Laptops weigh from 8 to 20 pounds and the dimensions are from 11.7 to 14 inches deep, 2.5 to 4 inches high. They can sit on the lap of a person traveling in a train or airplane and they function in the same way as the ordinary PC. They allow a traveler to operate computer programs exactly as if used in the office or at home.

Notebooks

Notebooks weigh 4 to 7.5 pounds and the dimensions are 11 to 11.7 inches long, 7.75 to 8.5 inches deep, 1.25 to 2.5 inches high. They can be used to take notes in libraries, business meetings, etc. Books properly prepared for computers can be read on a notebook computer which displays text and adds sound and motion for realism. Some notebooks have an electronic pen which is used to write on the screen. The script on the screen is converted to computer language inside the notebook. A few notebook computers have a built in radio for sending data to a main computer.

Sub-Notebooks

These are a smaller version of the notebook computers. They weigh 2.5 to 4 pounds and are 10 to 11 inches long, 5.9 to 7.5 inches deep and 1.5 inches or less high.

Hand-Helds

Hand-helds are the ultimate in size reduction of computers. The weight is 1 pound or less and the length is 6 to 7.75 inches long, 3.25 to 4 inches deep, and 1 inch high.

4.10 Computers Take Over the World

Education

Personal computers have become an integral part of the educational process in schools from kindergarten to graduate school. It starts with small toddlers drawing simple figures using the computer. Seymour Papert, an MIT educator, developed a programming language for children called LOGO. Children have embraced the computer world with more enthusiasm than their elders. Some colleges now require that entering students bring their own personal computers. Many universities have a central computer network that students can tap into with their own personal computers.

Business

Personal computers have also become pervasive throughout businesses from real estate, travel agencies, banks, accountants, and many others. Special computer programs such as Lotus l, 2, 3 were written to facilitate business operation. Secretaries in business offices in the last two decades of the 20th century had to turn in their typewriters for personal computers. Money now flows from business to bank and from bank to bank across national boundaries electronically controlled by computers.

Games

One of the earliest uses of personal computers was video games. Children and adults stared as Pac-Man ate his way through from one part of the screen to another. Video games were even considered by the United States Air Force as training young children who would be future combat pilots. Video games became a source of wealth for many companies such as Atari. They also became a sort of drug which prevented young children from doing their homework.

Music and Art

Music was composed on personal computers with special software. Computers became music synthesizers. Similarly personal computers are used by painters to create works of art. Computer programs provide a simulation of a painter's palette.

Word Processing

Word processing became one of the most popular uses of a personal computer. Writers can now handle text electronically, moving letters, words, and paragraphs. Formerly this usually meant retyping entire pages. Mail sent to thousands of readers can be sent all looking as if each is an original. Various programs such as Word Perfect became what the pen and the typewriter had been prior to 1975.

Engineering

Computer Aided Design (CAD) became an integral part of engineering. Computers took the place of rows of draftsmen working to place an engineering design on drawings. Engineers could now test design changes on a computer instead of actually building a trial model.

Architecture

Architects can now build a building on a computer. They are able to rotate the computer model and with virtual reality they can even move through the rooms and areas of a building long before it is built.

Language Translation

Computers are now used to translate from one language to another. Languages as different as Japanese and English can be translated from one to the other with the aid of computers.

Satellite and Spaceship Control

Satellites can be flown and controlled solely by computers. Small spaceships have been flown out to the limits of the solar system and beyond by computer.

Control of Communications

Mobile communications such as cellular phones and other two-way mobile systems are controlled by computers. Police and fire vehicles are expeditiously sent to where they are needed by computer aided dispatching. Vast telephone and data communications networks are completely controlled by computers. This is true in local as well as national and international networks.

Aircraft Navigation and Tower Control

Airplanes are now navigated with the aid of computers. Airplane control towers including radar displays depend on very sophisticated computers which can rapidly become out of date as rapid advances are made in both hardware and software.

Weather Forecasting

Weather on a five-day forecasting schedule routinely use computers. However, long range weather forecasting and climate modeling depends on the development of supercomputers that can handle billions and even trillions of computer operation steps. The book by Nebeker in the bibliography gives a history of the use of computers in weather forecasting.

4.11 Supercomputers

Supercomputer development started about 1972, subsidized by the United States in the cold war for code breaking and nuclear weapon design. The Cray-1 in 1976 was the world's fastest supercomputer, using silicon semiconductor emitter-follower technology and a single high-speed processor operating serially. The Cray Research, Inc. is based in Eagan, Minnesota, and is still considered the world's premier supercomputer maker although the technology is rapidly changing.

The Cray-2 and 4 parallel processors increased the speed obtained by the Cray-1. Other supercomputers soon came out with 16 parallel processors significantly increasing the speed of computing. Supercomputers fell into two main categories, the vector type using a small number of processors in parallel and the massively parallel processor (MPP) using hundreds or thousands of processors in parallel. An example of an MPP is the Thinking Machine CM-200 with 65,536 processors.

In 1982 Nobel physicist Kenneth Wilson suggested that researchers and their supercomputers could solve fundamental problems in science and technology. He made up a list that was called the "Grand Challenges" and has since grown into a long list some of which are shown below:

GRAND CHALLENGES

Climate modeling
Fluid turbulence modeling
Pollution dispersion modeling
Human genome
Ocean circulation modeling
Quantum chromodynamics
Semiconductor modeling
Superconductor modeling
Internal combustion system modeling
Vision and cognition
Pharmaceutical design
Aircraft design
Evaluating ozone depletion

As a result of these grand challenges, greatly improved supercomputers have been used to design automobiles that will better protect passengers in crashes. Internal combustion engines that burn fuel more efficiently, pharmaceuticals that are safer and more effective and integrated circuits of unprecedented complexity have been developed. There is, of course, much more that supercomputers can and will do.

The key to supercomputer development is speed. A measure of speed used is the FLOPS or floating point operations per second. The floating-point format is very similar to scientific notation. The number 6026.0 may be represented as 0.6026×10^4. In the floating-point method a zero always appears to the left of the decimal point. The number 0.6025 is called the mantissa and 4 is the exponent. Floating-point operations handle such numbers. An example is multiplying 0.6026×10^4 by 0.2×10^6. The product is found by multiplying the mantissas and adding the exponents to obtain 0.12052×10^{10}.

In supercomputers the speed is currently measured in gigaflops (billions of floating point operations per second) and in new supercomputers TERAFLOPS (trillions of floating point operations per second). These tremendous speeds account for the ability to handle very complex problems.

A race started between the United States and Japan to develop the fastest supercomputers in the TERAFLOPS range. In the United States there are four supercomputer centers: Cornell Theory Center, National Center for Supercomputing Applications at the University of Illinois, the San Diego Supercomputer Center and the Pittsburgh Supercomputing Center. Senior United States government officials, industrial, and academic scientists and engineers have cooperated in the Federal High Performance

Computing and Communications Initiatives (HPCCI). Various federal agencies funnel the funds to laboratories, private industries and academe, which then run the programs.

In Japan the Ministry of International Trade and Industry (MITI) pushed the development of supercomputers in different programs. One was the nine-year Superspeed project begun in 1981 with a goal of producing a scientific computer capable of 10 gigaflops. The MITI project concluded in 1990 with the demonstration of a Parallel Hierarchical Intelligent Computer which had four processors in parallel and achieved a peak-processing rate of over 10 gigaflops.

Hitachi Ltd. built a vector conventional supercomputer with a claimed peak speed of 32 gigaflops. It has 4 processors in parallel and a clock cycle of two billionths of a second.

MITI started a 10-year Real-World Computing Program (RWC) in 1992. The RWC has a budget of almost 500 million U.S. dollars. It will focus on R&D for intuitive information processing--the way human beings absorb information and make decisions. During the first phase various operating systems will be developed with as many as 10,000 processors in parallel. During the second phase a million processor system will be built.

RWC will also perform research in neural systems which humans use. The final goal is to develop a one million neuron network composed of 1,000 subnetworks of 1,000 neurons each, with a total processing speed of 10 teracups (10 trillion connection updates per second).

4.12 Problems of the Computer Age

Like most other new technologies the computer age brought unique problems which require their own particular solutions.

Software Error Disasters

Very small errors in programming can cause calamitous results. The omission of a comma caused a missile to go to a different destination than the one intended. Another mishap caused by a slight programming error was a series of shutdowns resulting in some 20 million telephone customers losing service in eight mysterious incidents of signaling equipment failure. These outages occurred between June 10 and July 2, 1991 in areas in northern and southern California, the Washington-Baltimore region and western Pennsylvania. At the time no one knew the cause but after a six months investigation the problem was found to be a slight error in the signal transfer point (STP) software.

STP's are digital switches that route telephone network control information. They are a key element in Signaling System 7, the internationally accepted protocol that searches for the best route under current traffic conditions. The protocol handles dial-800 toll-free calls, credit card calls and caller identification, as well as a host of other new services.

The STP software error was in one line of code among millions. The flaw involved only 3 bits, 0110 should have been 1101. 1^s and 0^s had been transposed. This small software error was enough to disable the STP's overload protection. Instead of discarding accumulated signaling messages when its storage buffers became full, an STP shut itself down. Signaling messages meant for it then were routed to an alternative STP, which itself became overloaded. That alternative unit, because of the 3-bit bug, shut itself down as well. In turn, the second alternative STP became overloaded and shut down. As time went on voice connections became deprived of the signaling connections they needed to set up calls and terminate them. Customers could not call within their region, except to phones connected to the same central office.

The error became apparent only in service under certain conditions. There were two conditions, one was heavy (but not excessive) voice traffic which generated a high signaling load for STPs and second a trigger mechanism which caused signaling messages to accumulate.

In one case the trigger was a hardware fault in a Baltimore STP. The STP automatically took some other components out of service to isolate the fault, and congestion and shutdown followed. This happened on June 26, 1991 and it disrupted telephone service for 5 million to 6 million customers for 6 hours in Washington, DC, Maryland, Virginia and parts of West Virginia.

Two hours later an error in routing data in a Los Angeles STP was the trigger and as messages accumulated the 3-bit bug went to work. Three million customers in Los Angeles and parts of Orange, Ventura, and San Diego counties lost their service for several hours.

It took 6 months of investigation before the 3-bit bug was discovered. A couple of patches to the STP were added to correct the malfunction. Previous testing of the original STP software did not reveal the error because of what are known as "What Ifs." There are so many of these that it is extremely difficult if not impossible to detect some software errors by repeated testing.

Another method of checking programs is a formal mathematical proof method. This is particularly used in Europe. Formal methods are actively being researched at the end of the 20th century in universities and corporate research centers in Europe. A working paper series of the University of Edinburgh by Cleland and MacKenzie in the bibliography gives an overview of this method as of 1994.

VIRUSES

A virus is a code embedded in an application or operating system program which will destroy the computer's ability to function. It causes a replication of a small program which continually repeats itself. The virus quickly takes over a computer system or network and destroys programs and files. It is often distributed through floppy discs. Individuals have inserted viruses in floppy discs and have wreaked havoc on computer networks.

A number of organizations have been formed to combat new viruses. CERT is the Computer Emergency Response Team located at Carnegie Mellon to resolve various software vulnerabilities. CIAC is the Computer Incident Advisory Capability located at Lawrence Livermore National Laboratory which is responsible for the immunity of Department of Energy computers. The FBI has formed CART, Computer Analysis and Response Team.

In addition, there are a number of programs that can detect specific known viruses such as *Stoned, Michelangelo,* etc. However, it is difficult to discriminate between all infected programs or an uninfected program when the virus is unknown.

COMPUTER OPERATOR PROBLEMS

Computer operators have suffered from a syndrome known as RSI or "repetitive stress injury." This formerly was mainly a problem for assembly-line workers. Many computer operators are being monitored by other computers to enhance efficiency. The result has been an outbreak of RSI which has crippled some computer operators. This situation has been aggravated by improper hand positions.

RSI in computer operators may be expressed in five different areas of symptoms. The first is *flexor tendinitis.* The friction in the sheath shrouding tendons associated with fingers may create pain and make movement difficult. The second is *Carpal Tunnel Syndrome.* Tendons, blood vessels and the median nerve pass through a narrow area in the wrists. Swelling in the wrist area may pinch a nerve. A third area is *Overloaded Arm Muscles and Fascia.* Incorrect hand positions may cause inflammation leading to shortening and scar-tissue formation. A fourth area is *DeQuervain's Disease.* Continued friction causes the sheath covering the thumb tendon to become inflamed and painful. *Extensor Tendinitis* is the fifth area. Here excessive friction may cause inflammation, leading to wrist pain.

There are a number of methods that are now used to avoid RSI for computer operators. In one newspaper in the United States, terminals now automatically remind computer operators to take a break every 50 minutes. In some businesses computer operators have adjustable-height desks, chairs and screens, footrests and wristrests are used to position the body correctly: elbows at right angles, knees slightly lower than hips, wrists relaxed in a neutral position.

Changes in the computer keyboard have been developed to reduce the strain of typing. However, all of this may be superseded by the development of new technology which will bypass the keyboard altogether. This is voice operated computer which at the time of this writing is still in its infancy.

A QUESTION OF VIDEO DATA TERMINAL HAZARDS

There have been some concerns as to whether VDT's (Video Data Terminals) cause miscarriages in pregnant women. These computer operators spend their working days in front of VDT's. There have been questions raised as to whether the pulsed magnetic field in the range of 15 kHz to 30 kHz might cause miscarriages.

The old type VDT's had magnetic fields of 0.5 microTeslas (or 5 milligauss). The European community specified a new type of VDT with a magnetic field of 0.05 microTeslas (or 0.5 milligauss). This is compared to the earth's magnetic field of 500 milligauss.

While some statistical studies seemed to show a correlation between working in front of a VDT and miscarriages there are some confounding factors which can skew the results. One is the psychological stress of working at a computer where the employee's performance is being constantly recorded and subject to review by management. Another confounding factor is posture where the employee sits in front of the screen for hours at a time.

One approach to reducing the problem is to use technology to lower the EMF from the VDT's the way the European community has done. In addition studies should be continued on effects of magnetic fields on membrane ion flow, hormone changes, protein synthesis and RNA transcription.

Also, pregnant women should be allowed to take breaks to relieve the psychological strain. Attention should be given to special chairs which will relieve this problem.

THE 21ST CENTURY TRANSITION HEADACHE

At the beginning of the computer age, computers were programmed to read year dates as just two numbers, not four.

For example, "00" stands for 1900 and 97 stands for 1997. So when 2000 arrives with the 21st century, present day computer systems, unless modified, will interpret "00" as 1900.

One example that could happen is a twenty minutes phone call starting just before midnight on December 31, 1999 and lasting into January 1, 2000. This could be interpreted by a computer as 20 minutes plus 99 years which would generate some phone bill.

Commercial giants like the telephone companies, utilities, and others are spending millions of dollars to correct the situation. In addition, software companies all over the world are coming up with solutions which may be quite expensive to implement.

BIBLIOGRAPHY

Babbage, Charles — *Passages From the Life of a Philosopher,* IEEE Computer Press, 1994

Cleland, George and Donald MacKenzie — "Inhibiting Factors, Market Structure and the Industrial Uptake of Formal Methods." *PICT*, University of Edinburgh, 1994

Garfinkel, Simon L. — *PGP: Pretty Good Privacy,* O'Reilly & Associates, 1995

Good, D.I. — "Toward a Man-Machine System for Proving Program Correctness," Ph.D. Thesis, University of Wisconsin, 1970

Kidwell, P.A. and Paul E. Ceruzzi — *Landmarks in Digital Computing, A Smithsonian Pictorial History,* Smithsonian Institution Press, 1994

Lee, John — *Computer Pioneers,* IEE Computer Society Press, 1994

MacKenzie, Donald — "The Automation by Proof: A Historical and Sociological Exploration," *IEEE Annals of the History of Computing,* Vol. 17, Number 3, Fall 1995

Nebeker, Frederick — *Calculating the Weather: Meteorology in the 20th Century,* Academic Press, 1995

Neumann, Peter G. — *Life in the Digital Breakdown Lane: Computer-Related Risks,* Addison-Wesley, 1995

Palfreman, John and Doran Swade — *The Dream Machine, Exploring the Computer Age,* BBC Publishers, 1991

Rutland, David — *Why Computers Are Computers: The SWAC and the PC,* Wren Publishers, 1995

CHAPTER 5

LET THERE BE PURE LIGHT

5.1 THE ROOTS OF THE LASER

Albert Einstein in November 1916 wrote to Michele Angelo Besso, a close personal friend, "A splendid light has dawned on me about the absorption and emission of radiation." He wrote two papers on this in 1916 and one in 1917. Einstein proposed that an atomic or molecular system exists in a sequence of quantum mechanical states of discrete energy and it can change its state from a lower level E_1 to a higher energy level E_2 by absorbing incoming electromagnetic radiation. The atomic or molecular system would then fall from the higher to the lower energy level, emitting radiation.

Einstein expressed this in an equation: $E_2 - E_1 = h f$, where h is a constant and f is the frequency of the electromagnetic radiation. When in a higher state the system can emit radiation by one of two methods. The first is spontaneous emission as in a fluorescent lamp. The energy is radiated in different directions at different times. The other method is stimulation which was described by Albert Einstein in 1917. He asked what would happen if while an atom or molecule was at a higher state of energy it absorbed one photon of radiation of the appropriate frequency. His answer, expressed mathematically was that the atom or molecule would then fall to its lower state emitting two photons of the same frequency, phase and direction. This stimulated emission became the theoretical basis of the MASER (Microwave Amplification by Stimulated Emission of Radiation) and later the LASER (Light Amplification by Stimulated Emission of Radiation).

The MASER was invented in the United States by Charles H. Townes and at the same time in the Soviet Union by Nikolai G. Basov and Aleksandr Prokhorov. The MASER operated at microwave frequencies well below optical frequencies. However, the MASER led directly to the idea of a LASER.

Townes, born July 28, 1915, was educated at Furman and Duke University in his home state of South Carolina, and at the California Institute of Technology where he obtained his Ph.D. in 1939. He worked at the Bell Telephone Laboratories (1939-47) before he joined Columbia University, New York, where he became a full professor in

1950. He moved to the Massachusetts Institute of Technology in 1961 and the University of California of Berkeley in 1967.

Basov served in the Soviet Army in World War II, following which he graduated from the Moscow Institute of Engineering Physics (1950). He studied at the Lebedev Institute of Physics of the Soviet Academy of Sciences in Moscow, earning his doctoral degree in 1956 and going on to become Deputy Director (1958), and later head of the Laboratory. Basov's colleague at the Institute, Aleksandr Prokhorov, was involved in the microwave spectroscopy of gases, with the aim of creating a precise frequency standard, for use in very accurate clocks and navigational systems. Basov and Prokhorov in 1955 developed a generator using a beam of excited ammonia molecules. This was the MASER developed simultaneously but independently in America by Charles Townes.

5.2 THE MASER

The Cold War was the incubator of the MASER both in the United States and the Soviet Union. The Soviet Union had exploded its first atomic bomb in 1949 and the Korean War had started in 1950. Townes and the Soviet Union scientists were well aware of Einstein's work on stimulated emission. In the United States in 1950 the Electronics Branch of the Office of Naval Research asked Townes to organize and lead an Advisory Committee on Millimeter Wave Generation. The United States government was anxious to develop millimeter waves for future military breakthroughs.

Townes knew that the conventional microwave sources developed during World War II were too large to generate millimeter waves. He decided to use stimulated emission to radiate at short wavelengths. One problem was that most of the atomic or molecular systems are in a lower state of energy. However, if an abnormal situation can be created in which there are more systems in a higher state, or population inversion, the incoming beam will itself be amplified but weakly.

In April 1951 Townes hit on the idea of using the regenerative oscillator discovered many years before by Armstrong and described in Chapter 2. Townes decided that the upper state systems could be enclosed in a resonator which would sustain the frequency that they emitted. Then the stimulated emissions would be reflected back into the system, inducing further emissions, increasing the radiation. When high enough levels were reached, the system would break into self-oscillation and become the source of millimeter waves.

It was well known that feedback oscillators in radio produced a single sharp wavelength with coherent radiation in phase. Townes thought this principle could be used to produce millimeter coherent waves of great strength similar to Armstrong's radio tube transmitters.

THE AMMONIA GAS MASER

Townes and a student at Columbia University, James P. Gordon, demonstrated the principle in early April 1954 using ordinary ammonia. This substance had an energy transition which produced radiation at 1.25 centimeters. Two basic components besides ammonia were necessary. First, a focuser to separate ammonia molecules in the upper state and second, a resonant cavity. The focuser sent the upper energy-state molecules of ammonia into the resonant cavity. The MASER used electric fields to separate the upper-state ammonia molecules in the focuser.

The MASER (Microwave Amplification by Stimulated Emission of Radiation) was both an amplifier and an oscillator. It produced oscillation with a continuous power output of about 0.01 microwatts. This was the first time continuous energy was obtained from molecular resonance. In 1955 Basov and Prokhorov in the Soviet Union independently demonstrated a generator using a beam of excited ammonia molecules.

THE SOLID STATE MASER

In 1955 Basov and Prokhorov proposed a so-called 'three-level' method of producing population inversion by pumping with a powerful auxiliary source of radiation. The next year Nicolaas Bloembergen of Harvard suggested the idea of using a molecular system with three unequally spaced energy levels. He proposed pumping the system from the lowest (level l) to the highest (level 3). The atoms at level 3 would fall to level 2 without emitting radiation. The number of systems in level 2 would now be greater than in level l. This would produce the desired population inversion. On September 7, 1956 Bloembergen negotiated an agreement with Bell Laboratories to patent the three-level MASER for him. A number of people started to build a three level solid state MASER.

Willow Run Laboratories, owned by the University of Michigan, began solid-state MASER work in 1956. Chihiro Kikuchi in January 1957 used pink ruby to investigate the possibility of a solid-state MASER. On December 20, 1957 Kikuchi and his co-workers demonstrated MASER action at 9,300 megahertz (3.2 centimeters). This three-level solid state MASER was an ultra-low-noise, tunable, continuously acting microwave amplifier of respectable gain.

5.3 THE DEVELOPMENT OF THE LASER

On September 14, 1957 Charles Townes put down his ideas in a notebook for a MASER operating at optical frequencies. He proposed a resonant cavity with sides about l centimeter long, much larger than the .00003 centimeter (30 microns) optical wave he had in mind. He proposed silvering the cavity to reflect the light and he also proposed making holes in the cavity through which radiation might be fed in or extracted. Townes at first thought of optically pumping the vapor of thallium with the irradiation of an incoherent light source.

Townes was a consultant to Bell Telephone Laboratories. His brother-in-law Arthur L. Schawlow was a member of the technical staff of Bell. Schawlow had worked for Townes in 1950-1951 as a research assistant. In October 1957 Townes discussed his ideas with Schawlow and they decided to collaborate on the development of the optical MASER at Bell Labs.

Townes tried to get information about optically pumping thallium vapor. He went to a 37-year-old graduate student, Gordon Gould, at Columbia University, who was doing a Ph.D. thesis under Professor Polykarp Kusch on the energy levels of excited thallium. Gould was using optical pumping to raise the thallium to higher energy levels. Gould became alarmed during these conversations because he too had been thinking about how to design an optical MASER.

Gould quickly put his own ideas down on 8 pages of notes and on November 13, 1957 had them notarized. He called the optical MASER, a LASER (Light Amplification by Stimulated Emission of Radiation). He proposed using a meter-long tube with reflecting plates at the ends. He mentioned the optical pumping of potassium vapor. He wrote down possible applications of the LASER, such as spectroscopy, interferometry, photochemistry, light amplification, radar and communications.

Gould went to a patent lawyer and discussed his ideas. However, the patent application itself was delayed until April 1959 but Schawlow and Townes had already filed a patent on behalf of Bell Laboratories in July 1958. Gould challenged the patent in court. At first the ruling went against him.

However, Gould persisted for many years against tremendous legal odds. He eventually received three patents on important aspects of laser technology. The last patent was issued in November 1987 some thirty years after Gould made the entries in his notebook on the laser. His persistence paid off in millions of dollars.

It should be pointed out that neither Gould or Schawlow and Townes had actually constructed a laser in 1959. In fact the latter were not able to build a proposed laser and TRG, a defense contractor, employing Gould, was able to reduce Gould's suggestions to practice only in 1962, two years after the first laser had been operated successfully.

By 1960 a race occurred in many laboratories to construct a working laser. In May 1960 Theodore H. Maiman of Hughes Laboratories built an experimental laser using red ruby and a pulsed flashlight to produce light with a wavelength of 694 nanometers (10^{-9} meters) at the long end of the red spectrum. Maiman placed a ruby cylinder inside a helical shaped xenon lamp. The ruby cylinder was l centimeter in diameter and 2 centimeters long, silvered on the ends to make it into a laser.

The red ruby laser as it was developed consisted of a ruby rod which chemically was AL_20_3 with 0.1% chromium. As shown in Figure 5-1 the rod is surrounded by a powerful

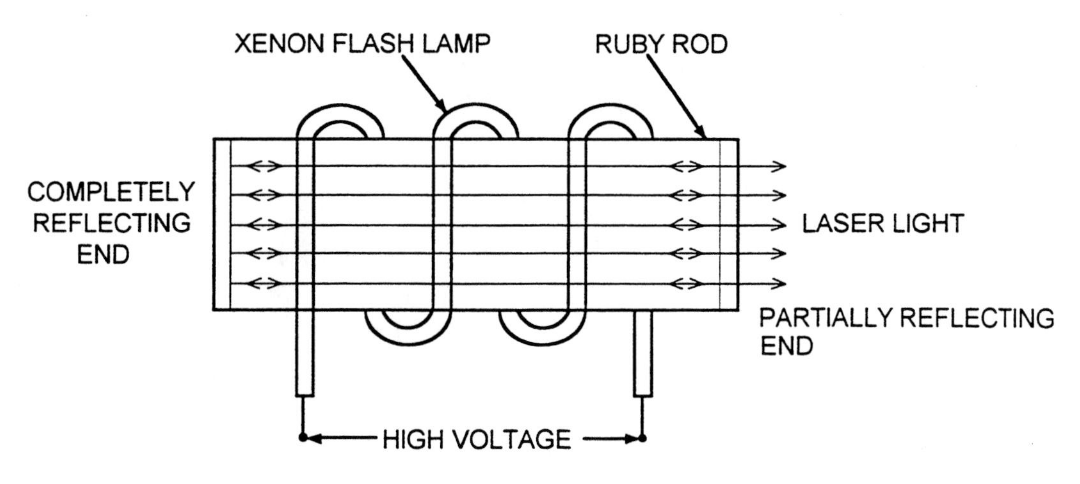

Figure 5-1. Ruby Laser

xenon flash lamp which supplies green light. When a chromium atom absorbs a photon from the flash lamp, electrons are raised to energy band E_3 as indicated in Figure 5-2. Because of interactions with other atoms, the electrons immediately drop into level E_2 where they stay temporarily. The purpose of the flash lamp is to increase greatly the number of atoms in the metastable state E_2 After the flash, the number of electrons in E_2 exceeds the number in E_1. This is the necessary population inversion.

To obtain a laser beam, all the electrons in level E_2 must be stimulated at practically the same instant to return to level E_1. This is the basic idea in laser operation:

$$hf_{12} = E_2 - E_1.$$

Most of the atoms in level E_2 interact with the red light and are in turn stimulated to emit photons of frequency f_{12}. Because the ends of the rod are polished and silvered the light is reflected back and forth in the rod and a standing wave rapidly builds up.

Laser light differs from ordinary light in three main ways. First, laser light is unusually intense. All the atoms emitting this light are stimulated to do so at practically the same instant by a standing wave in the laser rod. The rod is analogous to a vibrating string with two fixed ends. Second, laser light is very nearly of a single frequency. The light originates from only one kind of atomic transition. Finally laser light is coherent because the waves are all in phase. That is, the peaks and valleys of different waves are lined up at the same point.

In essence laser light consists of a single intense wave train confined to a very narrow cone. Such light has very little tendency to spread out and may consequently be

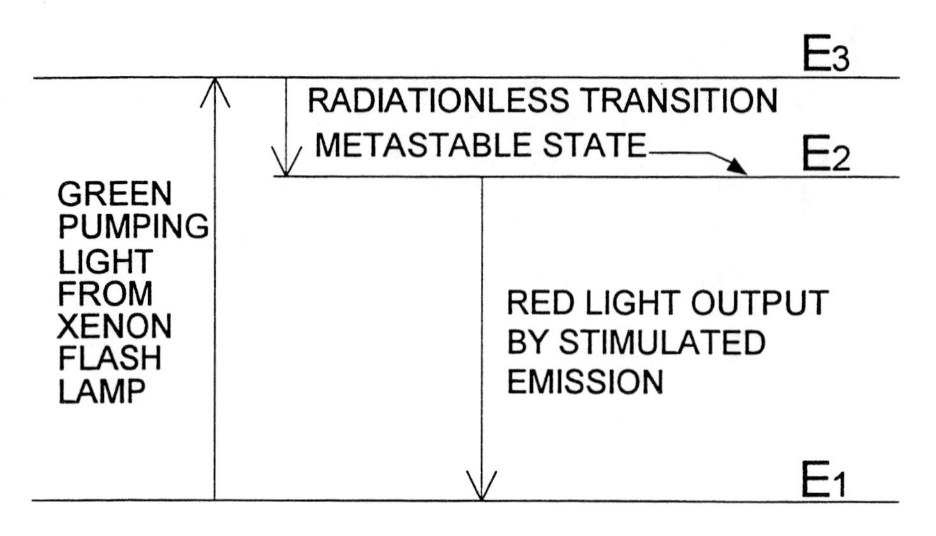

Figure 5-2. Energy Levels in a Chromium Atom

projected as a beam for hundreds of thousands of miles. In 1962 Louis Smullin and Giorgio Fiocco of MIT aimed a Raytheon ruby laser at the moon and recorded the reflected echo. To sum up, the ruby laser converts incoherent (out-of-phase) green pumping light into an intense beam of coherent red light by stimulating atoms to emit light in phase with a standing wave.

Very rapidly after the ruby pulsed three-level laser there were a number of different lasers that were produced. One was a four-level pulsed solid-state laser and another was a continuous gas laser. By 1962 there were about 400 companies that had laser research programs, many of them financed by the United States government. These companies had specific applications in mind. Hughes Aircraft built optical radars and American Optical had a research program on laser photocoagulator for retinal surgery. At the end of 1963, 20 to 30 United States firms had lasers on the market.

In Russia as early as 1957, Nikolai Basov of the Lebedev Institute started to work on a proposal to use impurity-doped semiconductors as lasers. His co-worker Prokhorov published a paper on masers and lasers in 1958.

In December 1964 Nikolai Basov, Aleksandr Prokhorov and Charles H. Townes were all awarded the Nobel Prize in physics. The award was for fundamental work in quantum electronics leading to maser and laser devices. Arthur Schawlow shared the Nobel Prize in physics in 1981 for research with lasers.

5.4 What Makes a Laser Lase?

After 1960 many different types of lasers were developed including various types of crystals, gases, liquids, and semiconductors. These all followed three basic principles which are necessary for laser operation. The first is the selection of materials that have at least two levels of molecular energy that satisfy Einstein's equation:

$$E_2 - E_1 = h f_{12}$$

where f_{12} falls in the optical region of the electromagnetic spectrum.

The second principle is population inversion. Normally most of the atoms are at the lower energy state E_1 and only a few are at the higher energy state E_2. For the laser to operate a mechanism must be supplied to invert this: that is to obtain more atoms in the E_2 state than are in the E_1 state. Population inversion may be produced in different ways. Some examples are optical pumping in ruby lasers, electric current injection in semiconductor lasers, and using very rapidly cooling of hot gases in the gas dynamic laser.

The third principle is the production of a standing wave of the desired frequency which will irradiate the lasing material producing stimulating radiation. All of the atoms at the E_2 energy level will fall down to the E_1 energy level simultaneously and coherent light of a single frequency will be radiated. One method of producing a standing wave is the use of mirrored surface at each of the two ends of the laser material as in the case of the ruby laser.

Many types of lasers that followed these basic principles were developed. Some were very large, measuring many meters of very high power. Some lasers were very small measuring in the centimeter range and emitting small amounts of power.

5.5 A Multiplicity of Lasers

There are over a hundred different types of lasers. A few of the more prominent types are described:

The Q-Switched High Power Ruby Laser

The ruby laser was described previously. There was a need to increase its power. In 1961, Robert W. Hellwarth suggested that if the reflectivity of the laser's end mirrors were suddenly switched from a value that was too low to permit lasing to one that was sufficient, the power output would be increased dramatically. This technique was called Q-switching. Hellwarth proposed inserting a Kerr cell together with a polarizer in front of one of the laser mirrors. A Kerr cell rotates the plane of polarized light when a voltage is applied to it. Applying voltage to the Kerr cell effectively stops lasing. Cutting off the voltage allows lasing. In early trials the peak output of a Q-switching ruby laser was 600 kilowatts. This was 100 times those of ordinary ruby lasers. Hellwarth calculated that peak powers as high as 12 megawatts could be obtained by Q-switching.

The Nd: YAG Laser

The neodymium doped yttrium aluminum garnet laser (Nd: YAG) was invented by Joseph E. Geusic of Bell Laboratories. A YAG crystal 0.25 centimeters in diameter and 3 centimeters long was used. First, tungsten filament lamps were used for optical pumping. They were soon replaced by Krypton pumping lamps. The wavelength of the Nd:YAG laser can generate high level power in the order of 700 watts. It is sometimes used in treating glaucoma.

The Argon Ion Laser

The argon ion laser was invented in February 1964 by William B. Bridges at Hughes Research Laboratories. It produced a few watts peak in 50 microsecond pulses. The wavelengths are in the 0.488 - 0.512-micron range. It was used in eye surgery.

The Carbon Dioxide Laser

C.K.N. Patel of Bell Laboratories invented the molecular carbon dioxide laser in 1964. Patel increased the efficiency of his CO_2 laser tenfold by exciting the CO_2 molecules by resonant transfer from excited nitrogen. The CO_2 laser's wavelength was at 10.6 microns and the peak power was in the kilowatt range.

The original CO_2 laser was very long for high power. For example, a discharge tube was 178 feet long for a power output of 2.3 kilowatts. A gas dynamic laser uses a different method of obtaining a population inversion. A heated mixture of CO_2, nitrogen and water vapor was used. The lower energy states have a larger population than the higher energy states. Cooling the gas fast, an upper energy level of CO_2 could be made to relax at a very slow rate. This is because different excited states relax at different rates. A lower energy state relaxes faster and is rapidly depopulated resulting in a population inversion.

The fastest way to lower the hot gas temperature is to expand the hot gas through supersonic nozzles. In a 1966 CO_2 laser the gases remained in a state of inverse population for about a meter from the nozzle. Laser mirrors placed with their normals perpendicular to the line of flow created a resonant cavity. The high rate of flow and high gas densities provided very large powers.

HELIUM-NEON LASER

This is one of the most popular gas lasers. Ali Javan, William Bennett, and Donald R. Herriott of Bell Laboratories proved in December 1960 that a mixture of helium and neon would be capable of laser action.

The helium in one of its excited stages has just the right energy to raise a neon atom from its ground state into an upper energy state. Originally a radio frequency exciter raised the helium to a higher energy state. Then the excited helium atoms raised neon atoms from the ground state resulting in a population inversion. Mirrors at each end supply the standing waves which stimulate the neon atoms resulting in coherent laser light. Later DC excitation was substituted for the radio frequency exciter.

DYE TUNED LASERS

Peter P. Sorokin and John R. Lankard of IBM invented the dye laser in 1965. Later it was found that it could be used as a tunable laser. Tunability of a laser was needed for monitoring chemical composition and processes. The dye tuned ruby laser using a group of dyes called metal phthalcyanines dissolved in liquid organic solvents was one answer. The molecular dyes were used to obtain stimulated emission using a ruby laser as a fast flash lamp. The dye beams had a broad spectral output because each molecular vibrational level has a rich substructure of rotational levels. The peak of this broad band can be varied by changing the concentration of dyes. The tunability of the dye became practical by using a rotating diffraction grating instead of one of the mirrors. The rotating diffraction grating functioned as a mirror reflecting a narrow band of frequencies around a frequency peak. This was invented by Bernard H. Soffer and Bill B. McFarland of Korad in 1967. Later improvements in 1970 made the dye tunable lasers capable of much better monochromatic outputs.

GLASS LASERS

In October 1961 Elias Snitzer of American Optical Company invented the glass laser. He used a rod of a barium crown glass doped with trivalent atoms of neodymium. Glass lasers were capable of megawatts of power. Later a set of neodymium glass slabs was used for laser fusion for hydrogen weapons simulation.

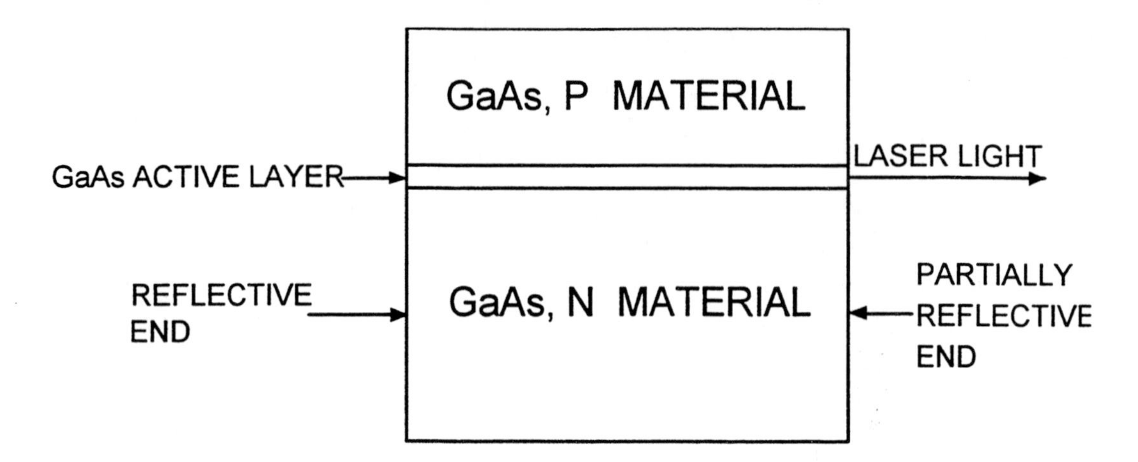

Figure 5-3. Homostructure Semiconductor Laser

Semiconductor Lasers

In 1947 the semiconductor transistor was invented. With the advent of the laser there was a race to develop a semiconductor laser. Pierre Aigran, a French scientist, had proposed using the transition of electrons from the higher energy conduction band of silicon to the lower energy valence band with the emission of a photon of light and the simultaneous release of a quantum of energy to the thermal vibration of the crystal as the basis for a laser. It was later found that gallium arsenide was better for lasing. Nikolai G. Basov of the Lebedev Institute of the Soviet Academy of Sciences proposed in1958 the idea of using semiconductors to achieve laser actions. In the fall of 1962 General Electric, IBM, and Lincoln Laboratory developed the semiconductor laser almost simultaneously.

The semiconductor lasers developed at the end of 1962 were compound semiconductor diodes of the gallium arsenic family. They measured in the order of 0.5 millimeter long and 0.1 millimeters wide. This compares with the centimeter length of ruby crystals and the meter size of gas lasers. They were excited by applying an electric current to the semiconductor instead of optical pumping. They could be easily and rapidly modulated by varying the strength of the current. The resonant cavity effect was obtained by polishing the ends of the diode so that they acted as mirrors.

A semiconductor laser must contain a minimum of three layers, an N-type layer, a separate active layer (either P or N) which emits the laser light and a P-type layer. In the first semiconductor lasers, the active layer and the surrounding layers were made of the same material. These were called homostructure junctions, as shown in Figure 5-3.

The homostructure semiconductor laser has the same index of refraction so light in the active layer can diffuse into surrounding layers. This meant that the first

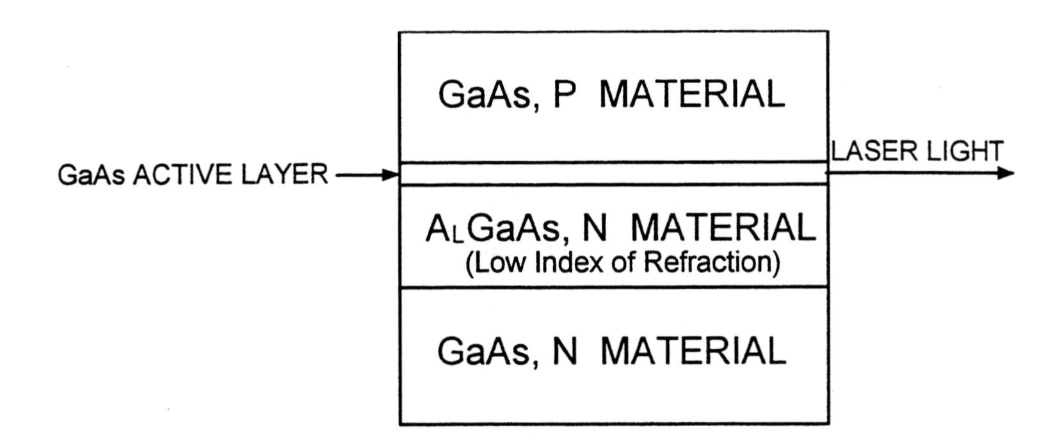

Figure 5-4. Double Heterostructure Semiconductor Laser

semiconductor lasers needed to be cooled with liquid nitrogen which made them impractical.

In order to overcome these difficulties, a double heterostructure laser was developed with the layers composed of different materials which have different refractive indexes. In a double heterostructure laser, the active layer has a higher refractive index than either of the two adjacent layers. This confines light generated in the active layer by creating a wave-guide effect. Figure 5-4 shows the double heterostructure laser.

The double-heterostructure is used in most semiconductor lasers to produce a continuous wave beam at room temperatures. However, there are some problems with this type of laser. One of these is that a number of adjacent frequencies are produced. Changes in temperature may cause a laser to hop from one dominant frequency mode to another. This becomes important in fiber-optic systems operating at the 1.55-micron wavelength for minimum loss. A small shift of the frequency of the laser will result in a loss in the signal transmission. This problem was alleviated by the development of the distributed feedback laser.

This is a diode laser design that includes methods of stabilizing the laser's output wavelength. A diffraction grating is used to scatter light back into the active layer. The feedback leads to interference effects which allow oscillation only at wavelengths that are reinforced by the feedback.

The conventional semiconductor laser described above is an edge-emitting laser in which the gain medium is grown epitaxially on a substrate. Epitaxy is the growth of a crystal by depositing atoms on a substrate a single layer at a time, by techniques employing molecular beams or gaseous chemicals. The two mirrors are formed by cleaving the resulting structure to expose reflective crystal facets. Light generated in the gain medium bounces horizontally between the facet mirrors and then bursts through one mirror to become the laser beam.

The Surface-Emitting Semiconductor Laser

In 1994 a new type of semiconductor laser was developed called a vertical surface-emitting laser. In this new class of devices photons of light bounce vertically between mirrors grown into the structure, and shoot up from their upper surface. These laser devices can be fabricated side by side on a wafer in great numbers. Millions of surface-emitting lasers can fit on a 3.9-inch diameter wafer. It is possible for them to be integrated on chips with transistors and other devices. The new lasers do not need to be wired individually to a circuit as in the conventional edge emitter semiconductor lasers. They have the advantage that the surface-emitter aperture can be shaped to give the beam the ideal circular cross-section. This results in a round collimated beam that couples its entire energy into an optical fiber. This contrasts with the conventional edge emitter which has an elliptical and divergent beam which is not as efficient for coupling to an optical fiber.

5.6 Wavelength Distribution of Lasers

This first maser, the predecessor of the laser, operated at a wavelength of 1.25 centimeters or 12,500 microns. The first laser operated at 0.76 microns. Newer lasers were developed through the different colors of the visible spectrum and then into the ultraviolet non-visible wavelength below 0.38 microns. The wavelength of newer lasers became smaller and smaller even into the X-ray range. However, it was difficult to sustain continuous wave lasing action, so that the shorter wave lasers, including the X-ray had to be pulsed.

5.7 Laser Applications

Drilling

One of the first applications of the laser was the drilling of holes in diamond dies at the end of December 1965 at the Western Electric, Buffalo, New York, plant. The diamond dies were used to manufacture fine wires for electrical connections. Previously, metal drills or diamond dust was used to pierce diamonds. This took many hours and many of the diamonds broke while being pierced. A ruby laser system of drilling took only a few minutes and proved extremely cost effective.

MICROMACHINING

Lasers were also used early in the trimming of microcircuit components such as resistors and capacitors. One example was trimming resistors encapsulated in glass. The lasers could machine through glass and other transparent materials. This became very useful in the manufacture of microelectronics which was taking off in the last half of the decade of the 1960's. The argon ion laser with an output of 1-10 watt range could be used for the micromachining of materials less than a centimeter thick but could not handle thicker material. Carbon dioxide lasers were soon available to micromachine slabs 2-3 centimeters thick. In 1968 Nd:YAG lasers were used for micromachining instead of carbon dioxide lasers because the new lasers were far more compact. The Nd:YAG laser had a wavelength one-tenth as long as the carbon dioxide laser.

OTHER LASER MANUFACTURING USES

The carbon dioxide laser was soon used in the cutting of cloth according to programmed patterns. The laser was also used to cut metals in conjunction with an oxygen jet. The laser heats the metal to ignition and the oxygen jet combines with the metal in a rapid exothermic reaction, while it simultaneously sweeps the oxides away from the workplace. The very small diameter of the laser saves material. Lasers in the early 1970's began to be used to machine aircraft materials such as titanium. Other manufacturing uses are in welding, heat treating and marking. Lasers are also used for measurement and control. In the microelectronic industry, lasers remove excessive pattern material, in micron-size features, from the photo masks used to fabricate integrated circuits. This makes possible the economical manufacture of chips which have become essential in electronic products.

THE LASER AND FIBER OPTICS IN COMMUNICATIONS

A very popular use of lasers is in communications. The GaAs/AlGaAs heterojunction laser diode is used with fiber optics to produce telephone and computer communications. Also used with fiber optics are the InGaAs and InP laser diodes. They emit in the 1.3 - 1.55-micron range. A micron is 10^{-6} meters or 10,000 angstroms. Pulse spreading or chromatic dispersion is minimal at 1.3 microns. Losses are minimum at 1.55 microns.

Fused silica optical fiber was produced in 1976 with transmission losses of only 0.5 decibels/kilometer. From 1980 to 1994 optical-fiber transmission capabilities have doubled every year while costs dropped exponentially. Figure 5-5 shows the basic cross-section structure of optical fibers. The core is the light transmission area of the fiber. The

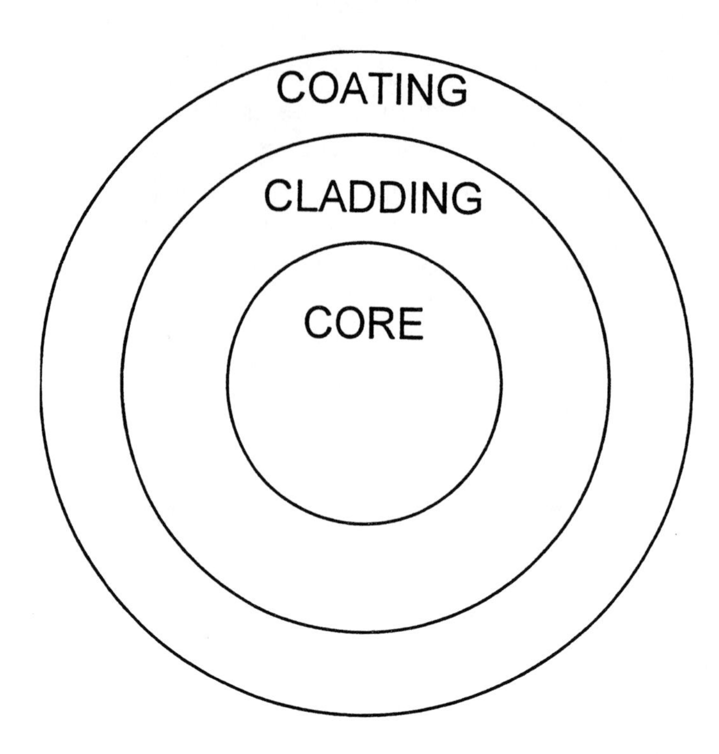

Figure 5-5. Cross Section of Optical Fiber

larger the core, the more light will be transmitted into the fiber. The function of the cladding is to provide a lower refractive index at the core interface in order to cause reflection within the core so that lightwaves are transmitted through the fiber. Coatings are usually multi-layers of plastic applied to preserve fiber strength, absorb shock and provide extra fiber protection. These buffer coatings are available from 250 microns to 900 microns.

The size of an optical fiber is commonly referred to by the outer diameter of its core, cladding, and coating. Example: 50/125/250 indicates a fiber with a core of 50 microns, cladding of 125 microns, and a coating of 250 microns. The coating is always removed when joined to connecting fibers.

There are three basic types of optical fibers: the step index, the graded index, and the single mode fiber. The first two use multimode transmission. Step index multimode has a sharp step-like difference in the refractive index of the core and cladding. In the more common graded index multimode fiber, the light rays are also guided down the fiber in multiple pathways. But unlike step index fiber, a graded index core contains many layers of glass, each with a lower index of refraction as the light goes outward from the axis. The effect of this grading is that light rays speed up in the outer layers, to match those rays going the shorter pathway directly down the axis.

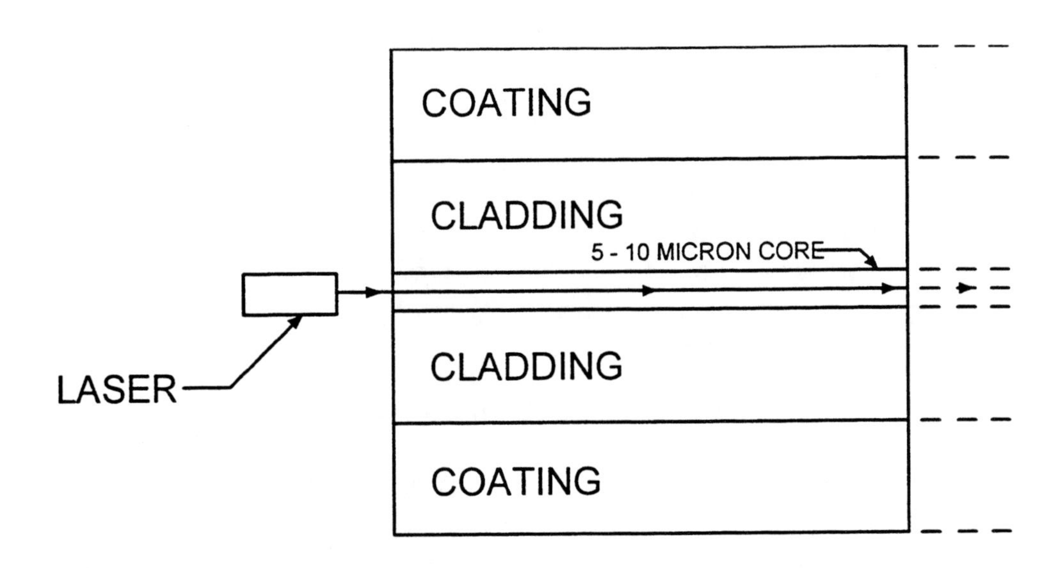

Figure 5-6. Laser Light Transmission in Single Mode Optical Fiber

The single mode fiber, Figure 5-6, has an extremely small diameter, approximately five to ten microns. The single mode fiber allows only a single light ray or mode to be transmitted down the core. This virtually eliminates any distortion due to the light pulses overlapping. The single mode is used for long distance transmission.

Fiber optic links are used for digital as well as analog transmission. In fiber optics, a digital pulse can be formed by turning the source "on" for a brief instant. The time of optical radiation from the laser is the pulse. The optical power on the fiber represents a binary "1" state. When the optical power is turned off, the binary "0" is represented. These two states represent binary signals. Digital signals consist of a series of bits that result in the laser being "on" or "off." In digital systems one parameter for system performance is bit error rate (BER). A BER of 1×10^{-6} means one error in 10^{6} bits.

Originally electronic repeaters converted the optical signal into an electric signal that was amplified to drive a semiconductor laser, which then generated the optical signal for the next stretch of fiber. The limitations of the electronic circuits involved present a restricted and finite signal bandwidth in the fiber.

Optical amplifiers can boost a wide range of wavelengths with no intervening photon-electron conversion. Optically pumping optical fiber with an external semiconductor laser yields amplifiers more than 50 km long. This distributed amplification offers the best band-to-end amplification and gives the best end-to-end signal-to-noise ratio. An erbium-doped fiber amplifier (EDFA) is used as a distributed optical amplifier. Erbium is a rare-earth element. When optically pumped with light from an outside semi-conductor laser, the erbium-doped fiber boosts the power of all signals traveling through it within a relatively wide range (25,000 GHz) of wavelengths centered on 1.55 microns. This is the wavelength for minimum attenuation of fiber optics.

Semi-conductor lasers, using EDFA with fiber optics, are capable of transmitting gigabits per second (one billion bits per second). This means that the entire Encyclopedia Britannica with one gigabit of information can be transmitted in one second. The technology for the information highway has been developed in the 20th century.

Special EDFA's can be used to transmit unique pulses called solitons thousands of kilometers. Solitons retain their shape because of the reciprocal effects of chromatic dispersion and an index of refraction that changes intensity. Practically speaking, soliton transmission would be effective over unlimited distances.

CONSUMER AND COMPUTER PRODUCTS

Another very important application of laser diodes is their use in audio and video compact disc players. The laser can produce digital stored information in the form of spots about l micrometer in diameter produced by local heating. The reflection of the laser beam from the spot creates readable information which may be either audio or video.

This technology is also used to store and read digital data on discs for computers. Optical discs provide much greater storage than conventional magnetic discs. Huge amounts of data can be stored on computer optical discs. The information of a number of books of an encyclopedia can be stored and read by lasers using one compact disc.

Computer printers using laser diodes encode information on photosensitive surfaces. The information is then transferred to paper. The quality is much better than that of typewriters or dot-matrix printers. The laser printers are much faster than mechanical printers and as the price goes down they will replace most other kinds of computer printers.

THE LASER AND SHOPPING

Another important use of lasers is in retail stores to scan bar codes on products. These employ the Universal Product Code (UPC) which uses a predetermined and reproducible pattern which corresponds to each number in a ten-digit code. A laser scanner detects the pattern and then decodes the number which is fed to a computer where the price is stored. The price is then sent back to the clerk where it is printed out.

Pioneers in this area in the early l970s were Shelley A. Harrison, Jerome Swartz and Susan Harrison, the principals of a company called Symbol Technology. They perfected "Gun Laser Scanners" that are used in all sorts of stores.

Another pioneer in the use of store laser scanners was Jerome H. Lemelson. In supermarkets the black bars on the product are read by a scanning helium-neon laser and a photodiode as the item is moved across a window. The identifying number is sent to a central computer where the price is stored. The price is then printed. The system also accommodates the automatic use of credit cards by insertion into a slot.

Scientific and Industrial Instrumentation

Lasers became the preferred light source for testing and measuring the quality of optical components. Lasers have been used in all types of spectroscopy. Tunable dye lasers are used for this purpose. Lasers are also used in material analysis and building construction. Visible laser beams are used with construction equipment to control ground leveling, pipe laying, and tunnel boring. Lasers are used in surveying and the control of the positioning of machine tools.

Laser Fusion Energy

Laboratory studies of laser fusion are being undertaken for work on the future development of fusion energy. Fusion energy in the 21st century may be a source of unlimited energy with relatively low radiation. In laser-driven fusion a target mixture of deuterium and tritium is irradiated by a laser pulse in the order of less than 10 nanoseconds. The target implodes, forming plasma dense and hot enough to cause a thermonuclear reaction. When the output power becomes greater than the input power, fusion energy will become practical. One type of laser used in this work is neodymium-doped glass.

Medical Applications

Lasers can deliver substantial amounts of energy to a small area, heating and destroying human tissue. This is very useful in certain surgical procedures. Different types of lasers are used for specific types of surgery.

The radiation wavelength from the argon ion laser is heavily absorbed by blood hemoglobin, rupturing blood cells, activating platelets, and damaging the lining of blood vessels. This promotes blood coagulation, which produces clotting in the vessel and cessation of bleeding.

The radiation from a carbon dioxide is strongly absorbed in the water in cells. The intense heat caused by absorbing laser radiation generates steam, which ruptures the cell. Impact depths of about 0.1 millimeter are typical, with injury to surrounding tissue limited to about 100 microns. The carbon dioxide laser has been used for the removal of nonvascular tissue, such as brain tissues and connective tissues. Solid state erbium lasers are beginning to replace the carbon dioxide laser for these operations.

Different types of lasers are employed in eye surgery. They are used to treat age-related macular degeneration, diabetic retinopathy and glaucoma. In panretinal photocoagulation a continuous wave argon laser burns about 500 microns in size in the retinal periphery. This induces regression of the abnormal blood cells, decreasing the chance of hemorrhage and retinal detachment. Glaucoma is treated by a laser-based irridectomy in which the flow of aqueous fluid is returned to normal by widening the opening in the iris. Argon ion lasers are widely used for this treatment. Also pulsed Nd:YAG lasers are used to treat glaucoma. Corneal sculpting with pulsed lasers is being used to correct the eye's ability to focus by altering the surface shape of the cornea. This could minimize the need for eyeglasses in some cases.

Lasers are also employed in dermatology to treat certain forms of skin cancer and to remove disfiguring marks. Lasers are used in gastroenterology to control bleeding from ulcers. In neurosurgery and urology, lasers remove certain tumors in otherwise inaccessible regions of the body and are also used to break up kidney stones. In cardiovascular medicine, lasers are used to remove obstructions in arteries. In oncology, lasers can optically activate cancerous cells previously tagged with hematoporphyrin dye, leading to the formation of chemically active molecules that selectively kill the cancerous cells.

Various other medical applications of lasers in the fields of pulmonology, urology, gynecology, orthopedic surgery and dentistry are being developed in the last decade of the 20th century.

Military Applications

Military applications can be divided into low-power and high-power lasers. Low-power lasers are employed as range finders in the United States Armed Forces. Nd:YAG lasers are used for this purpose and in target designation for missile guidance. Semi-conductor lasers are used in military fiber-optic communication links. Also laser diodes placed in air-to-air missiles act as proximity fuses. When light pulses from the laser are reflected from objects close by, the fuses are triggered.

Low-powered lasers are also used in simulated warfare for training. Weapons such as rifles or artillery are fitted with laser diodes. The personnel and vehicles participating in war games carry light sensors designed to sense the coded radiation from the lasers. A "hit" consists of the registration by a sensor of the laser light pulse. This has proven very effective in training exercises.

High-power lasers can be used to destroy enemy satellites and also to intercept incoming ballistic missiles. The chemical oxygen iodine laser at above 25 kilowatts is being developed as an antisatellite weapon. Electronically excited oxygen molecules are mixed with iodine vapor. A complex series of reactions produces laser emission at 1314 nanometers.

Another high powered laser for military applications is the free-electron laser (FEL). The FEL utilizes the coherent transfer of kinetic energy from a narrow beam of relativistic radiation in the presence of a periodic, static magnetic field. The FEL can be tuned by varying the voltage of the electron accelerator. The output power is expected to be scaleable in the megawatt region.

A U.S. Army weapon called a Dazer is a battery-powered, portable rifle-like weapon whose laser beam is hazardous to the eyes and skin. They can be used either to blind enemy forces or to disable battlefield surveillance systems such as thermal sights. During the Persian Gulf crisis in 1990, prototype Dazer weapons were sent to Saudi Arabia but they were not used. In 1995 the U.S. Army was working on three laser systems, one of which can fire a beam powerful enough to blind a person 1,000 yards away. In 1996 the Army responded to complaints by discontinuing work on the Dazer.

The military use of the laser is the dark side of this technology. Lasers like most technology can be used both for and against people. It is up to humanity to decide how and where to use technology.

One of the peaceful uses of the laser is in recording music, voice, video and data. This is a part of a revolution in recording technology developed in the 20th century and discussed in the next chapter.

5.8 THE PHASER

The phaser is a new method of generating coherent light. It does not use the population inversion of the conventional laser. In ordinary lasers more electrons are raised to a higher energy than remain in the lower-energy state. If this population inversion did not take place, the atoms in the lower-energy state would absorb the photons emitted by the excited atoms, instead of allowing them to take part in the laser beam. The coherent light emission would be quenched immediately.

The population inversion of the conventional laser takes a great deal of energy. Also the higher the frequency, the faster the atoms decay out of the excited state. The high frequency lasers, ultraviolet and X-rays, usually have to emit pulses instead of continuous beams. X-ray lasers are very large and limited to pulses.

A number of scientists in 1995 proposed a solution to the population inversion problem. These scientists are Olga Kocharovskaya and Yakov I. Khanin of the Institute of Applied Physics in Nizhniy Novgorod, Stephen E. Harris of Stanford University and Marlan O. Scully of the Center for Theoretical Physics at Texas A&M University in College Station.

The idea is to use quantum-mechanical waves to prevent photons from excited atoms of being absorbed by atoms in a lower-energy state. Scully's method which he has used successfully in gas experiments is to employ two beams on a cloud of electrons surrounding an atomic nucleus. One beam is a weak laser with randomly varying phases which boosts part of the electron population to a high energy level. The other beam deposits photons whose energy is equal to the difference between two lower electron energy levels of the atom in the gas. The beam links the two lower energy levels.

The electrons in the two linked lower energy states act like oscillating waves 180 degrees out of phase, canceling each other out. This means the lower energy electrons no longer are capable of absorbing photons. The electrons that are raised by the first beam to the excited state are not coupled to another state. They act like waves oscillating in phase. The medium can emit coherent light without absorbing its own emitted photons. This could lead to small continuous coherent wave ultraviolet and X-ray generators.

In the early 1990's an expensive large synchrotron was used by IBM and others to produce collimated X-rays to make experimental integrated circuits on chips. The phaser could possibly provide a much cheaper source of collimated, coherent X-rays.

BIBLIOGRAPHY

Bromberg, Joan Lisa — *The Laser in America,* 1950-1970, MIT Press, Cambridge, MA, 1991

Gourley, Paul L., Keven L. Lear, Richard P. Schneider, Jr. — "A Different Mirror." *IEEE Spectrum,* August 1994, pp. 31-37

Hecht, Jeff — *Understanding Lasers, An Entry-Level Guide.* IEEE Press, New York, 1992

Iga, Kenichi, Fumio Koyama, and Susumu Kinoshita — "Vertical Cavity Surface-Emitting Lasers." *Journal of Quantum Electronics*, Vol. 24, pp. 1845-1855

Jewell, Jack L. James P. Harbison, and Alex Scherer — "Microlasers," *Scientific American,* November 1991, pp. 86-94

Likourezos, George — "Jan. 28, 1958: A Laser is Born." *IEEE Spectrum,* May 1992, p. 43

Matthews, Larry L. — *Laser and Eye Safety in the Laboratory.* IEEE Press, New York, 1995

Schawlow, Leonard and Charles H. Townes — "Infrared and Optical Masers." *Physical Review,* Vol. 112, 1958, pp. 1940-49

Yamamoto, Yoshihisa and Robert E. Slusher — "Optical Processes in Microcavities." *Physics Today,* June 1993, pp. 66-73

CHAPTER 6

RECORDING TECHNOLOGY IN THE 20TH CENTURY

A number of different recording technologies were developed which included movies, methods of recording sound and video information, copiers, and superdense optical storage systems. Some of the technologies like motion pictures had roots in previous centuries.

6.1 THE MOTION PICTURE

The roots of the 20th century motion pictures were photography developed in the 19th century. The forerunner of the camera was the camera obscura, a dark room or chamber with a hole in one wall through which images of objects outside the room were projected on the opposite wall of the room. This was known by the ancient Greeks. Late in the 16th century Giambattista Della Porta, an Italian scientist, demonstrated the use of a camera obscura with a lens. In the 18th century artists commonly used various types of camera obscura to trace images from nature. In 1727 a German professor, Johann Heinrich Schulze, proved that light darkened silver salts. His discovery, in combination with the camera obscura, provided the basic technology necessary for photography which was developed by Daguerre and others in the 19th century.

Daguerre discovered in 1835 that light forms an image on a plate of iodized silver and that it can be "developed" and made visible by exposure to mercury vapor, which settles on the exposed parts of the image. By 1837 Daguerre was able to fix the image permanently by using a solution of table salt to dissolve the unexposed silver iodide. That year he produced a photograph of his studio on a silvered copper plate. Daguerre called the process Daguerreotype. Photography rapidly improved and changed and became widely used by the late 19th century.

In the period from 1884-1885 Eadweard Muybridge made a study of the motion of horses. He used a series of 12 to 24 cameras ranged side by side opposite a reflecting screen, with their shutters released by the breaking of threads as the horse dashed by. Muybridge secured sets of sequence photographs of successive phases of the movements.

In 1880 at San Francisco, Muybridge threw his photographs upon a screen one after the other with a lantern slide projector. This in effect was the world's first motion-picture presentation. A French physiologist, Etienne-Jules Marey, improved on Muybridge's work using a single camera.

On February 27, 1888, Professor Muybridge interviewed Thomas Alvin Edison as to the possibility of combining his zoopraxiscope projector with Edison's phonograph. Later Edison claimed he had already in 1887 thought of devising -- "an instrument which should do for the eye what the phonograph does for the ear and that by a combination of the two all motion and sound could be recorded and reproduced simultaneously."

Mr. William Kennedy Laurie Dickson who worked for Thomas Edison began in 1889 the development of a picture machine called a kinetoscope, a box into which one looked to see pictures in animation. A ribbon containing a series of pictures continuously passed between a small lamp and the eye of the observer. A rotating disc about a foot in diameter blocked out the picture except for a l/8th by one-inch radial slit. Through this flying slit the observer got a momentary sight of each picture frame as it came into position above the light. The frames passed at the rate of 46 per second. Edison patented the peephole kinetoscope on March 14, 1893. This was the first motion picture exhibiting machine employing a perforated film with equally spaced juxtaposed pictures. It was the first practical picture-exhibiting machine of any kind but the kinetoscope was not capable of projecting pictures on a screen. The first commercial showing of the kinetoscope was at the Holland Brothers Peep Show Parlour, 1155 Broadway, New York City, April 14, 1894. Later that year six Edison peep-show kinetoscopes were installed in old Broad Street in London by Robert W. Paul.

Edison in September 1902 patented the kind of film Dickson used. It was l 3/8 inch wide over all. It contained four perforations to each picture, the picture itself being l inch wide by 3/4 inch high. These dimensions became standard in the movie business.

Dickson also made the kinetograph which projected pictures on the screen but Edison abandoned it in favor of the kinetoscope which was able to be a commercial product in peep shows. The Edison kinetoscope appeared in a shop in Paris in 1894 and was seen by Louis Lumiere and his brother, owners of a factory making photographic equipment. They made a device which projected pictures on a screen. They called it the Lumiere Cinematograph. This was a combination camera, printer, and projector. As a projector it kept the film stationary for a time corresponding to the two-thirds of the total time, when an elementary image appeared exactly on the lens axis. The device allowed the frequency of 16 images per second and an illumination of 1/25 second per image.

On December 28, 1895 the Lumiere brothers opened a place in the basement of the Grand Cafe, on the Boulevard des Capucines, Paris. A small admission fee was charged for people who witnessed the following short films:

Men and Women Employees Leaving the Lumiere Factory
Arrival of a Train at the Station of La Ciotat
The Baby's Lunch

The Sprinkler Sprinkled
Boat Leaving the Harbor

This was the first time that an admission fee was collected for a motion picture projected on a screen. In a primitive way it was the start of "Going to the Movies," which later became the past time of millions of people in the 20th century after a series of inventions. Later the camera and projector were made into separate units.

The first practical motion projector, the Vitascope, was invented by Thomas Armat. This was the first projection machine employing a means of forming a loop. The Vitascope projected the picture on a large screen. Armat contracted with Edison to build 50 projectors. On the evening of April 23, 1896 Armat personally operated the machine at the first public exhibition of the Vitascope. It took place before a crowd at Koster and Bial's Music Hall in New York City. One of the scenes projected was of storm-tossed waves breaking over a pier on the beach of Dover, England. Edison's name was used in connection with the machine, partly for the commercial advantage of his name, and partly because he was the producer of and had patents pending covering the films, an essential part of the machine. This crowd called, "Edison, Edison, Speech, Speech."

The projector invented by Armat was bought in 1896 by people in many countries including England, South Africa, Denmark, Sweden, and Spain. Some of the projectionists were not trained well and some of the exhibitions were not always successful. For example, a projectionist in Barcelona failed in his first attempt and the disappointed audience threw knives at the screen and wrecked the theater. The projectionist was sentenced to serve a term in a Spanish jail.

The Vitascope was distributed in New York City by Norman Raff and Frank Gammon. At this time Edison was starting to make pictures in his Black Maria studio at West Orange, New Jersey. In order to overcome the difficulties at that studio, Raff and Gammon in 1896 built the first movie studio in New York City. They hired May Irwin and John Rice to reenact a kissing scene from the Broadway play, "The Widow Jones." Photographed on fifty feet of film, the May Irwin-John Rice kiss became a roaring hit. It made the Vitascope a worldwide sensation.

Thomas Armat also invented the star-wheel intermittent movement which became the "heart" of the motion picture projector. A small sprocket carrying the film could be given a gradually accelerated intermittent movement without film wear and tear and without jar to the mechanism. The star-wheel intermittent movement was patented on March 2, 1897.

Another essential invention for the motion projector was a device which frames the pictures while the machine is running. The Alfred E. Smith framing device was patented on April 30, 1901, patent number 673,329.

Later the number of frames was reduced from forty per second to twenty-four per second. The problem of flicker became more important with the reduction of frames per second. The John A. Press shutter was an important device for reducing flicker. It was patented on March 10, 1903, patent number 722,382.

In 1915 the U.S. Supreme Court legally designated the motion picture a mere "spectacle" and ruled in *Mutual Film Corporation vs. Industrial Commission of Ohio,* that the film was not entitled to constitutional protection under the first amendment, thus shackling the motion picture within the constraints of arbitrary censorship for the next 35 years.

The motion pictures became an important source of entertainment early in the 20th century. Two great improvements were the addition of sound and later of color.

SOUND

Strictly speaking, the silent film was never silent, since the very first pictures were shown with accompanying music. Furthermore, many tried from the start to have synchronized sound and picture using the phonograph method even though it gave mediocre quality and insufficient sound intensity. Only short films could be recorded with synchronized sound because in those days the running time of turntables and records was limited. When pictures increased in length, this first epoch of the short sound films ended and accompanying music gained more and more importance. In the early 20th century the showing of motion pictures was accompanied by a live piano player. It was the piano player's duty to adapt the music to the action.

There were two early basic methods of combining sound with motion pictures. One was photographic sound recording and the other was the synchronized electrically recorded disc. In 1926 the Vitaphone Corporation was founded to make sound motion pictures using synchronized electrically recorded discs. For synchronous recording the camera and the recording turntable were driven by selsyn motors. This driving system gives the equivalent of both being geared together and driven from one shaft. Starting marks on both film and disc are essential. For reproducing, the turntable and projector were mechanically geared together.

The first major sound picture by Vitaphone Corporation to be released was *Don Juan,* August 1926, which featured music by the New York Philharmonic Orchestra. Vitaphone then exhibited *The Jazz Singer* with Al Jolson, October 6, 1927. This is considered the first true modern talking movie.

The other method of synchronizing sound and pictures was photographic recording. The sound was converted to a beam of light, which was then recorded on the narrow soundtrack of the film. Professor Joseph T. Tykociner of the University of Illinois in June 1922 presented one of the first public exhibitions of sound on motion-picture film. He used a mercury vapor lamp whose light could be modulated by sound for recording on film. A photoelectric cell converted the variable density sound track on the film to an electric current which could then be converted to sound. Professor Tykociner never patented his device. It was very crude but it demonstrated the feasibility of placing the sound on motion picture film. Later a galvanometer with a revolving mirror became the RCA photophone standard for photographic recording.

Around 1925 a program of developing commercial sound-on film equipment was undertaken by General Electric Co. The first public entertainment picture to be shown with the G.E. developed sound on film system was *Wings,* a story of the Army Air Corps activities in World War I. The picture was produced by Paramount and the sound effects were added after the picture had been shot. *Wings* was exhibited in 1927 as a "Road Show" with a dozen sets of equipment having been supplied. The picture width was reduced from 1 inch to 7/8 inch to make room for a soundtrack.

Western Electric also developed a sound on film system for motion pictures. The commercial outlet for the Western Electric system was the Electrical Research Products, Inc. (ERPI). In 1928 the "Big Five" movie producers, Paramount, United Artists, M-G-M, First National, Universal, and several others signed agreements with ERI for licenses and recording equipment.

In the early theater installations, many projectors were equipped for both disc and film reproduction of sound. A speed of 90 ft./minute or 24 frames per second were chosen for both the sound on disc and the sound on film systems. At the end of 1927 there were some 157 theaters in the United States equipped for sound, of which 55 were for disc and film and 102 for disc only. At the end of 1928, of the 1046 ERPI theater installations, 1032 were for disc and film. Soon optical recording superseded the disc system.

It was discovered that distortion in the sound-track area of the film could be lowered by recording the light impulses as two separate but identical sets of patterns read simultaneously by a single photoelectric cell in the projector. Later when stereophonic motion pictures were introduced in 1976, the two optical tracks were easily converted into left and right channels which gave a more realistic sound reproduction of the musical score.

COLOR MOTION PICTURES

Still color photography was invented in 1907. The first color motion picture, *The Glorious Adventure,* was made in England around 1916. It used the Kenemacolor Process which photographed the color components by successive exposure. This led to a horse with two tails, one red and one green. It also led to visible fringes whenever there was rapid motion.

Technicolor Motion Picture Corporation in the United States attempted to improve the color process by using two simultaneous exposures from the same point of view. The process was two-color additive and the film was processed in a laboratory within a railway car. This car was completely equipped with a photochemical laboratory, darkrooms, fireproof safes and all the machinery and apparatus necessary to develop color film.

In 1917 the railroad car was rolled over the railway tracks from Boston, Massachusetts, where it was equipped, to Jacksonville, Florida, where the first Technicolor picture, *The Gulf Between*, was being made. The camera was the single-lens, beam-splitter. The color was introduced by projecting through two apertures, each with a color filter, bringing the two components into register on the screen by means of a thin adjusting glass element. This system demanded a projector operator who was a cross between a college professor and an acrobat. Technicolor decided after *The Gulf Between* to abandon additive color processes and special attachments on the projector to produce color.

In 1919-21 Technicolor developed a two component subtractive color process. The first color movie produced by the new process was *The Toll of the Sea* which had a general release in 1923. This picture was photographed in Hollywood by Metro-Goldwyn-Mayer and produced by Technicolor. The movie grossed more than $250,000, of which Technicolor received approximately $160,000. *The Toll of the Sea* was so successful that many moviemakers became interested in color. The Famous Players Lasky Corporation contracted with Technicolor to produce a picture in color, *The Wanderer of the Wasteland,* in 1923. It was shown in several thousand theaters over the country.

The two-color system was not completely satisfactory and Technicolor in 1932 developed an improved three component color system. The accuracy of tone and color reproduction was greatly improved and definition was markedly better. Gradually the movie evolved into excellent color reproduction combined with sound that became common in the late 1940's and beyond.

Problems With Early Films

One of the problems for many of the movies in the first half of the 20th century was the nitrocellulose-based film. Some of those early films were stored away in vaults. Much to the chagrin of researchers many of the films disintegrated into a pile of dust when the vaults were opened. Some of the movies were rescued in time and restored on new film but many were lost. Starting in 1951 a new type of more permanent film was used. The American Film Institute works with film professionals to preserve and restore old films, especially made before 1951 of flammable nitrate stock. Their goal is to convert all of these films to nonflammable stock before they are lost forever.

6.2 The Phonograph

The roots of the 20th century recording of sound go back to the summer of 1877. Edison was attempting to transcribe the human voice by making indentations on paper. The telephone at that time was not yet practical. In his experiments Edison used a stylus attached to a diaphragm. A person speaking into the diaphragm caused the stylus to vibrate up and down on a strip of parafinned paper creating indentations on the paper. Edison found that when the paper was pulled back beneath the stylus the sound of his

Figure 6-1 Edison and the First Phonograph, 1878
IEEE Center for History of Electrical Engineering

voice was vaguely reproduced. He decided to pursue this project vigorously.

In December 1877 Edison, Figure 6-1, improved on his discovery by replacing the strip of paper with a cylinder wrapped in tinfoil. This was the first machine that could both record and reproduce sound. The cylinder was rotated and the sound was recorded as variations in the depth of the groove cut by the stylus. For reproduction the groove was again run under the stylus tip with only enough pressure to maintain contact with the groove bottom. By the early 20th century the phonograph had been transformed by a number of inventions from a laboratory curiosity to a source of entertainment in many homes.

In 1885 two Americans, Chichester A. Bell and Charles Sumner, patented a machine that cut the groove in a cylindrical surface of wax with a sharp tool carried by the vibrating diaphragm. This gave much better sound reproduction than Edison's indented tinfoil.

Another inventor, the German-born Emil Berliner, made a radical departure using a flat disc instead of a cylinder. The disc was made of zinc and was coated with a thin layer of a fatty substance that protected it from the action of acid except where the scribe had traced its line. An acid bath then left a groove, the width and depth of which could be played on an appropriately designed reproducing mechanism which he called the Gramophone.

Berliner used the etched master to make a negative by electroforming. He made the master record surface electrically conductive by a wet-silvering method. He then plated it with copper (later nickel) creating a negative mold of the silvered-lacquer master. The electro-formed metal master disc was separated and used directly in a plastic molding press. He used this process to make records in a thermoplastic material, a procedure that subsequently became standard. Berliner also developed lateral recording in which vibrations are recorded as sidewise deflections of a groove of uniform depth rather than as variations in the depth of the groove. Starting in about 1900 lateral recording attained wide acceptance in the United States and in Europe.

Cylinder recording was still being improved and used. Methods of molding cylinder records of a relatively hard thermoplastic were introduced in1901, and in 1908 cylinders made of a new material called amberol enabled recording grooves to be spaced 200 to the inch increasing the playing time to four minutes per cylinder. Cylinders were replaced by the 78-RPM records by 1915. That record had a playing time of approximately 4 1/2 minutes per side.

The early phonographs used hand driven spring motors with friction disc governors. The friction increased rapidly if the speed exceeded the desired value. Extensive use of electric motors came in the 1930's when radio-phonograph combinations were introduced.

In early reproduction for the phonograph, performers huddled close to large recording acoustic horns. At the small end of the horn the sound waves were concentrated onto a diaphragm to which the cutting stylus was attached. During playback the acoustic output from the stylus-diaphragm was sent through the acoustic horns. The design of acoustical horns was improved in the early part of the 20th century.

In the early 1920's, a transition from acoustical to electrical methods took place in the phonograph business for both recording and playing. The development of the vacuum-tube amplifier made electrical methods practical. In recording, microphones collected the sound, which was then amplified electrically to the necessary level for recording with an electrical disc-cutting head. The vacuum tube was also applied to the replacement of the acoustical horn by the electric loudspeaker and to the development of the electrical phonograph cartridge.

A phonograph cartridge or pickup is a device that when actuated by the record groove produces a usable electrical signal proportional to the mechanical signal recorded in the groove. It is mounted in a tone arm which allows its stylus to track the spiral groove on the record. Steel was used for the stylus in early phonographs but they were displaced by diamond stylus tips which could be ground to precise dimensions. The stylus should retain contact with the two sides of the V-shaped record groove, and the point of the stylus should ride above the bottom of the V.

In order to generate an electrical signal, a number of techniques were developed such as the piezoelectric effect and the use of miniature magnets. The piezoelectric effect is the production of electricity by pressure on a crystal. The most popular use of high quality phono cartridges employs miniature magnets and coils. In order to produce an electrical audio signal, the stylus assembly either moves the magnet in relation to the fixed coils, or moves the coils in relation to the fixed coils. The purpose in each case is to cause a magnetic field to impinge on the coil that varies with the movement of the stylus, which in turn reflects the undulations in the record groove walls.

Twentieth century phonographs came of age in 1948 when Columbia Records introduced the long-playing (LP) record made of Vinylite. The new record was 11 7/8 inches in diameter, with a rotational speed of 33 1/3 rpm. LP's used fine-groove or microgroove recording. The old 78-RPM records had a playing time of approximately 4 1/2 minutes per side. The LP record extended the playing time to 30 minutes per side. Shortly afterward the RCA Victor Company introduced a microgroove seven-inch 45-rpm disc which played for eight minutes. The new records rapidly supplanted the old 78-rpm discs.

Stereophonic phonograph records, with two separate channels of information recorded in a single groove, was available commercially in 1958. This attempted to reproduce the quality of live musical performances where the sounds of various instruments come from different directions. Stereophonic systems, by reproducing sound with spatial perspective, produce a far greater sense of reality than single-channel or monophonic system. A stereophonic system requires two independent channels, with two separate microphones for recording and two separate loudspeakers for reproduction. On a stereo disc the outer groove wall contains the left-channel signal, the right channel being recorded on the inner groove wall. A single stylus is used with its axis of movement inclined at 45 degrees to the record surface. The pickup is designed to produce two practically independent signals in accordance with vertical and horizontal motion components. Operation at 45 degrees provides a symmetrical arrangement that offers advantages both in the design of disc cutters and reproducers and in quality of reproduction, because the electronic characteristics of the two channels are alike.

Two basic systems are used for motors and drive systems for the phonographs. One is a belt drive that couples a high-speed motor to the slow-moving turntable platter and the other is a direct-drive system employing a special semiconductor-controlled motor whose shaft rotates at the playing speed of the record. In the latter type of system the heavy metal platter that supports the record sits directly on the motor shaft. Precision machining and careful attention to design details are necessary to minimize "wow" and "flutter." "Wow" is a low frequency and "flutter" is a high frequency periodic change in signal frequency that results from momentary variations in turntable speed. "Rumble," a low-frequency noise picked up by the cartridge is caused primarily by vibrations from a phonograph motor that are transmitted through the motor supports or the drive system.

By 1940 many manufacturers were equipping their phonographs with record changers but the edges and center holes of the heavy 78-rpm records were damaged by the changer mechanisms. The introduction of 33 1/3 vinyl records in 1958 made changers practical since the vinyl records considerably reduced damage to the record by selector mechanism. The record changer used a small high-speed induction motor which communicated its rotation to the inner rim of the turntable platter through one or more rubber-edged idler wheels. When the tripping mechanism at the end of the record was triggered, the motor turned a series of gears connected to a mechanism that lifted the pick-up arm above the record just played, returned the arm to the side, dropped the next record down from the stack, and placed the tone arm down on the new record in its lead-in groove.

6.3 AUDIO MAGNETIC TAPE

Attempts in the early 20th century to record magnetically were hampered by problems of high distortion and noise and the absence of a medium on which to record. These problems were partially solved in 1927 with a patent by W.L. Carlson and G.W. Carpenter for the use of alternating-current bias to reduce distortion and increase the signal to noise ratio. In that same year the first United States patent was granted for magnetic tape made by drying a liquid containing magnetic particles on the surface of a strip of paper. In 1936 at the Berlin Radio Fair, the Magnetophon Company of Germany demonstrated the potential of tape recording by using a plastic-base magnetic tape to reproduce a performance by Sir Thomas Beecham leading the London Philharmonic Orchestra. During World War II Magnetophon machines were used for Nazi propaganda by the German government. Later the Magnetophon was imitated and improved upon all over the world.

After World War II, Marvin Camras of the Illinois Institute of Technology invented much of the new magnetic technology. Modern magnetic tapes are made up of a plastic-base film coated with a binder that holds a magnetic powder in place. The magnetic material most widely used in coating tape is a form of iron oxide, Fe_20_3, in very small needle shaped particles roughly 0.023 mils (thousands of inches) to 0.039 mils in length and one-tenth or less that in diameter. Other magnetic materials which were developed in the 1960's and 1970's are chromium dioxide and cobalt-adsorbed ferric oxide. Pure metal-particle (non-oxide) iron tape was introduced in 1978 by the 3M Company to produce better high-frequency response at high recording levels. The different magnetic materials are in the form of a powder.

The magnetic powder is mixed with a binder, a glue that holds the particles apart evenly dispersed throughout the coating. The wet mixture of binder and magnetic powder is precisely coated onto large rolls of the tape base. The magnetic coating for cassettes is about 0.24 mils.

Mylar and other polyester plastics are widely employed as the tape base. The most common thicknesses are 0.47 and 0.31 mils. The coated roll of film is passed over an intense unidirectional magnetic field to orient the needle-shaped particles uniformly and the coated film is then passed through drying ovens. The tape is pressed between sets of rollers to polish the surface and ensure a uniformly thick coating.

RECORDING ON MAGNETIC TAPE

The electrical signal to be recorded is applied to the tape by means of a "head." This consists of a coil wound around a core of magnetic iron. The core has a gap which allows the tape to move across its surface. The current in the coil magnetizes the particles in the tape. The tape receives a magnetic record containing direction (north or south pole), amplitude, and linear dimension (along the tape) corresponding to the direction (plus or minus), amplitude, and time of the original electrical signal.

In AC-biased recording the tape is passed across an erase head before it comes to the record head. Erasure removes any previous pattern in their north-south or south-north orientation, so that the distribution of north-seeking and south-seeking particles is random. After erasure the sum of the particles magnetic fields is zero. When the tape reaches the record-head gap an ultrasonic alternating bias current is added to the signal current. The bias current is much higher in amplitude than the signal current and several times greater than the frequency of the highest frequency signal to be recorded. While crossing the gap of the record head, the magnetic particles in the tape are subjected to many cycles of the bias current and leave the head with a magnetization that is proportional to the signal. The purpose of the AC bias is to assure low distortion and a high signal-to-noise ratio. Cassette heads for recording and playback are in the order of 50 millionths of an inch.

PLAYBACK

In cassette reproduction the tape is passed over the same head used in recording. The magnetized patterns on the tape passing over the gap in the playback cause the magnetic flux in the core to change, generating a voltage in the coil proportional to the speed of the tape and the amplitude and frequency of the recorded signal.

One of the advantages of magnetic recording is the ability to erase and reuse the recording medium thousands of time with little or no loss in the recording quality. Erasure is usually accomplished with the erase head.

THE CASSETTE

This was originally introduced around 1965 as a dictation device for business use by Philips of the Netherlands. However, it developed so remarkably into a high-fidelity medium that by the early1980's sales of prerecorded music cassettes exceeded those of phonograph records.

The tape is 0.15 inches wide and travels at a speed of 1 7/8 inches per second. After play is complete, the cassette can be inverted and played in the opposite direction. Cassettes are generally sold in C-60 and C-90 lengths (30 and 45 minutes per side) though some other lengths are available.

The most common drive mechanism found in the common tape cassette is the capstan drive. The tape is pulled forward by the friction between a pinch roller which consists of a cylindrical piece of rubber, and a motor-driven cylinder (capstan), whose rotational speed and diameter determine the speed of the tape.

THE DOLBY NOISE REDUCTION SYSTEM

The Dolby system is used in audio recording and transmission. In 1965 the Dolby A-type noise reduction system was developed by Ray Dolby with the first sales in 1965. As the music was being recorded a small amount of the recorded signal was fed back through a network of filters with a slight compression. Since this is a very low level it is mostly noise with a small amount of signal. This is then subtracted from the output of the recorder. The noise of the recording process is reduced but a small amount of signal is also reduced. To compensate for this an identical set of filters is placed at the input to the recorder. The output of these filters is added to the input of the recorder. This adds a small amount of signal to the input of the recorder to compensate for the small amount of signal lost at the output of the recorder.

The Dolby-A system was expensive and used for professional recording. In the years 1967-1969 the Dolby-B type consumer system for noise reduction was developed. It was a cheaper version of the Dolby-A system. The same principle of adding at the input and subtracting at the output was used although the filters used were much simpler and much less expensive.

In 1980 an improved version called Dolby-C was developed for the recording and playing of audio magnetic cassettes. This reduced the noise by 20 dB. It was used in over 20 million products by the end of 1986.

6.4 Copying with Ease

Up to about 1960 copying in business offices used stencils and ditto paper to make copies. This was laborious and scientists worked on new methods of reproducing a large number of copies using new technology. One of these was xerography.

Xerography

Chester F. Carlson, a patent attorney, became tired of the costly and slow process of producing multiple copies of patents. He decided to develop a new type of copying machine. Carlson enlisted the help of a refugee physicist, Otto Cornei. In 1938, Carlson made the first image using an electrostatic charge combined with a photoconductor, a material that becomes conductive when exposed to light. His patent #2,221,776 filed September 8, 1938 and issued November 19, 1940, described the invention which Carlson called electron photography. The process was further developed by the Batell Memorial Institute, a nonprofit institute.

In 1947 a manufacturer of photographic paper, the Haloid Company, acquired the commercial rights to Carlson's patents. The Haloid Company described the new process as xerography from the Greek words *xeros*, dry, and *graphein*, to write. Haloid first changed its name to Haloid Xerox, Inc. and later to Xerox Corporation. The first commercial xerographic copier for office use was marketed in 1960.

Xerography as it developed became very popular and spread to most business offices. In many machines a drum, belt, or plate coated with the photoconductive element selenium is charged with static electricity. Light reflected from the original document or illustration passes through a lens. The light strikes the photosensitive surface, forming on that surface a positively charged image corresponding to the dark areas of the original. The remainder of the surface loses its charge. Then negatively charged powdered ink (called toner) is dusted onto the surface. The negatively charged toner is attracted to the positively charged image. The inked image is transferred to positively charged paper and heated for an instant. The toner melts, creating a permanent copy. Some electrostatic copiers project the image from the original directly onto specially coated paper, instead of using a drum belt or plate.

Newer electronic copiers can generate over a hundred copies a minute. Special features were developed which included color reproduction, automatic document feeding and sorting, image reduction and enlargement and stapling.

As part of the digital revolution in the 1990's, a digital electrostatic copier was developed. The equipment converts the document into digital code which is stored in the copier's memory. The user can then edit the document, merging different parts of documents.

6.5 JOINING THE DIGITAL REVOLUTION

Most of the recording technology described previously is of the analogue type. In analogue recording there is a correspondence between the continuous variations of the sound waves and the methods by which they are stored. The electrical signals from the microphone, the undulations in the record groove, and the magnetic patterns recorded on tape are all analogues of the original sound. However, noises in the recording and playback such as from records are also reproduced in the analogue method.

In digital recording the sound waves are sampled at specific time intervals, and the amplitude value of each sample is converted into a binary number consisting of 0^s and 1^s Physically the ones are pulses and the lack of a pulse is a zero. The 0^s and 1^s are known as bits and are grouped into digital "words." In early audio digital recording 16 bit words were used for consumer application. A typical 16-bit word would look like this: 1010011101011000. This binary number represents the amplitude of a specific sample of an analogue of the original sound. Using such 16 bit words the dynamic range of the consumer digital system is far greater than the best available analogue recording and playing machines. The dynamic range is the difference between the highest amplitude signal and the amplitude of the noise. The dynamic range is 2^n where n is the number of bits in a digital "word." In commercial audio digital recording n is 16. The dynamic range is then 2^{16}. The dynamic range is expressed in decibels or the dynamic range = 20 $LOG_{10}\ 2^{16}$ = 96 dB. This is close to a live orchestra which has a dynamic range of 100 dB. By comparison the best figure of dynamic range for analogue disc and magnetic tapes is about 70 dB. The reason for using decibels is that the ear reacts to sound in a logarithmic manner.

The distortion using the digital recording system is also far less than the analogue method. The mechanical wow and flutter, which degrades analogue disc recording and playback systems, are eliminated by digital systems. The reduction in distortion and the great increase in dynamic range are the two main advantages of the digital method of recording.

The sampling rate for converting analogue to digital must be at least twice the highest frequency to be recorded. The highest signal that the human ear can hear is 20,000 Hz so that the sampling rate must be in excess of 40,000 Hz. A rate lower than 40,000 Hz would not pass frequencies of 20,000 Hz. Consumer digital audio uses a sampling rate of 44,100 Hz to allow a small margin of error.

A precursor of digital recording started in the 1970's in the form of digitally mastered recordings converted into conventional analogue records. By the early 1980's many master tapes of classical music were recorded digitally and then released in the form of analogue discs and tapes. This led to the invention of the Compact Audio Disc (CD) in 1982 by SONY in Japan. The CD could play up to 75 minutes of music or audio entertainment.

6.6 THE AUDIO COMPACT DISC

The audio CD is a system of audio recording in which an audio signal is digitally encoded on a rotating disc. This is accomplished by forming pits along spiral tracks on a transparent plastic disc, overlaying this with a reflective coating, and then covering this coating with a protective layer. A laser beam directed at the pits converts them into a series of 0^s and 1^s in digital form. The use of an optical pickup results in no wear since there is no mechanical contact between the pickup and the disc.

The focused laser spot size is extremely small so that a very great amount of information can be packed on the surface of the disc. Typically adjacent tracks of the spiral of pits are only 1.6 microns apart and 20,000 such tracks are available on a 120-millimeter (4.7-inch) audio CD. The pits vary in length from 0.9 to 3.3 microns. The track is optically scanned at a constant linear velocity of 1.25 meters per second or 4.1 feet per second.

The audio CD master is made from a glass disc covered by a uniform coating of photoresist. A laser is shone where the pits are to be formed. The light from the laser hardens the photoresist at these places. The photoresist is then developed and washed, leaving the master disc. A nickel mother is then derived from this master and is used to stamp out multiple copies of the disc in transparent plastic material. Each of these is then coated with a thin metallic reflecting layer, with a protective plastic coating on top of that.

The information that is read off the discs is in digital form as a sequence of 0^s and 1^s. These can be processed in many more ways than were possible with analogue systems. Information can be stored for as long as is desired, and then sent out at a rate that is controlled at the player's quartz-crystal oscillator clock. This eliminates the wow and flutter of conventional systems entirely.

The digital signal is converted back to analogue by special circuits. The conventional audio CD is a read only device. It only plays audio and does not allow for recording. Digital recording is performed on special tapes and on special discs.

The mini-disc is a digital compact disc about 2.5 inches in diameter. It was introduced in 1990 by SONY. The mini-disc plays and records digitally. They are

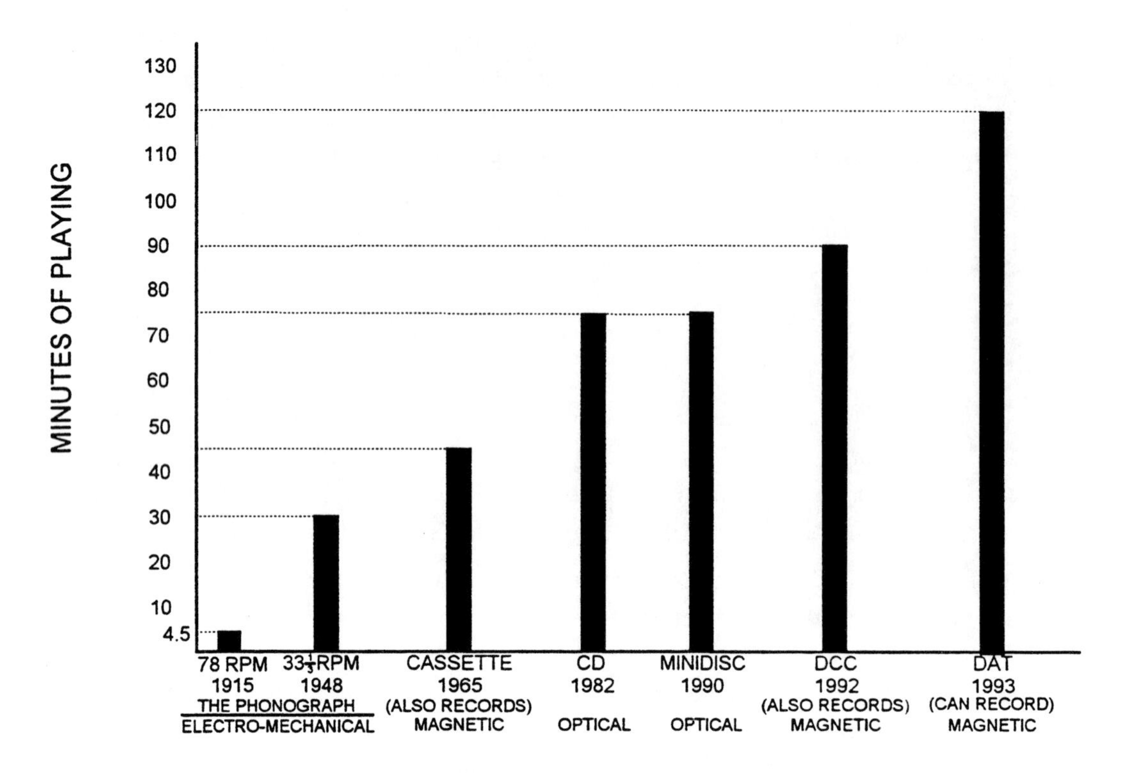

Figure 6-2 Audio Players in the 20th Century

portable and blank mini-discs are available for one-time recording. Figure 6-2 shows the development of audio players in the 20th century from the 78 RPM phonographs to the digital audiotape.

6.7 Audio Digital Recording Tapes

One type of digital tape is the Digital Compact Cassette (DCC) introduced in 1992. This offered digital recording and playback when used in new DCC playback decks and high quality analogue (non-digital) playback when used in older cassette decks. In size and shape the DCC is compatible with the analogue compact cassette.

Another format is the digital audiotape (DAT) which uses different sized cassettes, different machines, and has digital quality. The DAT equipment is not compatible with the conventional analogue compact cassette. The DAT cassettes are about two-thirds the size of a regular analogue cassette and need not be manually flipped over or reversed.

The SONY DAT can make a perfect copy of a CD but a Serial Copy Management System in non-professional DAT recorders prevents the user from making further digit-for-digit copies.

6.8 VIDEO RECORDING MAGNETIC TAPES

In the 1950's there was a problem in broadcasting television from New York studios to both the east and west coasts simultaneously. A program broadcast at seven PM in New York would be seen in California at four PM local time. This meant that workers on the West Coast would not be able to see these programs.

Ampex, a recording company in the United States, in 1956 built a video magnetic tape machine that would allow television programs in New York City to be rebroadcast later to the west coast thus overcoming the time difference between the two coasts. Engineers at Ampex soon realized that the large bulky videotape recorders could be redesigned for home use to record TV programs. However, the technology was sold by Ampex to Japan for $100,000.

In Japan, SONY Corporation introduced the first low cost videocassette in 1969. Around 1975 SONY developed the Betamax format. Matsushita Corporation came out with the VHS system in 1977. The two systems use a 0.5-inch wide tape system but they were mutually incompatible with each other. A cassette recorded on one system cannot be played back on the other system. In the United States a fierce battle occurred in the late 70's and early 80's between the two methods. The VHS format finally won out and became the dominant videocassette in the U.S.A.

A videocassette recorder can have from two to as many as seven tape heads that read and inscribe video and audio tracks on the magnetic tape. Most VCR's have fast-forward and reverse controls and a timer that enables television programs to be recorded automatically, and they can record a program on one television channel while a viewer watches a program on another channel of the same television set.

Color home movies can be made with the use of a hand-held camcorder. This consists of a videocassette recorder that is connected to a light and simple video camera. One camcorder system uses 8-millimeter videotape and adapters are available to use in a standard VCR. In August 1983 SONY published a description in the IEEE Transactions on Consumer Products of the development of an extremely small videotape recorder.

6.9 THE LASER VIDEO DISC

This is a system that incorporates a player and a prerecorded disc that allows very high-quality reproduction of video, audio, and data information on a television receiver. This technology is based on laser reading of prerecorded analogue video and digitally encoded audio information.

In recording, the analogue horizontal video amplitude modulated signal is first converted into a Frequency-Modulated (FM) signal with a constant amplitude. This FM video signal is converted into a pulse width encoding signal which modulates a laser beam shining on a master disc coated with photoresist similar to the making of a CD previously described. However, in videodisc recording the pit length is proportional to the pulse width of the signal. The pit length is in the order of a millionth of a meter or 0.000039 inches. A laser videodisc includes the analogue information or visual content described above and audio information digitally encoded as described in section 6.5.

The videodisc player system employs a laser. Usually a microscopic solid state diode red laser or a helium-neon green-blue gas laser is used. Power levels are in the order of a few milliwatts to ensure safety. The optical system including the lens, mirrors, and prisms focus the light and manipulate it in the head. Photodetectors detect pit location through reflected intensity.

The focusing servo system has motor-driven microactuators that move the lenses in response to the sharpness of the reflected image. This ensures that the laser spot stays in focus as the disc rotates. The movement servo system has actuators that move the head laterally to locate information while the tracking servo system has microactuators that tilt the focused spot to maintain it on a single track of data.

The motor and control system rotates the disc. Two methods are used, CAV (Constant Angular Velocity) and CLV (Constant Linear Velocity). The storage capacity of CLV is twice as high as that of CAV devices. The CLV has a storage capacity of 60 minutes per side for the NTSC system used in most of the Western Hemisphere and Japan. The CAV devices have faster access time and are used for interactivity.

There are a number of advantages of videodisc technology as compared to videotape recording. The videodisc has a very high audio quality since the signal-to-noise ratio is very great. Another advantage is the availability of a full television frame which is not available with VCR reproduction. Also VCR tapes will deteriorate with time while videodiscs do not. Videodiscs do not wear out with use since the laser does not touch the videodisc. The videodisc has the advantage of interfacing and interacting with computers.

A new type of videodisc, the Digital Video Disc (DVD) came out in 1996. The DVD is a two-sided, 4.7-inch optical disc which can hold 266 minutes of full-motion video and Dolby six-channel surround audio. A superior video and audio were obtained by digital compression.

The DVD has a language button which will change the original language of the film into actual conversations in one of eight languages. The DVD will also allow parents to show their youngsters a version of the film which leaves out sex and violence. This is accomplished by programming the player.

Figure 6-3 shows the development of video players and recorders from the Betamax to the Digital Video Disc.

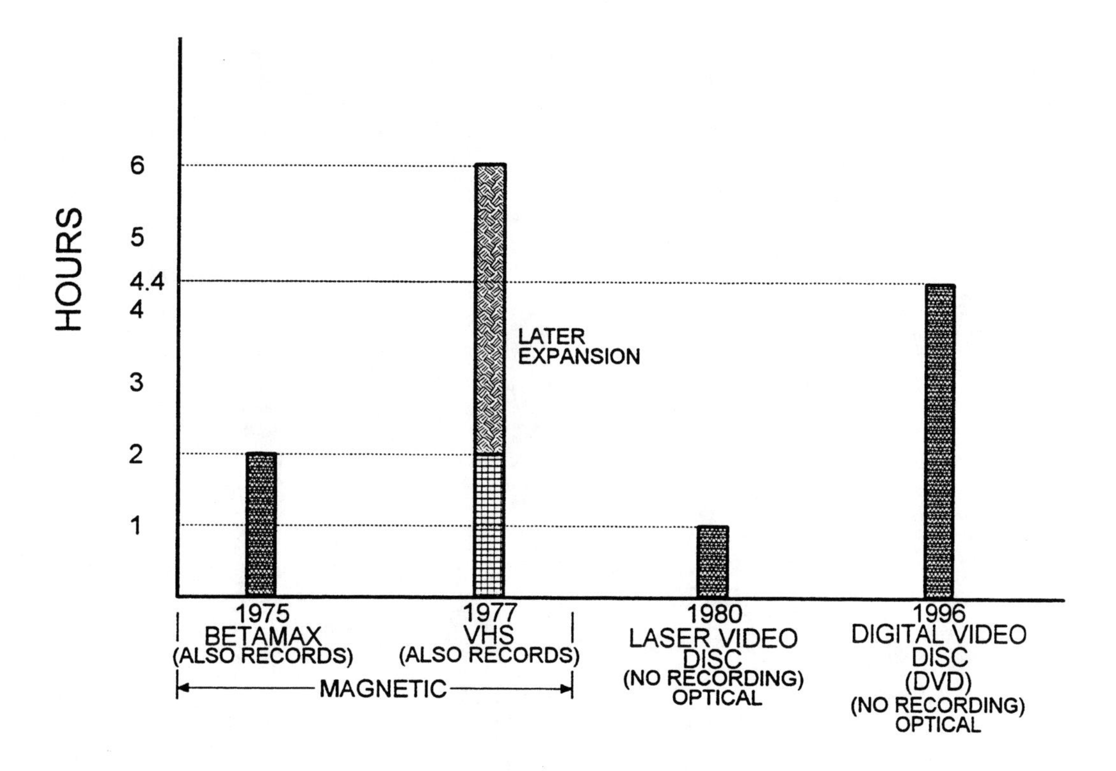

Figure 6-3 Video Players in the 20th Century

6.10 HIGH DENSITY OPTICAL STORAGE SYSTEMS

These use the laser to record and read digital information.

CD-ROM TYPES

In 1984 SONY came out with the CD-ROM which could store up to 680 megabytes of data including text and graphics. Whole encyclopedias could be stored on one disc. This could not be recorded or erased by the user.

In 1989 the CD-Recordable was developed. The user could record but not erase. The CD-Erasable in 1995 was capable of both recording and erasure by the user. This is expected as of 1999 to be a two-sided 150-millimeter disc which will be capable of storing 5 gigabits. It is intended to be a replacement for magnetic discs in computers. Figure 6-4 shows the development of the CD-ROM in the 20th century.

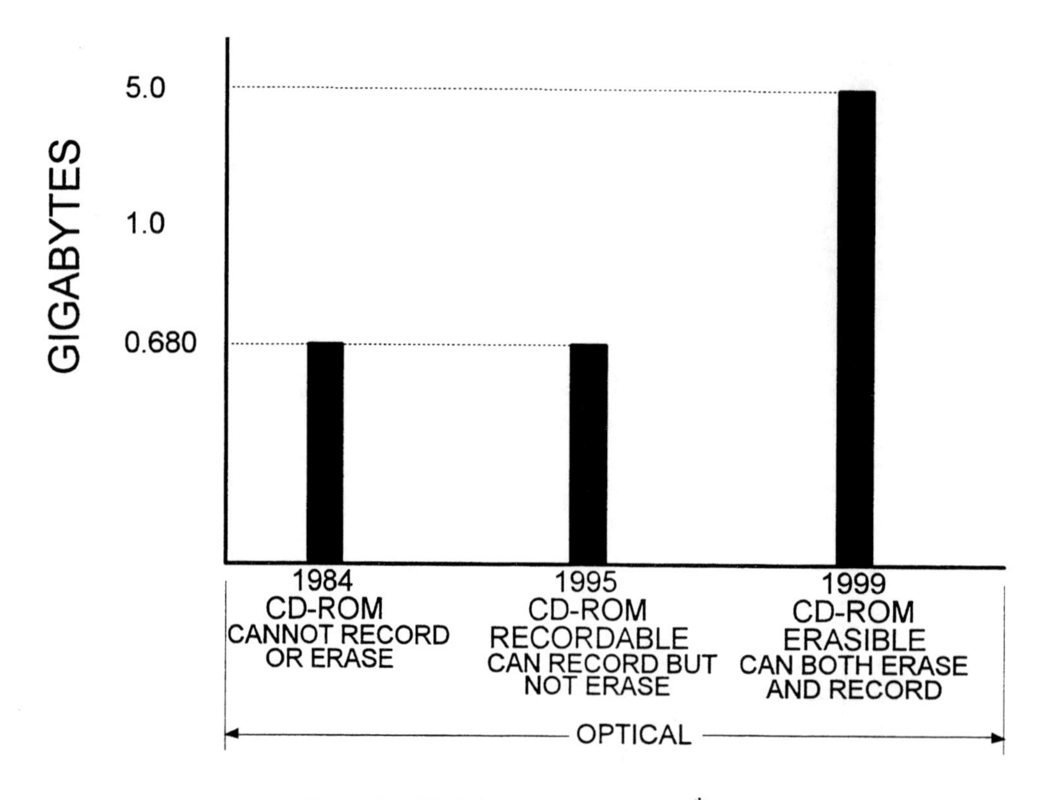

Figure 6-4 CD-ROM Players in the 20th Century

Optical Tape

Optical tape is a new technology using a laser that was developed in the last five years of the 20th century. Optical tape uses a polyethylene substrate on which an active layer and a dielectric film are sputtered. A back coat and a topcoat are added to improve handling and wear. It is expected that in mass production optical tape will be an extremely low cost medium. Data rates for optical tape are expected to reach 200 gigabytes by the end of the 20th century. Optical tapes have an advantage over magnetic tapes for long-term archiving. Magnetic tapes must be transferred every few years to new magnetic tapes to avoid deterioration.

6.11 PROBLEMS WITH 20TH CENTURY RECORDING

Some problems of 20th century recording were non-archival technologies and rapid obsolescence rendering collections unusable.

NON-ARCHIVAL TECHNOLOGIES

Starting with nitrocellular based film many of the technologies are not archival. That is they cannot be stored without deterioration. In the case of movies the solution was a different type of film and the copying of old films into the new differently chemical based film.

In magnetic tape such as cassettes or videotapes there is also deterioration over a relatively short time. The magnetic medium is degraded by creep, track deformation and print through. Print through is where in a tightly wound cartridge, information on one layer can magnetically affect information on neighboring layers. This problem of deterioration will be solved by the use of non-magnetic optical tape which uses the laser to record and read information.

OBSOLESCENT RECORDING TECHNOLOGIES

The change in speed of audio records posed some problems for collectors of recorded music in the first three-quarters of the 20th century. This was solved by machines that could play 78, 45, and 33 RPM records by a mechanical adjustment. Then came the different magnetic methods of recording such as the eight track and then the miniature cassette. This made collections of records obsolete and unplayable on new machines. Collections of eight tracks wound up first in flea markets and then in garbage cans. Miniature cassette collections were soon made obsolete by CD's. These in turn may become passé by optical tapes. Vast collections of music become junk as the players for those are displaced by new technologies.

Video recordings also are subject to changes as new formats and systems evolve. One example is the Betamax videotape which enjoyed a brief popularity and then was displaced completely. Obsolete collections of both audio and video information is one price to pay for new technologies.

6.12 From Recording Data to Satellites to Universal Communications

Recording data was used to control computers which was used to govern the development of satellites. The latter eventually led to the prospect of personal universal communications.

The next chapter covers the development of satellites. This leads to the following chapter combining many technologies into vast networks for universal personal communications.

Bibliography

Asthawa, Praveen and Blair Finkelstein — "Superdense Optical Storage." IEEE Spectrum, August 1995

Engel, F.K. — "Magnetic Tape," Journal of the Audio Engineering Society," Vol. 36, 1988, pp. 606-616

Fielding, Raymond, Editor — *A Technological History of Motion Pictures Television.* University of California Press, 1967

Hammar, Peter — "The Birth of Helical-Scan Videotape Recording." *Broadcast Engineering*, Vol. 27, May 1985, pp. 86-94.

Huber, David and Robert Runstein — *Modern Recording Techniques,* 4th Edition. Sams Publishing Co., 1995

Jorgensen, Finn — *The Complete Handbook of Magnetic Recording,* 4th Edition, McGraw Hill, 1996

Stanton, J.A. and M.J. Stanton — "Video Recording: A History." SMPTF Journal, March 1987, pp. 253-263

Taylor, Deems — *A Pictorial History of the Movies.* Simon and Schuster, 1943

CHAPTER 7

THE REVOLUTION OF THE ARTIFICIAL MOON

7.1 THE HISTORICAL BACKGROUND OF ROCKETS

The basic power for satellites in the 20th century has been rockets. The Chinese are credited with developing military rockets. Real progress was delayed until the early 19th century. William Congreve (1772-1828), a British scientist, published a paper in 1807, "*A Concise Account of the Origin and Progress of the Rocket System.*" In 1808 he invented the Congreve Rocket for use by the military. Congreve standardized the composition of gunpowder and added long flight-stabilizing guide sticks and built the first practical launching pad. Congreve's father, Colonel Commandant of the Royal Artillery and Superintendent of Military Machines, recommended to the War Office that the Congreve Rocket be employed for military purposes.

The Congreve Rocket was first used militarily in 1809 in an attempt to burn the French fleet in the Basque Roads. This was not as successful as had been expected but Congreve was rewarded and allowed to raise and organize two rocket companies in connection with the Corps of Royal Artillery. In the War of 1812 with the United States, the Congreve Rocket was used by the British fleet. The American national anthem written in 1812 refers to Congreve's Rockets *(The rockets red glare).* In 1813 Congreve was ordered with one of his rocket companies to fight Napoleon on the European continent. At the battle of Leipzig his rockets did not do much actual damage to the enemy but their noise and bright glare had a great effect in frightening the French and throwing them into confusion.

The next significant contributor to rocket power was William Hale (1797-1870), a British scientist and inventor. In 1844 he invented the rotating rocket which dispensed with the long guide stick of the Congreve Rocket. Hale's invention used a rocket's exhaust gases to rotate the projectile on its own axis. This gave the rocket stability through inertia and centrifugal force. In 1863 Hale wrote, Treatise on the "Comparative Merits of a Rifle, Gun and Rotary Rocket." This was one of the first works treating the exterior ballistics of spinning and nonspinning rockets. He also disproved the idea that rockets move because the exhaust gases push against the air. Hale correctly demonstrated rocket motion in terms of Newton's Third Law, *To Every Action There is Always Opposed an Equal Reaction.* This meant that rockets could operate in space where there is no air.

The next great advance to the development of rocketry was made by Konstantin Eduardovich Tsiolkovsky (1857-1935), a Russian scientist and inventor. Tsiolkovsky became interested in the possibility of interplanetary travel. In 1883 he wrote a paper about weightlessness in free space, "Svorodnoe Prostastvo (Free Space)." This paper included the principle of reactive motion for flight in a vacuum. This led to a simple plan for a spaceship. Tsiolkovsky studied the necessary conditions of life for plants and animals in space.

In 1896 Tsiolkovsky began to explore the possibility of interplanetary travel by means of rockets. The next year he developed the formulas which established the dependence between the velocity of a rocket at a given moment, the velocity of the expulsion of gas particles from the nozzle of the engine, the mass of the rocket, and the mass of the expended explosive material.

In *Issledovanie Mirovykh Prostranstv Reaktivnymi Priborami (A Study of Atmospheric Space Using Reactive Devices),* published in 1903, Tsiolkovsky set forth his theory of the motions of rockets, established the possibility of space travel by means of rockets and formulated the flight fundamentals

Tsiolkovsky contributed to the mechanics of bodies of changing mass. He postulated a theory of rocket flight which took into account the change of mass while in motion. He suggested the concept of gas-driven rudders for guiding a rocket in a vacuum.

Tsiolkovsky from 1903 to 1917 offered several plans for constructing rocket ships. He took into account principles of guiding a rocket in a vacuum, the use of a fuel component to cool the combustion chamber walls and many other factors.

Tsiolkovsky was treated with derision. He was considered to be a dreamer. During the years before the Revolution of 1917, Tsiolkovsky was embittered by the reception of his ideas. He received no money and he worked alone for many years under unfavorable conditions.

Life for Tsiolkovsky changed drastically with the October Revolution. In 1918 he became a member of the Academy, and in November 1921 he was allotted a personal pension which made it possible to devote himself completely to his scientific work.

Tsiolkovsky continued to work on rockets despite old age. In 1932 when he was 75 years old Tsiolkovsky wrote "Dostizhenie Stratosfery" (Reaching the Stratosphere). In this paper he formulated the requirements of explosive fuel in jet engines. In 1934 and 1935 Tsiolkovsky proposed using clusters of rockets in order to reach great speeds.

An American pioneer in rocket engineering was Robert Hutchings Goddard (1882-1945). On March 16, 1926 he designed, built and flew the first liquid-fuel rocket near Auburn, Massachusetts. Goddard received a Ph.D. in 1911 from Clark University. In 1913 he returned to Clark University to become a Professor of Physics.

Goddard explored the mathematical practicality of rocketry since 1906. He worked out the theory of rocket propulsion independently. Beginning in 1912 he investigated the experimental workability of reaction engines in laboratory vacuum tests. Goddard began to accumulate ideas for probing the earth's stratosphere. He received two patents in 1914 for a liquid-fuel rocket and a multistage step rocket.

In 1919 Goddard wrote a paper, "A Method of Reaching Extreme Altitude" in which he suggested that jet propulsion could be used to attain escape velocity and reach the moon. This received some bad publicity because of his moon rocket ideas. However, he continued to work on practical rockets. Almost alone, he designed, built, tested, and flew the first liquid-fuel rocket on March 16, 1926.

Goddard and a small staff moved to near Roswell, New Mexico, and continued his static and flight tests. He reached speeds of 700 miles per hour and altitudes above 8000 feet during the 1930's. His innovations included gyroscopic stabilization and control, instrumented payloads and recovery systems, fuel-injection systems, regenerative cooling of combustion chambers, gimbaled and cluster engines, and aluminum fuel and oxidizer pumps.

Goddard worked mostly alone or with a very small staff. His developments were soon taken up and furthered by large teams particularly in Germany from 1936 and through World War II. After his death in 1945 the U.S. government awarded his estate one million dollars for all rights to his collection of over 200 patents. Early in the 1960's the National Aeronautics and Space Administration (NASA) named its first new physical facility at Greenbelt, Maryland, in honor of Goddard.

Another pioneer in space research was Hermann Julius Oberth (1894-1989). He was born in Nagyszeben, Transylvania (then part of Austria-Hungary). It is now called Sibui, Romania. He studied medicine in Munich but his education was interrupted by service in the Austro-Hungarian Army during World War I. He was wounded and transferred to a Field Ambulance Unit. Later he said, "This was a piece of good luck for it was here I found that I should probably not have made a good doctor."

After the war he had difficulty in returning to Germany because he had become a citizen of Romania when that country annexed Transylvania. He finally was able to go back to Germany where he studied mathematics and physics at several universities. He wrote a doctoral thesis on space travel which was rejected by Munich University but it was accepted for publication by a Munich book publisher in 1923. It was entitled, *Die Rakete Zu Den Planetenraumen* (By Rocket to Interplanetary Space). It set forth with mathematical explanations, certain of Oberth's ideas on rocket travel. Oberth stated that machines could be built which would reach such speeds that they could pass beyond the limits of the earth's atmosphere and would escape the earth's gravity. He proposed that such vehicles could be constructed so as to allow human beings to travel in them with safety.

Until 1933 Oberth was unfamiliar with the work of Robert Goddard and until 1925 with that of Konstantin Tsiolkovsky in the Soviet Union. After corresponding with both men, he acknowledged their precedence in deriving the equations associated with space flight.

In 1929 Oberth wrote a book, *Wege zur Raumschiffart* (Ways to Spaceflight), which won the first annual Robert Esnault-Pelterie-Andre Hirsch Prize of 10,000 francs. This enabled him to finance his research on liquid propellant rocket motors. In 1931 Oberth received a patent for a liquid propellant rocket from the Romanian Patent Office. He launched a liquid propellant rocket in May 1931 near Berlin.

According to one version Oberth was an anti-Nazi at one time and tried at the beginning of World War II to leave Germany and return to Romania. He was offered the choice of German citizenship or concentration camp. He became a German citizen in 1940. In 1941 he went to the German rocket development in Peenemunde where he worked for Wernher von Braun, his former assistant.

At the end of World War II he was placed in an internment camp and interrogated a number of times. He was released and he went to his country home at Feucht, Germany, where he was unable to find employment. Later he went to Switzerland where he continued his writing on astronautics. Then from 1950 to 1953 he worked in Italy for the Italian Navy on solid propellant rockets. He then returned to Germany where he wrote *Menschen im Weltraun* (Men into Space) in 1954.

In 1955 Wernher von Braun asked Oberth to join him to work on rockets in the United States. He did advanced space research for the Army until he retired to West Germany in1958. He spent his retirement until his death in 1989 on theoretical studies.

Wernher von Braun (1912-1977 was a German (later American) engineer who played a prominent role in all aspects of rocketry and space exploration, first in Germany and after World War II in the United States.

Braun did not do well in school in physics and mathematics. However, in 1925 he acquired a copy of Oberth's book, *Die Rakete Zu Den Planetenraumen.* Frustrated by his inability to understand the mathematics he applied himself at school until he led his class. He enrolled in the Berlin Institute of Technology. He joined the German Society for Space Travel in 1930. In his spare time he assisted Oberth in liquid-fuel rocket motor tests. Braun graduated from the Technical Institute with a B.S. degree in Mechanical Engineering in 1932 and he entered the Berlin University to study physics.

Captain Walter R. Dornberger in the fall of 1932 was in charge of solid-fuel rocket research and development in the Ordnance Department of the German Army. He arranged a research grant for Braun on liquid fueled rockets. In 1934 Braun received a Ph.D. in Physics from the University of Berlin. His thesis contained the theoretical investigation and developmental experiments on 300 and 660 pound thrust rocket engines.

A large military rocket development facility was erected at the village of Peenemunde in northeastern Germany on the Baltic Sea with Dornberger as the military commander and Braun as the technical director. The A-4 long-range ballistic missile was developed at Peenemunde. The Nazi Propaganda Ministry designated the A-4 as V-2, meaning Vengeance Weapon 2.

According to von Braun in his book, *History of Rocketry & Space Travel,* he was arrested in February 1944 by the Gestapo and imprisoned for two weeks. He was accused of not really being interested in war rockets, but was working on space exploration. The court said he was opposed to the use of V-2's against England, and he was about to escape to Britain in a small plane, taking vital rocket secrets with him. Dornberger went directly to Hitler and said that without von Braun there would be no V-2. Von Braun was released.

The V-2 assault began in September 1944 and ended on March 27, 1945. More than 5,000 V-2's were built before the war ended. They were fired against Great Britain and targets on the Continent. There was no defense against the V-2 since it dropped down on its target at 3,500 miles per hour just five minutes after taking off. More than 1,500 V-2's were reported to have landed in southern England or just off its shores. They were responsible for more than 2,500 deaths and great property damage.

Much of the V-2 rockets were produced at Nordhausen, two hundred miles to the southwest of Peenemunde. When the first American troops arrived at Nordhausen they found thousands of bodies of slave laborers stacked. An average of 150 workers had died per day towards the end of the war. Von Braun many years later said about this, "I did not know what was going on, but suspected. And in my position I could have found out, but I didn't and I despise myself for it." Von Braun, his younger brother Magnus, Dornberger and the entire German rocket development team surrendered to U.S. troops in May 1945. Within a few months Braun and about 100 members of his group were at the U.S. Army Ordnance Corps test site at White Sands, New Mexico. Here Braun and team assembled, and supervised the launching of captured V-2's for high altitude research purposes.

Moving to Huntsville, Alabama, in 1952, Braun became Technical Director (later Chief of the U.S. Army ballistic weapon program). Under his leadership, the Redstone, Jupiter-C, Juno and Pershing missiles were developed. In 1955 he became a U.S. citizen. Later after the National Aeronautics and Space Administration (NASA) was formed in 1958, Braun and his group were transferred from the Army to that organization.

One of the first Russians to visit Peenemunde after World War II was a 38-year old engineer named Sergei Korolev. He was a former test pilot who had become the Soviet's most important designer of jet aircraft and rockets. He had long dreamed of space exploration.

Korolev graduated from the elite Moscow Higher Technical School. He then worked at the Group for the Study of Rocket Propulsion System (GIRD). In 1932 he was placed in charge of GIRD's Design and Production Department, and in 1933 he launched the Soviet's first liquid fuel rocket, the 09.

Korolev became a victim of Stalin's purges in the 1930's. He was arrested and sentenced to hard labor in the Siberian Gold Mines in Kolyma. A year after his arrest Korolev was saved by an old colleague, Andrei Tupolev, who requested that he be allowed to join his program.

When Korolev visited Peenemunde, although he was still technically a prisoner, he was placed in charge of reestablishing V-2 testing and production. Korolev recruited German rocket scientists and delivered them to Russia on October 28, 1946.

Stalin died in 1953 clearing the way for a nonmilitary, orbital satellite launch. In 1954 Korolev unveiled the R-7 booster, the Semyoka, the world's first true Intercontinental Ballistic Missile (IBM). Korolev determined to use the R-7 to launch a small non-military orbital satellite. Korolev enlisted the help of the Politburo, a group of Soviet leaders, who had taken over after Stalin's death. The leader of the politburo saw Korolev's satellite as a political weapon which could be used without a war.

7.2 THE EARLY SATELLITES

In 1954 a U.S. secret Army-Navy program, Project Orbiter, had been prepared by Wernher von Braun to put a satellite in orbit. On August 3, 1954 von Braun published a paper for the Orbiter Group, "The Minimum Satellite Vehicle Based Upon Components Available From Missile Development of the Army Ordnance Corps." In this paper von Braun demonstrated that a satellite could be orbited with existing hardware. The Defense Department on September 9, 1955 killed Orbiter by voting to try Vanguard, the alternate Naval Research Laboratory proposal.

On October 4, 1957 the United States public was startled by a small Soviet Union satellite that circled the earth about every 90 minutes in an elliptical orbit that ranged in altitude from about 140 to 560 miles. Sputnik l remained in orbit, gradually losing altitude, until January 4, 1958, when it disintegrated upon reentering the denser portion of the atmosphere.

Sputnik 1 was launched by a single R7 intercontinental Ballistic Missile from the Baikonur Cosmodrome in the steppes of Kazakstan in the Soviet Union. Although the Soviets had announced the launch prior to October 4, 1957 and had even published the radio frequencies, it still came as a shock to the American public. Korolev's name was left out of the publicity by the Soviet. For a number of years until his death in 1965 he was referred to as the Chief Designer.

On November 3 the Russians launched another Sputnik but this one weighed 1,120 pounds and carried a dog, Laika. Laika was the first living thing to completely orbit the earth. The air eventually gave out and the dog died before the satellite disintegrated upon reentering the denser portion of the atmosphere.

On December 6, 1957 the U.S. attempted to launch its first artificial satellite, Vanguard. It was to be carried into space by a three-stage rocket. The first stage used liquid oxygen and kerosene. The second stage engine burned white fumic nitric acid and unsymmetrical dimethyl hydrazine. The third stage used a solid propellant to deliver the Vanguard into an orbit. The three-stage rocket rose four feet, wavered, toppled over, and exploded in flames. Although later Vanguard satellites did orbit the earth, this failure of the first Vanguard right after the success of Soviet's Sputnik caused a psychological shock.

The press called Vanguard "Rearguard, " "Kaputnik" and "Stayputnik." Great apprehension spread throughout the United States. There was the feeling that the Soviet Union could dominate the world from the heights of space using this new technology. Von Braun and his group were instructed to proceed with Project Orbiter. On January 31, 1958 the first U.S. satellite, Explorer I, was successfully launched by a Jupiter-C rocket. This was a modified Army Redstone Liquid Propellant Medium-Range Ballistic Missile to which was added three upper stages of solid-propellant rockets. The race between the United States and the Soviet Union was under way. The satellite itself was 80 inches long, 6 inches in diameter, and weighed only18 pounds.

In 1960 and 1961 the Soviet Union orbited the Korabl Sputniks, large satellites containing dogs. Capsules from three of them were recovered with live dogs aboard. This was proof that animals could return safely to earth after a number of orbits. The Korabl Sputniks were the test vehicles for the manned Vostok (East) spacecraft.

On April 1, 1961 Vostok 1 carried the first man in orbit, Major Yuri A. Gagarin. He made one trip around the earth. This was the first of six manned Vostok flights. In June 1963 Vostoks 5 and 6 flew together around the earth, at one point traveling within three miles of each other.

The second generation manned Soviet satellite was Voskhod (Ascent), which could carry three cosmonauts, were launched in 1964-1965. A third generation manned Soviet satellite Soyuz (Union), first developed in 1967, became a space workhorse into the 1990's.

7.3 THE RACE TO THE MOON

President John F. Kennedy announced on May 25, 1961 that the U.S. intended to land on the moon within the decade. NASA had previously announced a $20 billion Apollo program. This was now accelerated by Kennedy's speech.

A huge Saturn 5-carrier vehicle was proposed for this mission. The vehicle was to stand 363 feet tall and weigh more than 6 million pounds. The Apollo spacecraft launched by Saturn 5 was to carry three astronauts. The Apollo was to first orbit the earth and then to proceed to the moon and go into orbit around it. A detachable lunar module would carry two of the three astronauts down to the moon. After a suitable stay on the moon the lunar module would blast off and rejoin the orbiting Apollo. The spaceship would then return to the earth.

This was a very ambitious program and it had many critics including prominent scientists. However, with the presidential authority behind it, NASA proceeded with an extensive series of preliminary unmanned and manned space flights. There was fear that the Soviets might win the race since they succeeded in launching unmanned lunar probes which soft landed on the moon and sent back the first pictures. The Americans determined to proceed with their program.

Project Mercury was the United States' equivalent of the Soviet Vostok series. Marine Colonel John H. Glenn (later U.S. Senator) was the first American to orbit the earth in a Mercury capsule in February 1962. The next step was to orbit the earth with more than one occupant in the spacecraft.

The U.S. Gemini program was developed to launch two astronauts in space. The first of the Gemini flights occurred on March 23, 1965. However, the Soviets were again first

Figure 7-1 Astronaut Aldrin Deploying a Scientific Package on the Moon
Photograph by Neil A. Armstrong Courtesy of NASA

with a space walk. They launched three cosmonauts in the first Voskhod spacecraft. Then on March 18, 1965, Lt. Colonel Aleksei A. Leonov left his Voskhod 2 craft for the first "walk" in space.

The Americans were the first to dock in space which occurred in 1966 when Gemini 8 hooked up with an Agena target vehicle. The last of the Gemini, No. 12, was orbited in November 1966. In April 1967 the Soviet Union inaugurated the Soyuz spaceship that had been designed by Korolev for the moon program.

A series of Apollo earth orbital flights by the U.S. without incident gave the Americans the lead in the race to the moon. On July 20, 1969, astronauts Neil A. Armstrong and Edwin E. Aldrin, Jr. left Apollo 11 in a lunar module and landed on the moon, Figure 7-1, while Michael Collins remained in orbit above in the command service module. The lunar module returned from the moon to the command service module in lunar orbit, Figure 7-2.

The Russian equivalent of the Saturn 5 which blasted the Apollo to the moon was the N-l. It was 310 feet long, powered by thirty engines. Its construction was delayed by Korolev's death. Finally in 1969, the same year Armstrong landed on the moon, the first N-1 was ready for testing. It blew up on the launch pad, as did the next three. The Americans had the moon to themselves.

Apollos 12 through 17 continued the manned lunar exploration. The only failure was the Apollo 13 but the crippled spacecraft was able to circumnavigate the moon and return to earth. The later Apollos surveyed the moon in a vehicle and gathered lunar rocks for study back on the earth.

7.4 THE SPACE SHUTTLE

In 1967 in the U.S., the President's Science Advisory Committee had recommended that "Studies should be made of more economical ferry systems, presumably involving partial or total recovery and use." In March 1970, President Nixon approved development of a shuttle vehicle that could launch and repair satellites and perform scientific studies, and transfer passengers to a space station.

NASA during 1970 and 1971 designed a totally reusable two-stage vehicle. The first stage with a crew of two would provide the initial thrust out of the atmosphere and then fly back to a runway, while the second stage continued to orbit. The second stage with a crew of two, plus room for as many as 12 passengers, destined for the space station. The cargo bay was to be big enough to accommodate space station modules. This shuttle would have cost over $10 billion to develop. However, the Budget Bureau objected vigorously to the cost and NASA scrapped the project.

NASA redesigned the shuttle so that the development would be approximately half the cost of the original design. The new shuttle called the Space Transportation System had three stages. The first stage consisted of two giant solid rocket boosters (SRB's). The liquid hydrogen and oxygen fuel for the second stage's ascent to orbit would be contained in an external tank (ET) attached to the belly of the manned vehicle, a delta-winged craft known as the orbiter. When empty the SRB's would be parachuted into the ocean and recovered for refurbishment and reuse. The ET would be jettisoned shortly before reaching orbit and disintegrate during reentry, its remnants would fall into the ocean. The orbiter with three liquid-fueled main engines would continue alone into orbit where it could remain for as long as 30 days. The orbiter would touch down without engine power at over 200 miles per hour.

These modifications to the original concept resulted in an STS that was not totally

Figure 7-2 Apollo 11 Lunar Module Leaving the Moon
The Earth is in the Background.
Courtesy of NASA

reusable, and one that had higher operating costs per flight. Although the development costs had been reduced by $6 billion dollars the Budget Bureau in 1976 cut NASA's request to just over $3 billion. The managerial responsibility within NASA was divided among a number of organizations. The Johnson Space Center in Houston, Texas, was responsible for the orbiter, the Marshall Space Flight Center at Huntsville, Alabama, was in charge of the orbiter's three liquid-fueled main engines, the ET and the SRB's. The Kennedy Space Center at Cape Canaveral would assemble the components, check them out, and conduct the initial launches.

The first space shuttle, STS-1, was launched in 1981 from the Kennedy Space Center in Florida, Figure 7-3. A remote camera at Launch Pad 39A took this picture of the opening of a new era in space. Astronauts John Young and Robert Crippen orbited the Earth for 54 hours. The mission ended with an unpowered landing, Figure 7-4, at

Figure 7-3 Maiden Flight of First Space Shuttle
Courtesy of NASA

Edwards Air Force. This flight started a series of successful shuttle flights for five years.

The orbiter itself is protected from entry heating by some 30,000 ceramic tiles glued to its skin. The tiles are composed of silica fibers. Those on the orbiter's belly are exposed to temperatures of nearly 2000° F. At the rear of the orbiter are five rocket

Figure 7-4 Landing of First Space Shuttle
Courtesy of NASA

motors, two small and three large. The two small motors burn hydrazine and nitrogen tetroixide and are used to provide the final burst of speed to reach orbit, to change orbits, and to deorbit. The three main engines are liquid fueled.

The orbiter itself does not carry fuel for the three main engines but routes liquid hydrogen and oxygen to them through 17-inch diameter lines coming from the ET. When the ET runs dry, shortly before reaching orbit it is jettisoned and disintegrates as it enters the atmosphere.

The SRB's each contain over a million pounds of a rubbery solid propellant, a mixture of four ingredients. The ingredients are mixed in 600-gallon bowls, using the equivalent of a giant eggbeater. For ease of fabrication, handling, and transportation, the propellant is poured from these bowls at the Thiokol factory in Utah into four segments that are shipped individually by rail and assembled at Cape Canaveral. The three solid fuel motors of the orbiter are each composed of four segments pinned together and sealed with O-rings. This design later caused a terrible accident that killed seven astronauts on January 28, 1986, Figure 7-5.

For 25 years before that the NASA had successfully flown 133 men and women in space without an in-flight loss of life. The Challenger, a space shuttle, was launched on January 28, 1986. The flight lasted 1 minute and 13 seconds. Lift-off was normal but less than a second later cameras recorded a puff of gray smoke coming out of the side of the Right Solid Rocket Booster. Three seconds later the smoke disappeared and all appeared well until 58 seconds after lift-off, when a flickering flame appeared from the same part of the Right Solid Rocket Booster. The flame quickly grew into a large plume that impinged on the surface of the External Tank. At 65 seconds a change in flame color indicated that hydrogen from the ET was mixing with the fire, which continued to grow until 72 seconds. At that point massive amounts of hydrogen and oxygen burned enveloping the orbiter in a fireball. The forward fuselage containing the crew of seven broke out of the orbiter and hit the Atlantic Ocean 2 minutes and 45 seconds later, at a speed of about 200 miles an hour. Seven astronauts died, including the shuttle pilot commander, Michael J. Smith, in the worst space disaster.

A commission under William P. Rogers, a Washington lawyer, was formed to investigate the Challenger explosion. One of the members was the famous physicist, Dr. Richard Feynman. During the investigation pieces of the right SRB corroborated the fact that a failure had occurred between the two lower segments. Portions of the two O-ring seals designed to prevent hot gases from leaking through the joint were destroyed, allowing a fiery jet to escape and to impinge on the adjacent surface of the ET.

Feynman demonstrated that the O-rings lost their elasticity at low temperatures. He did it by placing them in his refrigerator and showing how brittle they had become. The temperature at the launch of the Challenger was 26° F, a temperature below freezing. It was the coldest launch temperature during the shuttle series.

NASA had the joints and O-rings of the Solid Rocket Boosters redesigned and revised its operating procedure. After a 3 year long hiatus the shuttle program resumed and in early1994 astronauts boosted into orbit by a shuttle walked in space to repair the Hubble spacecraft, a scientific telescope satellite orbiting the earth.

The space shuttle Challenger disaster had an immediate effect on the launching of satellites into orbit. NASA decided not to use the space shuttle for lifting commercial satellites into space. In addition there was a three-year hiatus in shuttle flights after the accident. This encouraged the use of unmanned rocket boosters. They included the Atlas in the United States, the French Ariane, and the Russian Proton.

In 1996 NASA signed a contract with Lockheed Martin to develop the prototype X-33 for the next generation space shuttle. This will lead to a completely reusable vehicle for the 21st century.

7.5 SPACE STATIONS

Space stations are orbiting platforms which have docking facilities for other spacecraft. They are in effect scientific laboratories and observation posts. The Soviets were pioneers in the construction and long term operation of space stations.

Figure 7-5 Challenger Disaster
Courtesy of NASA

SALYUT SPACE LAB

On April 19, 1971 the Soviet Union launched the first space station Salyut l. On April 24th three cosmonauts in Soyus 10 docked with Salyut l but did not board.

In June three cosmonauts docked with Salyut for about 22 days to conduct engineering proving trials and experiments in biology and space medicine. This was the first manned orbiting scientific laboratory.

The crew died while returning to earth when a pressure equalization valve opened.

Skylab

On May 14, 1973 America's first embryonic space station, Skylab, was orbited. A two stage Saturn V launched Skylab from the Kennedy Space Station at Cape Canaveral. Skylab was damaged by air pressure during ascent, which ripped away the meteoroid shield and one solar "wing." The other solar wing was left in closed position. Figure 7-6 shows Skylab in space. The solar panel on the left side was lost on launch day.

Three successive visits by three-man crews repaired Skylab and performed a wide range of studies and experiments. The Skylab astronauts spent more than 170 hours in space and traveled some 70,500,000 miles.

Skylab's orbit, 275 miles above the earth's surface was too low to last for a long time. A scattering of molecules from the upper atmosphere, as well as particles coming from the sun, would bump into it, and little by little reduce its speed and start it plunging into the lower atmosphere and burning up. This happened on July 11, 1979 and pieces came down in western Australia. There were reports of spectacular visual effects and loud noises but no one was hurt. At least one chunk of material was found.

MIR Space Station

On February 19, 1986 the Soviet Union launched the core of space station MIR (Peace) into orbit. This core was developed from Salyut. The overall length was 42.6 feet with a maximum diameter of 13.6 feet. The mass of the core was 21 tons. Provisions were made for six docking ports. Other modules were sent up later to dock with the core to become a space station. The Soviets also sent up a series of Progress freighters which docked with MIR supplying propellants, air, food, and water. Engines of the Progress were used to raise the orbit of the station complex to prolong its life. Soyuz and later space shuttles exchanged crews with the space station. Figure 7-7 is a photo of MIR.

The Soviets used the MIR to study the effects of long term zero gravity on cosmonauts. Several problems were found including the loss of calcium in the bones and the loss of muscle tone. Various procedures were initiated to minimize these and other problems.

The Freedom Space Station Proposal

In 1984 President Reagan proposed a new space station, Freedom, orbiting earth at 240 miles. Its mission would include satellite repair, assembly of spacecraft outbound

Figure 7-6 Skylab in Space
Courtesy of NASA

from earth, life sciences, environmental studies, astronomy and the production of specialized materials.

However, the end of the Cold War and budgetary problems drastically scaled back the plans. It was decided to join other nations in an international program leading to a space station which would serve as a jumping off place for a manned expedition to Mars.

THE INTERNATIONAL SPACE STATION

The successor to the Freedom station proposal is a space station which is truly international. The United States, Russia, Japan, Canada and the European Space Agency are all to take part in the construction and operation. The European Space Agency itself

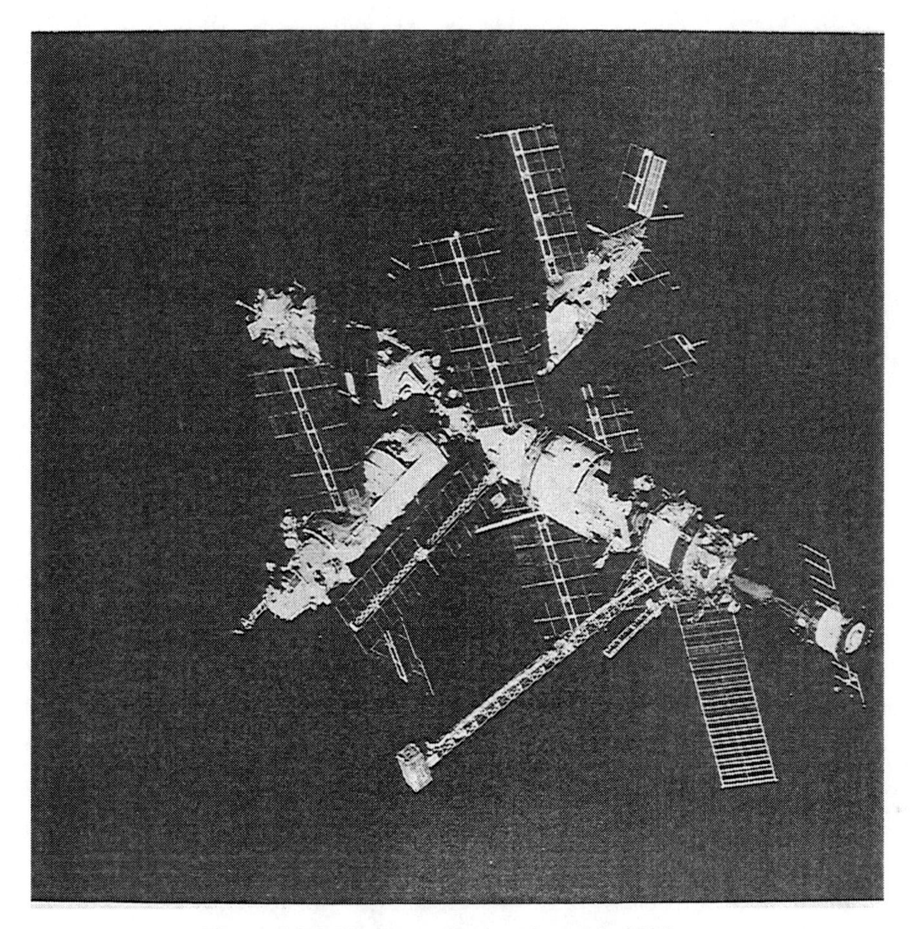

Figure 7-7 MIR Space Station, June 29, 1995
Courtesy of NASA

consists of a large number of countries including Austria, Belgium, Denmark, France, Germany, Ireland, Italy, the Netherlands, Norway, Spain, Sweden, and the United Kingdom.

In 1995 the construction of major parts of the International Space Station (ISS) was begun on earth. The first of these components will be launched in late 1997. Completion is scheduled for 2002. The ISS will be the largest man-made object in space. The length will be 109.1 meters and the weight will be 423,000 kilograms. The six-person crews will be able to float through inter-connected modules which give them a large working area. The ISS will allow scientists to work on fundamental research in materials and life sciences for periods as long as 10 years. This will set the stage for manned missions to Mars.

The ISS will orbit the earth at 240 miles. This is considered to be a low earth orbit.

7.6 Types of Orbits

There are a number of different types of orbits of artificial satellites around the earth pending on applications.

Low Earth Orbits (LEO)

Satellites in low earth orbits are located several hundred miles above the surface of the earth. Some have typical revolution times of about 90 minutes. They are used for space stations, military functions, and for certain types of communications.

Geosynchronous Orbits

Geosynchronous orbits have a period, or time of revolution around the earth, that exactly matches the earth's rotation. The geosynchronous orbit may have an inclination to the earth's equatorial plane. The satellite appears to hover at a fixed longitude over the earth but it moves up and down in latitude in a narrow figure-eight motion. It takes one day to complete the figure. The inclination establishes the maximum north or south latitudes that the satellite reaches in its travel. The height for geosynchronous satellites is 22,238 miles above the planet, revolving at a velocity of slightly over 6,900 miles per hour.

Geostationary Orbits (GSO)

These are a special case of the geosynchronous orbit in which the satellite plane lies in the equatorial plane of the earth. The geostationary appears to stay directly over the equator at one particular longitude. It appears to an observer on the earth that the satellite does not move. This means that antennas pointed at the satellite do not have to be reoriented with time. The geostationary satellite can see about one-third of the earth. The coverage extends from 83°S to 83°N latitude and plus or minus 83° in longitude from the subsatellite point (the point on the equator below the satellite). Three satellites are required for GSO systems.

Sun-Synchronous Orbit

This is a near polar orbit which has its plane directed at the sun. The altitude is usually around 512 miles above the earth. As observed from a position on the sun, the satellite orbit plane remains in the same apparent orientation throughout the year. This means that the shadows on the earth and the illumination of clouds remain constant throughout the year. This is valuable for making weather observations. The typical number of revolutions of the satellite is 15 or 16 times around the earth in one day.

7.7 WEATHER AND ENVIRONMENTAL SATELLITES

TIROS

In April 1960 the United States launched its first Television and InfraRed Observation Satellite (TIROS) for weather observation. It circled the earth in a polar sun synchronous orbit taking 19,000 photographs of clouds. Later models transmitted more than 10,000 pictures each. A total of 10 of the first series of TIROS were launched from 1960 to 1965 and established the feasibility of weather observation from above. Each of the satellites carried two miniature cameras, a scanning infrared radiometer and an earth budget instrument. This last monitored the amount of radiation received and re-radiated by the earth.

The first operational generation of TIROS was the nine pairs of Environmental Science Service Administration (ESSA) satellites launched from 1966 to 1969. The first satellite of each pair provided global weather data to the U.S. Department of Commerce's stations in Wallops Island, Virginia, and Fairbanks, Alaska. This data was relayed to the National Environmental Satellite Service at Suitland, Maryland for processing and forwarding to the major forecasting centers all over the world. The second of ESSA pair of satellites provided high quality pictures of the earth's cloud cover every 24 hours.

The second generation of TIROS operational satellites was the Improved TIROS Operational Satellite (ITOS) first launched on January 23, 1970. ITOS used only one satellite instead of the pair employed in the ESSA system. ITOS supplied observation of the earth's cloud cover every 12 hours. Included were temperature soundings of the atmosphere. The spacecraft was operated by the National Oceanic and Atmospheric Administration successor to ESSA.

The third generation operational TIROS were the TIROS-N first launched in 1978. This series obtained more accurate sea-surface temperature mapping and identification of snow and ice. A data collection system (DCS) received environmental data from fixed or moving platforms such as buoys or balloons and retained it for transmission to ground stations.

The fourth generation operational TIROS were designated Advanced TIROS-N, Figure 7-8, launched from the late 1980's into the early 1990's. They were designed to fulfill the USA's requirements for polar-orbiting civilian weather satellites to the end of the 20th century.

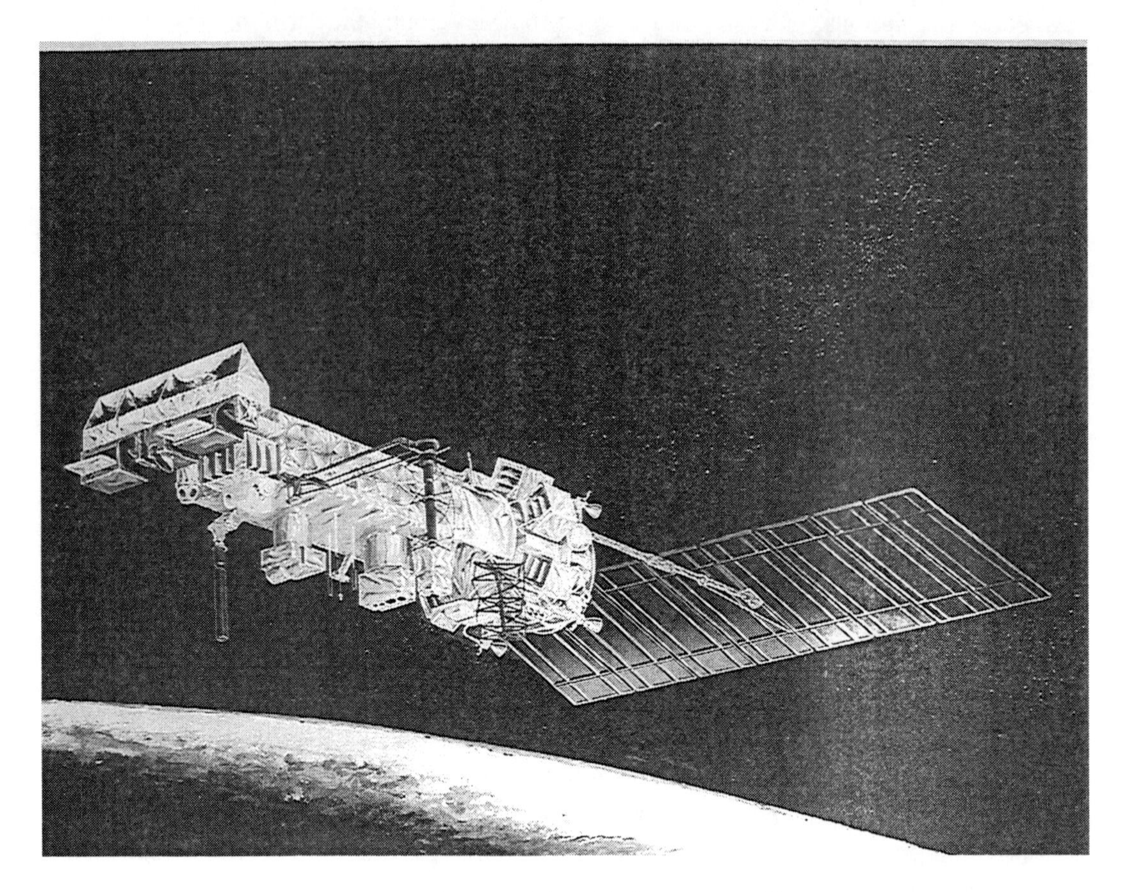

Figure 7-8 Advance Tiros-N Weather Satellite
Courtesy of NASA

Advanced TIROS-N operates in a sun-synchronous near-polar circular orbit at an altitude of either 518 or 541 miles. In the operational configuration, two satellites are used with a nominal orbit plane separation of 90°.

The weather instrument payload of the Advanced TIROS-N includes the following:

1. A number of infrared detectors that produce stratospheric* and tropospheric** temperature profiles. One of them also produces tropospheric moisture profiles.
2. A Microwave Sounding Unit that produces temperature profiles in the presence of clouds.
3. A Data Collection System (DCS) for the collection of meteorological data from buoys, balloons and remote weather stations.

---

* Stratosphere--Upper portion of the atmosphere extending from approximately 7 miles to about 31 miles above the earth.

** Troposphere--The portion of the atmoshere below the stratosphere.

4. Special instruments to measure the earth's ozone distribution. This is vital to measure the depletion of the ozone layer due to CFC's.
5. Special instrumentation to determine radiation gains and losses to and from the earth.

The Advanced TIROS-N at the end of the 20th century in its polar orbit covers large northern and southern areas. It is complemented by a series of satellites which cover the earth in geostationary orbits.

GOES

A second type of weather satellite used by the United States is the Geostationary Operational Environmental Satellite (GOES). This unlike TIROS uses a geostationary orbit at 22,300 miles above the equator which allows a continuous view of the earth's cloud cover. The first operational version was GOES-1 launched on October 16, 1975. This was followed by a series of GOES with some significant accidents. A GOES satellite blew up during launch in 1986 and another failed in orbit in 1989.

The GOES are operated by the National Oceanic and Atmospheric Administration (NOAA). Normally to ensure full coverage of the United States (including Alaska and Hawaii) the GOES must work in pairs, one each off the east and west coasts at the equator. However, by 1993 there was only the GOES-7, launched on February 26, 1987 over the Pacific Coast. It was out of maneuvering fuel and drifting. Occupying the second spot over the Atlantic in 1993 was an aging spare on loan from Europe. The GOES system was in serious trouble. The United States had relied on GOES for almost two decades for continuous weather monitoring--especially to track life-threatening storms such as hurricanes, torrential downpours and ice storms.

To replace the ailing GOES-7 and its aged European counterpart, NOAA has developed a new improved series of five GOES which have been designated GOES-i, j, k, l, m. The first of the new series, a yellow cube, seven feet on a side and packed with instruments, Figure 7-9, was launched aboard an Atlas rocket on April 13, 1994. This new series of satellites will provide improved weather coverage of the United States to the end of the 20th century and beyond. One of the two main instruments is an improved scanning television camera to provide better depth of contrast, more detail and sharper images in both the visible and infrared wavelengths. The other main instrument is an improved infrared detector called a sounder, which is designed to penetrate vertical columns of atmosphere to measure the temperature and humidity at various altitudes. This will enable the satellite to look deep inside storms. The two instruments can work simultaneously, unlike the previous GOES' series.

Figure 7-9 GOES-i Weather Station
Courtesy of NASA

If the two key instruments work as planned, they will enable meteorologists to pinpoint hurricanes and other storms as much as 24 hours in advance. Forecasters hope to predict variations in rainfall and snowfall amounts from one city to the next.

The precision of the new GOES series is also the result of a basic design change. The original GOES satellites spun constantly in order to maintain stability and used special spin-scan cameras. They could pinpoint storm paths only to within a range of 6 to 12 miles. Each new craft is to fly in a fixed attitude--stabilized by a system of gyroscopes and other controls. The new GOES can pinpoint storm paths to within a mile.

The cost of the development and launching of the new GOES series is estimated at $2 billion. However, proponents claim that it will produce vast benefits for farmers, marine and aviation navigation and others. Also taxpayers may save $1 million in evacuation and related costs for every mile it brings the forecasts closer to the actual point of a hurricane's landfall.

Mission to Planet Earth

This is a NASA program to launch a number of environmental satellites to monitor the atmosphere. The launch of EOS-AM, the first major component of this effort, is scheduled for 1998. It will be placed in a 512 miles high sun-synchronous orbit.

The problems to be addressed include the greenhouse effect and holes in the ozone layer. The El Niño phenomenon that alters drought and monsoon cycles will also be studied.

Four polar-orbit platforms will comprise the Earth Observing System. The instruments will be grouped aboard platforms in complementary orbits so that measurements can be taken at the same times and thus compared on an equal footing.

7.8 Earth Resources Satellites

It was discovered early in the space program that minute variations in infrared radiation would give information on earth resources. It was soon found that multi-spectral sensors could identify features on the earth's surface by the energy that emit or reflect from the sun. For example, the spectral signature for vegetation is different from that of rock, soil, or water.

These differences are registered by the sensing equipment aboard the satellite. They can then be resolved into different false colors which give a picture of earth's resources. With this technology, it is possible to recognize specific crops under cultivation. Changes in spectral signature indicate such information as poor soil condition, moisture content of the soil as well as crops affected by disease or insect infestation.

Landsat

General Electric developed a series of satellites called Landsat for NASA starting in the 1970's. They were designed to take pictures of Earth's resources from low earth orbit. They were used to correct and update maps. Satellite imagery charted underwater features such as coral reefs which are potentially dangerous.

Landsat in the mid-1970's demonstrated how a satellite system could help forecast the field of important world crops. Total acreage for a particular crop could be calculated from Landsat surveys. This was then compared with the potential yield per acre based on knowledge of past meteorological data between crop harvests. Landsat can pick out and identify many different types of crop in one area such as corn, soybeans, sorghum, oats, grasses, lettuce, mustard, tomatoes, carrots, and onions. Landsat can become a global food watch which could help humankind avoid disastrous food shortages. Landsat can also achieve better management of crop and timber resources.

French Spot Satellites

A French program, System Probatoire d'Observation de la Terre (SPOT) was announced in 1978. The aims of this project are to maintain an inventory of non-renewable and slowly renewable resources, agriculture and the atmosphere. SPOT wants to identify, predict and control some of the processes relating to oceanography, climatology, soil erosion and water pollution as well as keeping track of potentially dangerous natural phenomena such as floods, droughts, storms, earthquakes and volcanoes. SPOT 1 was launched by Ariane 1 in February 1986 and achieved a sun-synchronous polar orbit. Very good picture data was transmitted to earth.

Russian Earth Resources Program

This uses a battery of six multi-zonal cameras installed in satellites and space stations. Each camera had its own film and lens filter registering different information. One brings out details of the soil structure, including moisture content and rock composition. Another captures information on the types of vegetation. One camera concentrates on water quality in lakes and oceans and the extent of pollution.

7.9 Communication Satellites

Arthur C. Clarke wrote a memorandum on May 25, 1945 to the Council of the British Interplanetary Society (BIS) proposing the use of satellites for worldwide communications. Clarke pointed out that a satellite in a circular orbit 22,300 miles above the earth would take 24 hours to complete one circuit. This is the same period as one rotation of the earth. A satellite whose orbit was in the plane of the earth's equator would stay over the same spot continuously. This is called a geostationary orbit.

Clarke proposed in his memorandum to the BIS that three satellites spaced at equal intervals above the equator at 22,300 miles could cover radio communications worldwide. The BIS recommended that Clarke publicize his idea. He agreed and published an article on satellite communications using geostable orbits in "Wireless World" in October 1946. This opened the way to a number of experimental systems which led to worldwide communication satellites.

ECHO

One of the first experiments was Echo. In the early 1960's long range experiments were made by bouncing radio signals from large balloon satellites. The balloons, Echo 1 and Echo 2, were made of aluminized mylar. They were used on an experimental basis to relay signals between two ground stations.

TELESTAR

Echo was an engineering test but a true communication satellite, Telestar 1, was launched on July 10, 1962. Telestar was owned by AT&T and it was the first commercially funded satellite. Telestar weighed 170 pounds and had a low earth orbit.

The ground stations were located at Andover, Maine, in the United States, Goonhilly Downs in the United Kingdom, and Pleumeur-Bodou in France. The first TV broadcast, 15 hours after launch, relayed a picture of the American flag at Andover to Britain and France. Two weeks later, millions of people on both sides of the Atlantic watched and listened to a two-way conversation across the Atlantic. Some historians date the birth of the "Global Village" from that video conversation.

RELAY

Relay was another low-orbit experimental communication satellite which was designed to discover the limits of satellite performance. Relay 1 was launched on December 13, 1962 and Relay 2 on January 21, 1964. The Relay and Telestar experiments laid the groundwork for working communication satellites.

SYNCOM

Syncom was a series of communication satellites in synchronous orbits. Syncom 3 was placed in a geostationary orbit over the equator near the international dateline on August 19, 1964. The satellite broadcast live the opening ceremonies of the Olympics in Japan. This was the true beginning of communication satellites and led to international corporations using this technology.

INTELSAT

In August 1964, the International Telecommunications Satellite Organization (INTELSAT) was formed. It is owned by member nations prorated by their share of the traffic. The 800 ground stations remain the property of the 165 user countries. The consortium's operational arm is the Communications Satellite Corporation (COMSAT) in Washington, DC.

INTELSAT 1, known as Early Bird, went into regular service with 240 telephone circuits on June 28, 1965. The satellite was a cylinder 2.36 feet wide and 1.93 feet high with a mass of 86 pounds. Solar cells wrapped around the satellite provided 40 watts of power. It stayed in service for four years.

Several generations of geostationary communication satellites were launched by INTELSAT. Each handled more telephone and television circuits and had an increased power. In April 1982 a contract for an INTELSAT 6-series was awarded to Hughes Aircraft. It carried 120,000 telephone circuits and three TV channels. INTELSAT 6 has 48 transponders and the solar power output is 2,600 watts. In space with its solar array skirt and antennas deployed the INTELSAT 6 is 38.4 feet tall. Each INTELSAT 6 was designed to last 14 years. However, the launching was delayed by the Challenger disaster. It was decided to take all the INTELSAT 6 series off the shuttle. The first one was launched in 1989, three years late, by the Ariane 4 rocket. A later INTELSAT 7 series with newer communication equipment was designed by Ford Aerospace. The first launch was in the summer of 1992.

INMARSAT

The International Maritime Satellite Organization (INMARSAT) is an organization of countries similar to INTELSAT which provides communications for ships and offshore oil platforms. INMARSAT was established in 1979 by a consortium of maritime nations and is the equivalent of INTELSAT, with headquarters in London, England. The INMARSAT uses three geostationary satellites over the three main ocean regions with global beam coverage of each zone. A number of shore-based large antenna stations provide interconnection with the international telephone system. The eventual goal is to service 10,000 ships. By the 1990's almost all communications to and from ocean-going ships was via INMARSAT.

BIG LEO AND LITTLE LEO

LEO stands for low earth satellites which circle the earth only about 200 or so miles above its surface. Big LEOs operate at frequencies above 1 GHz and are used to relay mobile voice communications. One example is the Motorola project Iridium which uses

66 satellites. Little LEOs with frequencies below 1 GHz use clusters of satellites to handle brief digital messages to provide electronic mail and paging to portable and mobile devices.

Satellite TV Broadcasting

Originally very large satellite dishes were used to receive TV broadcast from satellites. However, with the development of higher powered satellite TV transmitters it became possible to make the ground antennas much smaller. The 18-inch diameter satellite dish made it possible to readily install TV satellite receivers in private homes.

7.10 Satellite Navigation Systems

GPS

In the early 1960's the United States Department of Defense realized that satellites could be used for extremely accurate targeting of intercontinental missiles. Over the years this idea developed into the Global Positioning System (GPS). Although it was originally built for the military at a cost of $12 billion dollars, GPS is also used for civilian navigation and accurate determination of position.

The basic principle is that radio waves travel at 186,000 miles per second. A measured time difference between a transmitted signal from a satellite and a received signal on the earth can be used to determine the distance between them. For example, if the time difference was one tenth of a second, the distance would be 186,000 x 0.1 or 18,600 miles. The receiver on earth would be on an imaginary sphere whose radius was 18,600 miles and whose center was the position of the satellite at that time. Three satellites would theoretically give the location of the receiver where all three spheres coincide.

However, there are some practical considerations which require a fourth satellite. The atomic clocks in the satellites are too heavy and expensive to be used in the receiver whose location is to be determined. Instead a small, relatively accurate time device is used at the receiver. This time measurement will introduce an error called the offset in the calculations when three satellites are used. However, a fourth satellite will set up 4 equations with 4 unknowns. A small computer in the receiver will solve the equations for the offset and give an accurate position for the receiver.

This makes practical a small hand-held receiver. The GPS accuracy in civilian use is better than ten feet. In military use the GPS is even considerably more accurate. GPS can be used for locating a ship, a plane or a car. It is also a very accurate navigation tool.

The 24 GPS satellites in use travel at 11,000 miles above the earth and take 12 hours to complete an orbit. They pass over U.S. Department of Defense bases which precisely locate the satellites. A special code is used by the military for great accuracy. This is not available for civilians. However, the accuracy for the latter is quite adequate for most position location and navigation

7.11 Scientific Satellites

Satellites from the beginning were used to make scientific discoveries.

Explorer 1

Explorer 1 was the first American scientific satellite launched on January 31, 1958. It was designed by James A. Van Allen of the State University of Iowa. The satellite was responsible for the discovery of the Van Allen radiation belts around the earth. These are created when the earth's magnetic field captures high-energy protons and electrons from the sun. It is the phenomenon that produces the auroras in both the northern and southern polar areas where the magnetic field is the most concentrated.

Orbiting Geophysical Observatory (OGO)

OGO satellites weighed upwards of 1,000 pounds. OGO-1 was launched on September 4, 1964 in an elongated orbit that carried it as far as 90,000 miles away from the earth, as close as 175 miles above the earth. OGO's were orbited annually in the 1966-1969 period.

Orbiting Solar Observatories (OSO)

OSO satellites launched from 1965-1969 supplied data on solar flares.

Orbiting Astronomical Observatories (OAO)

OAO-2 weighing 4,446 pounds was launched on December 7, 1968 into a circular orbit. It contained eleven telescopes and it was the first time astronomers were able to make long-term observations unhampered by the disturbing effects of the earth's atmosphere.

Solar Maximum Mission Satellite

This was launched on February 14, 1980 to study the sun during the most active part of an International Solar Maximum Year. It observed solar flares in ultraviolet, X-ray and gamma regions of the spectrum and measured the sun's total radiation with great accuracy.

Infra-Red Astronomical Satellite (IRAS)

This was launched on June 25, 1983. It was the first satellite to conduct an all-sky survey to search for astronomical objects emitting infrared radiation. IRAS was an international venture involving the UK, US, and the Netherlands.

Hubble Space Telescope

This should have been launched in 1986 but the Challenger left it earthbound until it was launched on April 25, 1990, Figure 7-10, by the space shuttle Discovery. Hubble was equipped with a 94 inch diameter mirror which in space should be able to observe objects fifty times fainter than those seen by the 200 inch Mount Palomar telescope. However, the primary mirror had a major defect. It was ground to a curvature that was too shallow by a very small amount. Light rays that hit the outer edges of the mirror did not focus to the same point as rays from the center of the mirror. The difference between the two focal points was an inch. The result was that pictures were often fuzzy.

In 1993 a team of astronauts aboard a space shuttle installed corrective lenses which acted like eyeglasses to correct the faulty curvature of Hubble's primary mirror. Other repairs and equipment replacement were performed to bring the Hubble space telescope into tip-top operation, Figure 7-11.

Solar and Heliospheric Observatory (SOHO)

On December 2, 1995 SOHO was launched on an unmanned Atlas rocket and headed for a point nearly 1 million miles from earth and 92 million miles from the sun. Here the gravitational pulls of earth and sun cancel each other. SOHO was jointly sponsored by NASA and the European Space Agency to study the sun's deep interior, the Corona, its outer atmosphere and the solar wind which can affect radio communications on earth.

The solar wind is a stream of electrically charged particles going outward from the sun's atmosphere. They pass the earth at a speed of 250 miles per second. The earth's magnetic field is compressed by the solar wind on the side facing the sun.

The above represents only a few of the scientific satellites. Many more have been built

Figure 7-10 Hubble Space Telescope Being Launched April 25, 1990
Courtesy of NASA

and launched by different nations. In addition to scientific satellites there have been a number of planetary probes in the last half of the 20th century that explored the far reaches of the solar system.

7.12 Planetary Probes

Pioneer 5, the first interplanetary probe, was launched into an elliptical orbit around the sun on March 11, 1960. It studied interplanetary space.

Pioneer 10 was launched in 1972 to fly by Jupiter. It was powered by nuclear thermoelectric generators because it went where the sun's light was too weak to energize the spacecraft. Pioneer 10 sent back pictures of Jupiter and its moons. Pioneer 11 was launched a year later where it also sent back pictures but then went on to pass by Saturn.

Figure 7-11 Hubble Space Telescope Being Repaired, December 4, 1993
Courtesy of NASA

Both Pioneer 10 and 11 left the solar system with plaques carrying a message from Earth to some form of intelligent life somewhere in the universe, perhaps millions of years from now. The plaques included naked figures of a man and a woman. This caused a fuss at the time of launch. Figure 7-12 is artist's concept of Pioneer at Jupiter.

In 1992, 20 years after launch, Pioneer was still sending back data from five billion miles away even though the 570 pound craft was designed to operate for a minimum of 21 months.

MARINER

On December 14, 1962, a U.S. planetary probe, Mariner 2 performed a flyby of the planet Venus at a distance of 216,400 miles. Mariner 2 found no appreciable magnetic

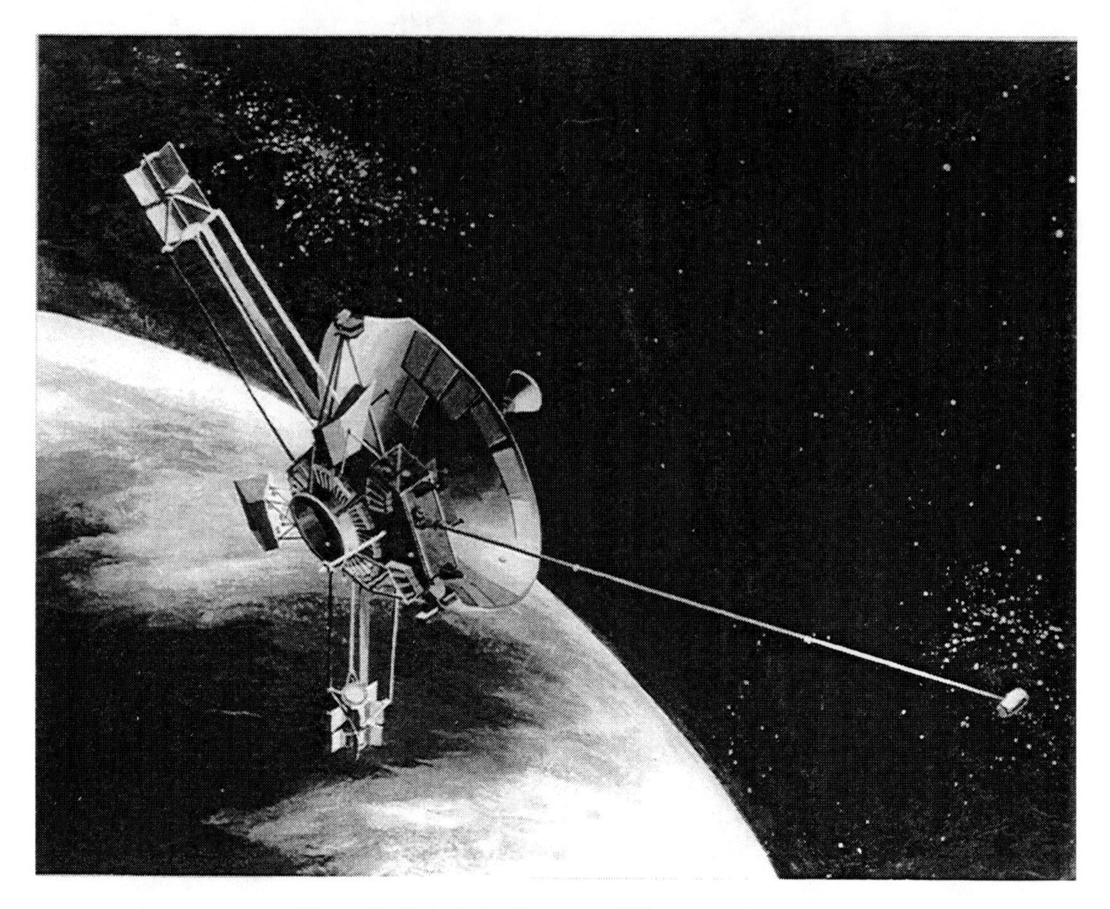

Figure 7-12 Artist's Concept of Pioneer at Jupiter
Courtesy of NASA

field or radiation belts such as those that surround the Earth. It also found that the surface was dry and extremely hot at 425°C. The surface atmospheric pressure was found to be at least 20 times sea-level pressure on Earth.

On July 14, 1965, Mariner 4 swept past Mars at a distance of 6,000 miles. Mars like Venus had no magnetic fields or radiation belts. The probe sent back pictures of the Martian surface similar to the moon.

Mariner 5 flew within 2,500 miles of Venus in 1967 and added to the accumulating store of knowledge about the planet which is the next closest to the sun from Earth.

Mariner 9 was launched in 1971 and became the first spacecraft to orbit another planet. It circled Mars and sent back over 7,000 pictures from its orbit around Mars. Some of these showed a huge rift stretching one-fifth of the way around the planet. It was 155 miles at the widest point and 4 miles deep.

VOYAGER

In 1979 NASA launched Voyager 1 and Voyager 2 to take detailed pictures of Jupiter, Saturn and their moons. Voyager 2 then made a flyby of the planet Uranus in 1986. Pictures were sent back of rings and tiny moons.

VIKING

The United States in 1975 began a search for life on Mars by launching Viking 1 and Viking 2. Viking 1 orbited Mars and dispatched a lander vehicle to a soft landing on July 1976. Several weeks later Viking Lander 2 set down on the surface of Mars to begin a coordinated search for life. Pictures of the landing site did not show any evidence of life on Mars.

GALILEO

On October 18, 1989 NASA launched a 2-1/2 ton spacecraft to Jupiter from a space shuttle. On December 7, 1995 Galileo arrived and swung into an orbit around the giant planet. This was the first time that any spacecraft orbited Jupiter.

On the same day, a 746 pound capsule, which detached from Galileo on July 12, 1995, plunged through the gases of the planet. It sent back data to Galileo after deploying a parachute.

Jupiter itself is composed of 89 percent hydrogen and 10 percent of helium. It is the largest planet, 318 times more massive than the earth.

Galileo was designed to study Jupiter and its moons for two years. The scientist Galileo discovered four moons of Jupiter in 1620. Galileo the spaceship explored the Jupiter moons as well as Jupiter itself.

MAGELLAN

The NASA Magellan probe orbited Venus in 1990 and increased the knowledge about the planet.

MARS PATHFINDER

Mars pathfinder used a Delta II rocket to blast off from Cape Canaveral on December 4, 1996. It landed on July 4, 1997. Pathfinder consisted of a lander and a rover called Sojourner. Sojourner was equipped to analyze rocks and continue the search for evidence of life on Mars.

SOVIET PLANETARY PROBES

The Soviets launched a series of spacecraft, Mars 1 through 6, to explore the planet Mars.

The Soviets also explored the planet Venus with a series of spacecraft named Venera. The Soviets in 1967 launched Venera 4 which released a spherical capsule into the atmosphere of Venus. This showed that the atmosphere was almost entirely carbon dioxide. Venera 7 in 1970 reached the surface of Venus. It reported back that the pressure was 90 times that of earth and that the surface temperature was hot enough to melt lead.

Venera 9 and 10 became the first spacecraft to orbit Venus in 1975. Both sent probes that landed on the surface of Venus. They were fitted with TV cameras which sent back the first pictures of the surfaces of Venus.

7.13 SPACE TRASH

Since Sputnik became earth's artificial satellite in 1957, space trash included dead satellites, spent rocket stages, solar panels, dropped tools, discarded clothes, and garbage bags. By 1991 there were an estimated 7,000 objects larger than four inches in diameter orbiting the earth. These can be detected by radar.

Objects smaller than four inches evaded detection. They include the nuts, bolts, springs, brackets, switches and other assorted remnants of collided objects and upper stage rockets that exploded due to leftover fuel. It was estimated that this number is at least 14,000. In addition, there are trillions of minute particles of aluminum oxide from rocket fuel exhaust.

Even the smallest piece of orbiting debris, traveling at speeds up to 17,500 miles per hour, pack a wallop that can threaten a spacecraft and astronauts. In 1983 a paint fleck collided head-on with the space shuttle Challenger embedding itself in a windshield. Had the fleck been any larger, it might have shattered the windshield.

All objects eventually fall from orbit depending on their orbit. In low earth orbit 150 miles up, a piece of space garbage stays up for a few months or years; in geosynchronous orbit, 22,300 miles above the earth's equator, the same object will remain aloft for centuries.

On July 11, 1979 pieces of Skylab came down in Western Australia. Fortunately the area was sparsely populated and no one was hurt. Parts of a Soviet nuclear powered satellite came down in the Arctic without harming people. On October 28, 1993 a 4,000-pound Chinese satellite malfunctioned and crashed west of Baja California, in the Pacific Ocean. In November 1996, a Russian nuclear powered Mars probe also malfunctioned and crashed in the Pacific. Again fortunately no one was hurt. Eventually, this may not always be the case.

BIBLIOGRAPHY

Braun, von Wernher and Frederick Ordway, Ill — *History of Rocketry and Space Travel,* 3rd Edition, Thomas Y. Crowell Company, New York, 1975

Collins, Michael — *Liftoff, The Story of America's Adventure in Space,* Grove Press, New York 1988

Dooling, David — "Research Outpost Beyond the Sky." *IEEE Spectrum*, October 1995

Gatland, Kenneth — *The Illustrated Encyclopedia of Space Technology,* 2nd Edition, Salamander Books, Limited, 1989

Hum, Jeff — *GPS, A Guide to the Next Utility,* Trimble Navigation, 1989

Pratt, Timothy and Charles W. Bostian — *Satellite Communications,* John Wiley & Sons, Inc. 1986

Walter, William J. — *Space Age,* QED Communications, 1992

CHAPTER 8

MOVING TOWARDS PERSONAL UNIVERSAL COMMUNICATIONS

The twentieth century saw the development of personal universal communications starting from the roots in the mid nineteenth century, the International Telegraph to the future Public Land Mobile Telecommunication System at the end of the century. This last includes small pocket wireless transceivers that can transmit and receive all kinds of information worldwide through a vast network of satellites.

8.1 LOOKING BACK AT THE ROOTS, THE TELEGRAPH

Around 1810 William Sturgeon of England (1783-1850) used Ampere's discovery of electromagnetism to construct electromagnets. He wound copper around an insulated iron core and produced magnetism by sending an electric current through it.

Joseph Henry (1797-1878), an American scientist, visited Sturgeon in England in 1811 and observed his electromagnet. Henry returned to America and built an improved version. He insulated the copper wire and produced a stronger electromagnet by winding many more turns on it. Henry was intrigued by the fact that the electromagnet, unlike a permanent magnet, could be switched on and off by making or breaking the current. This led him in 1832 to construct an experimental telegraph. An electromagnet activated by a battery, repulsed a permanent magnet, causing it to strike a bell.

In the same year that Henry constructed his experimental telegraph, an American artist, Samuel F.B. Morse (1791-1872) was returning to America aboard the packet "Sully" after studying art in Paris. He had a luncheon with Dr. Charles Jackson of Philadelphia. Jackson told Morse that Benjamin Franklin had sent electricity through many miles of wire almost instantaneously. Jackson was wrong but his statement gave Morse the idea for the telegraph. He drew rows of dots and dashes made on moving tape in response to electrical signals sent down a wire. These dots and dashes could represent letters and numbers.

Morse knew nothing of Henry's experimental telegraph until 1836. In the meanwhile Morse became a Professor of Art at New York University where he experimented with his own idea of the telegraph. He built a sending instrument and a receiver connected by wires. In one type of sender he used a metal lever which contacted a disc on which were raised metal conductors with varying spaces and lengths. This produced long (dashes) and short electrical impulses (dots) on a long wire connected to the disc. The long wire led to an electromagnet in the receiver. Here a pencil controlled by the electromagnet produced a wavy line whose output could be interpreted as dots and dashes. However, Morse's telegraph could work only over a short distance since he was using one cell battery and a weak electromagnet.

In 1836 Morse read Henry's 1832 article about his experimental telegraph. Henry had pointed out that a many celled battery and a large number of turns of the electromagnet were both necessary. Morse then rebuilt his apparatus and in November 1837 sent a signal through 10 miles of wire wound on a reel.

In the meanwhile, Wheatstone and other Europeans were working on telegraph equipment and Morse looked for capital to press his patent claims. He demonstrated his telegraph to a wealthy art student of his at New York University, Alfred Vail. Vail's family owned the Speedwell Iron Works in Morristown, New Jersey. The young man became interested and in 1837 brought Morse to see his father, Judge Stephen Vail in Morristown, New Jersey. The Judge scribbled a message for Morse to send to Alfred inside the barn. The message read, "Patient waiters are no losers." Alfred Vail was able to read the message which was sent over a wire which entered the barn. Judge Vail then supplied Morse with money to build a demonstration model and apply for a preliminary patent called a caveat.

Alfred Vail worked to convert the visual system into an audible one. It was known that an electromagnet when activated by an electrical current could attract an iron lever down to meet the iron core of the magnet. This produced an audible click. When the current was turned off, a spring would pull the lever back producing two successive clicks. A short time interval between clicks was a dot and a long interval was a dash. A telegraph key at the transmitter controlled by a skillful operator translated numbers and letters into an alphabet of dots and dashes. At the receiver a telegrapher translated the dots and dashes back into numbers and letters.

A demonstration telegraph line was built from Washington to Baltimore in 1844. A company, the Magnetic Telegraph Company, was formed which extended the line in 1846 and 1847 from Philadelphia to Newark.

Ezra Cornell in 1847 built a telegraph line from Piedmont, New York, parallel to the Erie Railroad being built by Eleazer Lord. It was soon found that the telegraph could be used by an Erie Railroad superintendent to check ahead on the oncoming train. Erie bought Cornell's telegraph line and it was used to dispatch trains by telegraph. In 1855 the Erie Railroad Telegraph opened to commercial businesses in the cities served by the Erie. Later, Erie Railroad Telegraph became Western Union.

Trained telegraph operators were in great demand as the new technology spread over the United States. One of these was Thomas Alva Edison. When he was 15 years old, he saved the life of a child who had been playing on a railroad track. The grateful child's father, a telegraph operator, gave Edison lessons in telegraphy.

For the next five years, Edison worked as a roving telegrapher mostly on the night shift so that he could devote his time during the day to performing scientific experiments and developing inventions. Many of his early inventions were in the field of telegraphy which fascinated him. Later he named two of his children Dot and Dash.

Western Union planned a worldwide telegraph network that would cross the United States, Canada, Alaska, and Siberia to European capitals. In 1860 the Russian link was begun by the Russian Telegraph Administration to install a 7,000-mile telegraph line from Moscow through Siberia to Vladivostok. The Great Northern Telegraph Company, a Danish firm, worked with the Russian Telegraph Administration.

Great Northern built telegraph lines in China and laid a submarine cable between Vladivostok and Nagasaki, Japan. Great Northern opened a European-Far East service in 1872.

The Western Union plan for an Alaska-Siberia telegraph line was superseded by a transatlantic cable laid in 1865 by Cyrus Field. The U.S.-European underwater cable went into operation in 1867.

Great Britain was anxious to establish a telegraph service between London and the growing Empire in India. In 1867 the German company Siemens offered a plan to pass through the Prussian-Persian lines to connect, at Teheran, with a system administered by the British-India Authority. The Indo-European Telegraph Company, Ltd. was formed to construct and operate the line from London to India. Automatic repeaters were used throughout the system. The overhead line was more than a quarter of the earth's circumference. It opened for business in 1870 and provided service for almost 60 years with the exception of World War I.

The telegraph was the first of several methods of obtaining a rapid worldwide communication network. Attempts to transmit telegraph messages faster and to send a number of messages simultaneously led in the 20th century to new inventions such as the teleprinter, the public switched telephone network, the facsimile, and computer networks.

8.2 THE TELEPRINTER

In the 20th century the dots and dashes of the telegraph was replaced by the teleprinter. The prototype of the modern teleprinter was devised by Charles Krumm of the United States in 1907. The teleprinter, also called teletypewriter or teletype, is any one of various telegraph instruments that transmit and receive written messages and data via long-distance telephone cables or radio relay systems.

The teleprinter consists of a typewriter-like keyboard and a printer. A message is sent by typing on the keyboard. Each keystroke generates coded electrical pulses. The pulses are routed by an electric switching system over an appropriate transmission line to the destination. There a receiving teleprinter decodes the incoming pulses and prints out the message on paper.

Teleprinters use either of two keyboard coding schemes for message transmission. Some units use a variation of the Baudot code (Murray code) in which a letter, number, punctuation mark, or symbol is represented by a combination of five "on" and "off" pulses. Depressing a key of the teleprinter keyboard activates a set of five code bars that produces a pattern of pulses corresponding to the code combination of the particular character selected. Each character is preceded by a "start" pulse and terminated by a "stop" pulse.

Other teleprinters use a keyboard code called the American Standard Code for Information Interchange (ASCII) that provides for seven digits. An eighth digit is used as a parity check.

Internationally there was a service called Telex which consisted of a network of teleprinters. Telex systems were initiated in the 1930's in several European countries. After World War II, Telex developed into an international service. Telex used a modified Baudot (Murray) code and sent messages at 67 words per minute using both wire and radio circuits.

In 1931 AT&T developed the Teletypewriter Exchange Service (TWX). TWX teleprinters used ASCII and transmitted at speeds up to 150 words per minute. In 1970 Western Union acquired TWX for operation in the United States. Computers can transform ASCII into the Murray code and vice versa.

8.3 THE PUBLIC SWITCHED TELEPHONE NETWORKS

The public switched telephone network (PSTN) using digital transmission for voice, video, data, and facsimile was a development of the 20th century. However, the telephone itself was invented in the last quarter of the 19th century. This invention originated in work by Alexander Graham Bell (1847-1922) to develop a multiple telegraph which could send several messages simultaneously over a single wire.

Alexander came from a family of teachers of speech in Edinburgh, Scotland. His grandfather had taught corrective speech and his father had invented a universal phonetic alphabet. Alexander Graham Bell had two brothers who both died of tuberculosis. The remaining family fled to Ontario, Canada, in 1870.

Alexander became a teacher of the deaf. In the fall of 1872 he went to Boston and opened his own private school for the deaf. While there Bell began experimenting with the invention of a multiple telegraph. His idea was to transmit a number of messages, each at a different pitch. At the receiver there were a number of reeds. Each reed responded to a specific pitch.

In January 1875 Bell hired a young electrical worker, Thomas A. Watson, as his assistant to develop the multiple telegraph. During their experiments one of the receivers stuck to its electromagnet. The batteries were disconnected and Watson plucked the reed free. In the next room Bell was startled to see the reed in one of his receivers vibrate very strongly even though the batteries were all disconnected. With only the strength of residual magnetism, the plucked reed induced a current strong enough to travel over the wire and excite the receiver's electromagnet in the next room and make it vibrate.

Bell had Watson repeat his plucking of the receiver reed a number of times. The inventor in the next room pressed his ear to a reed in a receiver and heard a musical tone of the pitch of Watson's reed. Bell found that the vibrations of Watson's reed was reproduced in both pitch and loudness in another reed in the next room. Sound itself could be converted into an electric current and transmitted.

Bell realized that this phenomenon could be used to transmit speech. He discarded the idea of the multiple telegraph and concentrated on the telephone. Bell sketched a diaphragm transmitter and a light vibrating reed which caused an electric current to vary with the variations of the voice. Watson later built a diaphragm receiver much like the transmitter.

On February 14, 1876 Bell's invention was filed with the U.S. Patent Office. On March 3, 1876 the Patent Office Examiners approved Bell's Patent No. 174,465, the basic telephone patent.

Bell knew that this transmitter could not be used for practical voice transmission. He decided to make a liquid variable resistance transmitter. His idea was to use the varying audio voice to vary the conductivity in an electric current. This would cause the electrical current from the transmitter to follow the variations of the voice. On March 9, 1876 Bell sketched a liquid transmitter which included this principle.

The transmitter was a chamber filled with a diluted acid. The varying pressure of the voice on the diaphragm of the chamber caused the conductivity of the acid to vary. A wire from the liquid carried a varying current to a receiver.

On March 10, 1876 Bell yelled into the mouthpiece of a liquid transmitter, "Mr. Watson -- come here -- I want you." This was the first practical voice transmission and the telephone was born.

In the spring of 1876 Bell invented another form of receiver which became known as the "Centennial Iron-Box Receiver." He placed an electromagnet in a hollow iron cylinder with one end closed. A sheet-iron lid was fastened to the open end and the electromagnet was adjusted so that it was close to but not touching the lid. The lid acted as a diaphragm, responding to the electromagnetic fluctuations induced by the undulatory current.

Bell exhibited his telephone at the Centennial Exposition in Philadelphia on June 25, 1876 to an audience which included Emperor Pedro II of Brazil and his Empress. The demonstration was a great success.

Like most inventions there was a great patent dispute with inventors like Elisha Gray and others. The litigation went on for years until Bell was awarded the fundamental patent. The Bell Telephone Company was founded in July 1877.

In 1876 Edison invented the carbon-button transmitter which was a pressure sensitive solid conductor. Later in 1880 Francis Blake invented the Blake carbon transmitter which improved the quality of speech. The pressure of the air from the speaker would cause the resistance of the carbon to change which in turn varied the electric current. A carbon transmitter working with the Centennial Iron-Box Receiver became the standard telephone.

After the 1876 Centennial Exhibition in Philadelphia, Bell offered to sell his telephone patent to Western Union for $100,000. The giant telegraph company turned him down and instead paid Elisha Gray a huge sum for his multiple telegraph.

SWITCHBOARDS

The first switchboard in the world was opened in1878. A push-button signaled the operator. Later a handcrank was used to alert the operator. The caller then told the lady at the switchboard who he or she wanted to talk to. The operator placed a plug in the appropriate jack. This made the connection between the two parties.

The invention of the first automatic telephone switch in 1892 has an interesting legend behind it. Almon B. Strowger was a Kansas City undertaker who was losing business to a rival undertaker whose wife was the switchboard operator for the local telephone system. According to Strowger the operator was connecting all undertaker calls to her husband. Strowger decided to invent a device to eliminate the switchboard operator.

His invention was automatic switching gear which allowed the customer to make a connection without the help of an operator. Strowger's invention was a ten-position rotary selector switch with a pivoting central arm that could rotate to connect with any of ten electrical contacts. The pivoting arm was moved by an arrangement of electromagnets, springs and ratchets. Each time the electromagnet received a pulse of current, it advanced the arm by one position. Originally the customer operated the switch by a button. If the caller wants to send a 5, he or she presses a button five times, thereby sending five pulses of current to the electromagnet, driving the selection arm. The button was replaced by the dial phone which came into service in 1919.

The early 20th century saw a gradual increase of telephones. This led to an elaboration of the Strowbridge switch to interconnect more and more subscribers. A single Strowbridge switch could interconnect ten subscribers. When a caller picked up the receiver, the line would be connecting to the central selector arm. Dialing a one-digit number would then ring one of the other nine telephones.

Adding a second stage of switching could expand the service to a hundred subscribers. Now the original switch, instead of being connected directly to ten subscriber lines, would be linked to a bank of ten more identical switches. Each subscriber would be identified by a two-digit telephone number. When the subscriber dialed the first digit, for example a 4, the first selector would connect the line to the selector arm of the switch leading to line 40 through 49. Dialing a second digit would move the selector arm of the second switch to the appropriate contact. This idea could be expanded to 1,000 and even to 10,000 subscribers but there is a limit to the size of the mechanical switching gear.

The telephone company decided on a solution. It set up central switching offices with as many as 10,000 subscribers, then provided trunk calls between central offices. A three-digit number selected the desired central office. The three-digit number was followed by a four-digit number which made the desired connection on that central office.

The Strowger switch was replaced by the cross bar switch in telephone central offices. The cross bar switch had many vertical and many horizontal paths. Electromagnets mechanically connect any of the horizontal paths to any of the vertical paths. Some cross bar switches were used until the 1990's.

In the second half of the 20th century there were a number of improvements using the development of solid state electronics. In 1951 direct long distance dialing started. Switching changed from mechanical to electronic. In 1963 the touch-tone telephone came into service. Each number and symbol on a touch-tone pad activates two simultaneous audio tones, one from a high frequency group and one from a low frequency group.

Later in 1973 the touch-a-matic set was the first telephone with a solid state memory. A touch of a single button would dial any of 31 pre-recorded numbers. In the 1990's phone number selection could be accomplished directly by voice instead of by touch-tone.

DIGITAL NETWORKS

The invention of tiny lasers and the development of fiber optics in the 1970's led to the replacement of copper wire cables with much smaller glass fiber optics. This allowed the development of a new type of public switched telephone network using digital instead of analog transmission. The human voice was digitized and became part of a stream of digital information which included data from computers, facsimile and video.

The first digital system to do this was called the Integrated-Services Digital Network (ISDN). This is a wholly digital public telecommunications network capable of handling both voice and relatively slow data traffic starting at 144 kb/s.

ISDN is being replaced by broadband ISDN (B-ISDN). The minimal rate envisioned for B-ISDN is 156-M b/s. The telecommunication industry standard way of transmitting on B-ISDN is called Asynchronous Transmission Mode (ATM). ATM is the heart of the B-ISDN and is a standard packet of 48 bytes of information and five bytes of address. Video, voice, and data can all be transmitted by this standard packet at very high speeds. ATM is expected to be the switches for the B-ISDN by the end of the 20th century.

The International Telephone Networks

In the 20th century telephone networks spread all over the world. The United Nations agency that regulates international telecommunications divides the world into nine zones. World Zone 1 includes the United States and Canada and about a dozen Caribbean nations. Zone 2 is in Africa; 3 and 4 cover Europe; 5 is Central and South America; 6 is the Pacific; 7 is the territory of the former USSR; 8 is Asia; and 9 is the Middle East.

Many of these international telephone networks go through the AT&T center at Bedminister, New Jersey. Here computers route the information throughout the United States and the world. Forty large TV screens give a picture through CNN of happenings in each area over the world. Network managers can re-route messages if natural disasters such as earthquakes appear on one of the TV screens.

FLAG (Fiber-Optic Link Around The Globe)

FLAG is a complex undersea optical fiber cable that spans more than two-thirds of the earth's circumference. It links Great Britain to Japan. FLAG goes through the Atlantic Ocean, the Mediterranean and the Red Sea, the Indian Ocean, and the Pacific Ocean. FLAG was completed in 1997. FLAG can carry 120,000 circuits at 64 kb/s on two fiber pairs.

All-optical-technology amplifiers are embedded approximately 30 miles or so in the cable. Electro-optical repeaters at the landings regenerate the signal at higher power levels for the next series of undersea amplifiers.

FLAG starts in Cornwall in the United Kingdom and terminates at Miura on Tokyo Bay. On the way it connects to landing points in Spain, Italy, Egypt, United Arab Emirates, Malaysia, Thailand, Hong Kong, People's Republic of China, and the Republic of Korea.

Semi-conductor lasers at the transmitter generate an optical signal at 1.558 microns which is transmitted at 5.3 Gb/s through the undersea plant to the next receiving equipment on shore. FLAG is the longest man-made structure ever assembled in the world's most ambitious undersea lightwave communications system.

8.4 Security on the Public Switched Telephone Network

In 1977 a Data Encryption Standard (DES) was introduced by the United States Bureau of Standards* This was to be the cryptographic standard for commercial use for

*This is now the National Institute of Standards and Technology

the 1980's. However, it was known that with the advent of greater computer power, DES would not be enough for the last few years of the 20th century.

In the early 1990's the United States National Institute of Standards and Technology (NIST) selected a classified computer algorithm called Skipjack incorporated into licensed hardware named the Clipper Chip. The Clipper Chip incorporates among other things a unit key. The ultimate objective is to have each telephone and all communication devices equipped with Clipper hardware.

The unit key is unique to each chip. The key is split into two numbers which when combined will produce the unit key. One part of the key will be held by the U.S. Treasury Department and one part by the NIST. Federal law enforcement agencies by court order can obtain the keys for domestic telephones. They do not need a court order to tap international phone conversations.

The United States government requires all companies doing business with them to use Clipper. It is the only cryptographic hardware approved for export without review. This means that any U.S. multinational company desiring secure communications must use products based on the Clipper standard.

There has been a great deal of opposition to the mandated use of the Clipper security system. In May 1994 the U.S. Congress held hearings to discuss Clipper and the broader issue of cryptography policy. Many people interested in secure communications opposed the use of the Clipper Chip and have developed their own systems using the concept of the public key.

PUBLIC AND PRIVATE KEYS FOR SECURITY

Traditional encryption methods are based on the private key method. This is a large integer number to encrypt information. Only a recipient who knows the private key or a code breaker with a supercomputer can decode the information. The private key must be delivered from the transmitter site to the receiver site. This is the weak link in this system. The author in 1945 had to deliver a private key to American forces trying to take the City of Myitkyina in Burma from the Japanese. The private key was in the form of rotor wheels for a cipher machine. An armed guard provided security to prevent any attempt at seizing the private key.

In order to solve this problem, modern encryption systems use a technology called public key, a derivative of private key. The user announces publicly that a certain large integer number should be used as a public key to encrypt messages. The public key is derived from another large integer number called the secret key. The recipient uses it to decrypt the message with the aid of digital computers. The public key technology avoids the distribution problems of private key encryption. Public key was developed originally in 1975 by Whitefield Diffie and a Stanford University Professor, Martin Hellman. This has been improved by a number of computer programmers and others.

THE PGP (PRETTY GOOD PRIVACY) SYSTEM

Phil Zimmerman developed a security program called PGP*. He had distributed this system free. The U.S. Government does not want PGP to proliferate overseas. In 1995 Zimmerman was awarded the Pioneer Award from the Electronic Frontier Foundation. However, at the same time a federal prosecutor in San Jose, California, investigated him for violating export regulations in distributing PGP.

8.5 FACSIMILE

The invention of the telegraph in 1837 not only led to the telephone but even earlier the facsimile machine. Shortly after the advent of the telegraph, Alexander Bain thought up a crude method of sending pictures over telegraph wire. He was a well known Scotch clockmaker who was very familiar with the operation of pendulums. His idea was to use characters of metal at the sending side. A pendulum tipped with a metal stylus swung back and forth over the character. At the receiving end of the telegraph line, a similar pendulum swung back and forth across chemically treated paper.

Each time the sending pendulum struck the metal, it completed an electrical circuit and sent an electric pulse through the electric wire. Arriving at the swinging pendulum at the receiver, the electricity discolored the chemically treated paper, leaving a trace on the paper each time contact was made. Clocks on both ends advanced the metal characters and the paper a fraction of an inch simultaneously. In practice the timing and synchronization of the transmitting and receiving ends proved impractical and this machine was never developed.

The next step was made by Frederick Bakewell who came up with a rotating drum. The message was written on the drum with a nonconducting ink. A metal sheet was wound around a revolving cylinder. A metal stylus was set on the cylinder and a screw mechanism kept the stylus moving along at a uniform rate.

As the stylus moved along the bare metal, it emitted an electric current. When the stylus touched the insulated ink, the current was stopped. The electric current was sent on the telegraph wires to a receiving rotating drum covered with a sheet of chemically treated paper. The electrical impulse discolored the paper and produced a reproduction of the original. Bakewell patented his facsimile machine but synchronization remained a problem and it was never commercialized.

It was up to an Italian Priest, Giovanni Caselli, to make the first practical facsimile machine. He used Bain's idea of swinging pendulums but he employed only one clock in the transmitting end that synchronized the pendulum at each end perfectly. Also Caselli worked with ordinary ink. He used very high quality recording paper soaked in potassium cyanide which changed color each time electricity passed through it. Caselli

* See both Garfinkel and Stallins in the Bibliograpghy for details about PGP

patented his machine called the Pantelegraph in the United States in1863. In 1865 Caselli's machines began operating on the existing Paris-Lyons telegraph line. Two years later a leg to Marseilles was added.

The 1870 war between France and Prussia and the siege of Paris interrupted the operation of the facsimile link. After the war ended, the line was not restored. It was not until the early 20th century that a new type of facsimile using light took over from swinging pendulums.

G.R. Carey of Boston in 1875 built a mosaic of selenium photo cells. An image was focused on the mosaic and each cell gave off an amount of electricity in proportion to the light falling on it. At the receiving end, a mosaic made of shutters reproduced a crude copy of the transmitted picture. Arthur Korn, a German physicist, in 1902 built a rotating cylinder of glass. He positioned a selenium cell inside and wrapped the printed matter or picture to be transmitted around the cylinder. Korn shone a light through the picture and onto the selenium where the picture was converted into electrical signals. At the receiver end, the signal was converted back to light and was shone on photographic film making a permanent copy. Korn's invention was used by newspapers to transmit photographs.

A Frenchman, Edward Belin, built a facsimile machine using the same principles that Korn used but simpler and smaller. Belin's machine, the Belino, built in 1913, was the world's first portable fax. It could be hooked up to telegraph or telephone lines. In 1914 the Belino sent back the world's first remote fax photo/news report.

In World War II both sides used fax machines to transmit photographs, weather data, and other information. From 1948 to 1958 Western Union sold some 50,000 desk-fax units so customers could send and receive telegrams. The fax machine was further developed in the mid 1960's at Magnavox and Xerox Corp. They developed general purpose analog equipment which took 4 to 6 minutes to transmit a photocell scanned page.

In 1968 the U.S. government took a decisive step in the development of fax systems. It allowed fax machines access to the public switched telephone network. In 1972 Japan did the same.

The initiative in new fax developments passed to Japan because the Japanese language was at a definite disadvantage in existing telecommunication systems. That language used over 2,000 Kanji characters. Even the Japanese phonetic alphabet has 48 symbols compared to the 26 letters of the English alphabet. The Japanese language does not adapt easily to Western telegraph and telex systems. For this reason the Japanese embarked on digital fax research. By the 1970's the Japanese domestic telephone company operated the world's largest facsimile research laboratory.

In 1980 Group 3 fax standards were set up by the United Nations. This used digital signals to be sent over regular telephone lines in one minute or less. The pictures or text to be transmitted are converted to zeros and ones. At the receiver end the fax reconstructs the image from the digital code. Group 3 fax became worldwide in the late 1980's and early 1990's using Group 3 standards.

In the modern facsimile an image is formed by a lens in a way similar to that of an ordinary camera. A linear array of small photodiodes is substituted in the fax transmitter for the film. The portion of the image falling on the linear-diode array is a thin line, 0.005 inches across the top of the page being transmitted. Usually, 1,728 diodes are used for a page 8 1/2 inches wide. Each of the 1,728 diodes is checked in sequence to read across the page. Then the original page is stepped the height of this thin line, and the next line is read. The step-and-read process is repeated until the whole page has been scanned.

In the fax receiver the signals are converted into a copy of the original. The most commonly used system is the thermal recorder. It has 1,728 very fine wires positioned in a row across the recording paper. These wires touch the paper and produce very small hot spots as current passes through them. These hot-spot sections of the wires form a straight line across the page at a resolution of 200 dots per minute. Each of the 1,728 recording wires in the receiver corresponds to one of the 1,728 photodiodes in the transmitter. Thermally sensitive coated paper is used to produce the image, but ordinary paper is used in some more expensive systems.

Fax transmission over good quality telephone networks is usually at 9,600 bits per second. If the telephone-line quality in some countries is not good enough for transmission at this speed, then the transmitter automatically steps down to 7,200 bits per second. If the telephone quality is still not good enough, then the process of lowering the speed continues down to 4,800 and even 2,400 bits per second.

Computers can be used to transmit and receive facsimile. A fax board is plugged into a vacant slot in the computer, and a software program converts computer language files into Group 3 facsimile files. A standard facsimile modem of the fax board makes the signals and the communication protocol identical with Group 3. Documents can be viewed on the computer display screen or a hard copy can be made on the computer's printer.

8.6 INTERNET

Internet is a network of interconnected computer networks that spans the world, connecting many millions of people on their computers.

ORIGINS

One of the first computer networks in the United States was ARPANET which was founded in 1969 by the Advanced Research Projects Agency of the Defense Department. ARPANET initially linked researchers for the Department of Defense with remote computer centers. The network allowed the researchers to share the hardware and software.

Later ARPA became DARPA, the Defense Advanced Research Projects Agency. Soon other experimental networks using radio and satellites were connected to ARPANET using an internetwork technology sponsored by DARPA. In 1980 an unclassified military network (MILNET) split off from ARPANET but connections made between the networks allowed communications to continue. This interconnection of networks was first called the DARPA Internet which later was shortened to Internet.

In the early 1980's other networks such as the Computer Science Network (CSNET) and Because It's Time Network (BITNET) began providing nationwide networks to the academic and research communities. These originally were not part of the Internet but soon connections were made between the various communities.

In 1986 the next great advance in the Internet occurred with the establishment of the National Science Foundation Network (NSFNET) which linked researchers with supercomputer centers. One of the founders of NSFNET was Dennis Jennings of the University College, Dublin. In 1985 he was offered the job of NSF Program Director. One problem was that he had to be a citizen of the United States or one of the allies. Ireland did not qualify. However, Jennings, an Irish citizen, was born in England and this allowed the NSF to employ him. Jennings in August 1985 proposed a NSFNET linking four supercomputers and the National Center for Atmospheric Research (NCAR).

Steve Wolff, Jenning's replacement at NSF, in 1986 expanded the NSFNET and funded international links and encouraged the growth of regional and campus networks. NSFNET had no directly attached users but was connected to regional systems. The NSFNET began to replace the ARPANET for research networking. The ARPANET was dismantled in 1991. The NSFNET had the fastest speed on the Internet. It is capable of sending about 5,000 typescript pages per second. Other networks on the Internet have slower speeds.

NSFNET is an example of a high-speed central network known as a backbone. They accept traffic from and deliver it to the mid-level networks. Mid-level networks take traffic from the backbones and distribute it to their own member networks usually at a lower speed.

In December 1991 a law was passed by the United States Congress combining the computer networks of all academic research institutions into one high capacity, high-speed network. This is known as the National Research and Education Network (NREN).

International Aspects of Internet

International connections of the Internet spread to networks all over the world including Japan. In Japan there was at first some reluctance to join Internet because of the Japanese use of Kanji characters instead of the Roman alphabet. This was solved by Jun Murai of Keio University. He developed special software which allows Kanji users to automatically transmit and receive Romanized Japanese. In Japan, Nifty Serve is a network that acts as the licensee to the U.S. network CompuServe. The software was rewritten to support Kanji characters.

In 1992 the European Backbone (EBONE) was formed. This was similar to NSFET in the United States. It too had no directly attached users but was connected to regional systems which in turn would be connected to users. The Backbone has five major hubs. They are in London, Montpellier (France), CERN (the high-energy particle physics center in Geneva), Amsterdam, and Stockholm. EBONE also has three links to the United States. Although targeted for the benefit of academic and research use, commercial traffic can pass over the Backbone.

Internet also spread to small countries around the world such as Singapore. In 1992 the National University of Singapore put in a connection to Princeton University and joined the Internet. Ireland used BITNET to connect various research laboratories and universities and then made connections to the Internet. Areas like New Zealand, Hong Kong, Australia, Greenland, Easter Island are linked on the NSI (NASA Science Internet). This is one of the core Backbones for the Internet and links scientists in universities and research laboratories all over the world.

COMMERCIAL NETWORKS

The Internet and its predecessors served as federally funded research networks for catching the attention of the business community and the general public. In the early 1990's, the nongovernment sections of the Internet were opened to commercial options. These ranged from individual access through networks at a low monthly cost to a full feature, privacy-enhanced connections for corporations. Internet has since grown into a network connecting many other networks.

By 1996 there were two general methods in the United States for getting on the Internet, Commercial On-Line Services and Internet Access Providers.

Commercial On-Line Services, such as America Online, Prodigy, CompuServe, Microsoft Network, and WOW!, all offer access to the Internet and, in addition, they have their own information centers, forums and chat rooms. Their customers log on by obtaining an account number and password. They all offer free software.

Large Internet Access Providers, such as AT&T, Sprint, Southern New England Telephone, MCI Corp., Netcom, Pipeline, PSINET and UUNET Technologies provide direct connection. Many Internet providers give away free software.

In 1996, a simplified inexpensive network computer, the NC, was introduced for use on the Internet. The NC basically uses software stored on the network. The NC is smaller and simpler to operate than the conventional PC.

USES OF THE INTERNET

Although one of the purposes of the Internet is still the linking of scientists and laboratories and universities, many other uses have developed. These include electronic mail, bulletin boards, public discussions, remote log-in and file transfer.

E-MAIL

Electronic mail, known as e-mail is the most commonly available and most frequently used on the Internet. Electronic mail is controlled by e-mail program software in the computer. Received messages are stored in the computer and may be read later. E-mail computer programs let the sender compose and send e-mail and read, organize, and store messages. The e-mail program prompts the sender for pertinent information such as the recipient's e-mail address. Internet e-mail addresses consist of a username and a host part separated by an @ sign in this manner: *user name @ host name*. The username is the name that is logged into the computer. The host part is a series of words which describe the organization. A made up example is: *esmith @ dax.-com,* where the suffix *com* stands for commercial. Other suffixes are *org* for not-for-profit organization, *edu* for educational, *gov* for government, *mil* for military and *net* for network.

REMOTE LOG-IN

This is an interactive tool on the Internet. It allows the user to access the programs and applications available on another computer. The user can interactively query the computer on the Internet for needed information.

FILE TRANSFER

File transfer allows files to be transferred from one computer to another. This includes documents, graphics, software, and spreadsheets.

WORLD WIDE WEB

The Internet's World Wide Web (WWW) is a dynamic information source. Globally WWW represents all the computers (servers) that offer users (clients) access to hypermedia based information and documents. WWW uses a concept called *hypertext.* Every piece of information has links which allow the user to follow other pertinent documents no matter where they are located anywhere in the world.

The person credited with the invention of the World Wide Web is Tim Berners-Lee, a consulting engineer at CERN, the European particle-physics laboratory in Geneva. He developed a system that provided easy-to-follow links between documents stored on a number of different computer systems created by various groups at CERN. Berners-Lee expanded the system and made it available on the Internet in 1991.

The best way to cruise the World Wide Web is with a computer program called a Web browser. There are a number of browsers in use. One of these is called Mosaic, developed originally at the University of Illinois' National Center for Supercomputer

Application. Mosaic was further developed by a private company, Spyglass, Inc. Microsoft included a Mosaic based browser in their operating system, Windows 95.

Another browser is the Netscape Navigator from Netscape Communications, Inc. In 1995 the Navigator captured a large part of the WWW browser market and World Wide Web home pages became popular for businesses and others to advertise. The address of a home page has the prefix, HTTP://. HT stands for hypertext, the second T for transfer and the P is for Protocol. The Web home page is specified by giving its computer address or URL (Uniform Resource Locator).

World Wide Web home pages can be constructed by Hypertext Markup Language (HTML). Software has been developed which converts documents into HTML.

Up to December 1996, the only way to surf the World Wide Web was with a conventional computer. Then Sony and Philips introduced WebTV which allows surfing the Web on a regular TV set without a PC. It automatically connects to a WebTV Network for a monthly fee. The technology was developed by WebTV Networks, Inc.

The Internet Terminal, which is book size, hooks up to any TV. It contains a modem which is connected to a telephone line. World Wide Web home pages can be seen and heard on the television set by means of a built in WebTV browser. There is a connector for use with a printer.

A small remote unit controls the TV and also enables the surfing of the web. The remote has a thumb control that acts like a mouse in a conventional computer.

In addition to surfing the Web, there is an optional keyboard for e-mail use. This is an example of combining television and the computer.

Protocols of the Internet

The various networks of the Internet have to be connected by a common protocol. A protocol is a group of procedures and conventions for allowing communications between specific endpoints of a network. The Internet Activities Board (IAB) oversees development of the Internet suite of protocols. The protocol most in use in the Internet is the Transmission Control Protocol/Interface Protocol (TCP/IP).

TCP/IP breaks up information into chunks called packets. Each packet contains several hundred characters plus tags such as the addresses of the sending and receiving computers. The Internet is a packet switched network. The switches are computers called routers. They are programmed to figure out the best packet routes. Figure 8-1 shows the Internet connections.

Security Access Problems on the Internet

Many classified laboratories and government agencies are connected to the Internet. There are some supposedly secret passwords that act as keys to the Internet. Starting in the summer of 1993 there was a rash of break-ins to the classified portions of the Internet and to connecting computer networks. Many secret passwords were stolen. On February

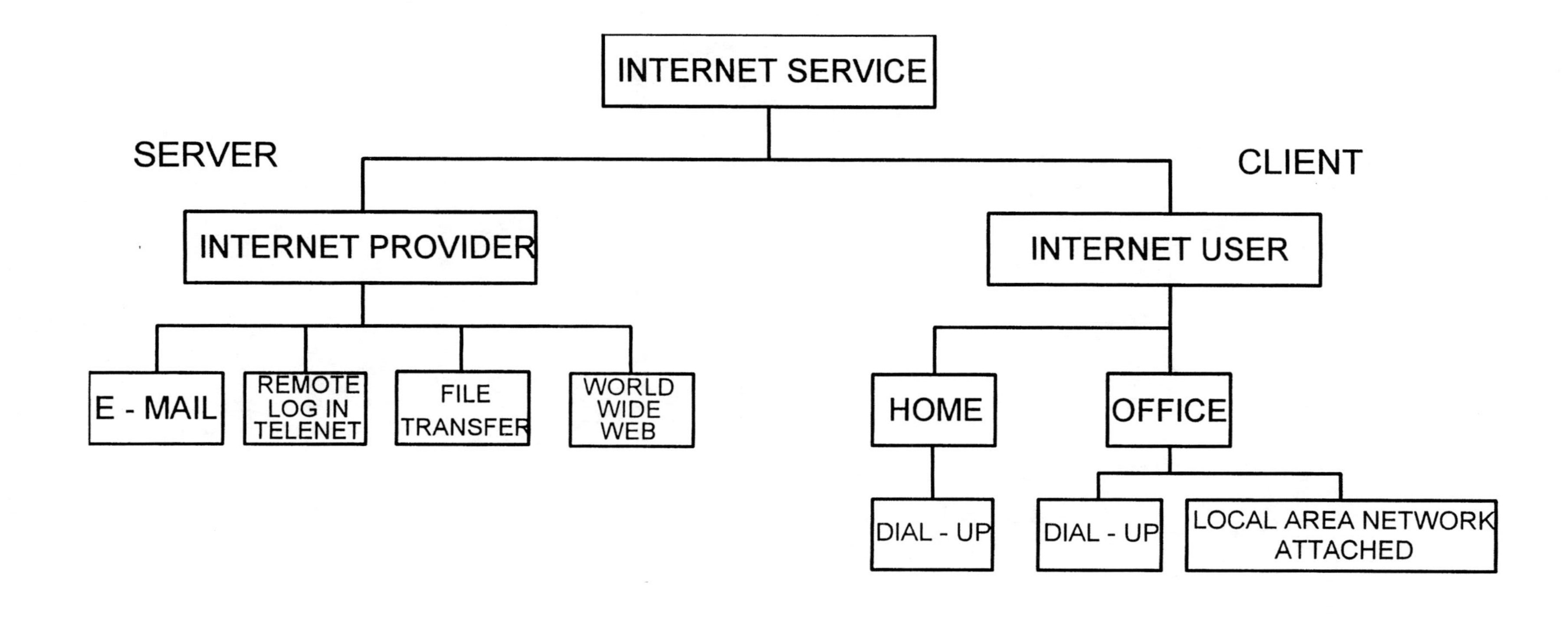

Figure 8-1 Internet Connections

3, 1994 the Internet and other connecting computer networks were alerted to the problem.

Unauthorized people who get into the Internet can destroy very valuable computer files. Individuals who eavesdrop on electronic conversations can snatch passwords, called snarfing, and obtain unauthorized access to confidential files.

One of a number of proposed methods to protect computer networks is the Kerberos security system developed at M.I.T. under project Athena. In Greek mythology Kerberos was a three-headed dog, Figure 8-2, who guarded the gates of Hades. In computer network security Kerberos has three main components, database, authentication server, and ticket-granting server. The Kerberos architecture eliminates the need for clear unencrypted text passwords to be passed over the network. The procedure is as follows:

1. A scientist, Dr. Jones, wants to enter the Naval Research Laboratory (NRL) database. He uses an authentication server, Alpha network, as a gateway via the Internet to the NRL computer database. Dr. Jones contacts Alpha and gets a special password called a "secret key."
2. When Dr. Jones wants to enter the NRL database he sends a message to Alpha's authentication computer.
3. The authentication computer sends Dr. Jones an encrypted response that can be read only with Dr. Jones' "secret key."
4. The encrypted response contains a temporary key good for this session only and an encrypted ticket that will expire if he doesn't use it soon.
5. Dr. Jones sends two things back to the Alpha computer, the encrypted ticket and an encrypted message coded with the temporary key.
6. Alpha then sends Dr. Jones a second encrypted ticket. Dr. Jones decodes it with his temporary key.
7. Dr. Jones then sends this final ticket over the Internet to NRL's computer database requesting the desired information.
8. The NRL computer queries the Alpha computer to verify that Dr. Jones has a valid ticket. Alpha confirms that Dr. Jones is authorized and the NRL computer opens its gates. Dr. Jones' secret key at no time goes over the Internet to risk interception or "snarfing."

THE DIRT ROAD OF THE INFORMATION HIGHWAY

The Internet has been used to spread racial and religious hatred on hyperspace bulletin boards.

Figure 8-2 Kerberos, a Proposed Password Guardian on the Internet.
Illustration by: Dianna Gabay

The Internet has also been used to transmit pictures showing child pornography. Legislation has been passed by the U.S. Congress. This has been opposed by various Internet users as interfering with free speech.

In addition, children have been seduced over the Internet by impostors posing as fellow children.

All of the above has raised constitutional questions which will probably be fought out in the courts.

8.7 Cellular Radio

The ancestor of the portable cellular phone was the walkie-talkie. Early models from 1936 to 1950 are shown in Figure 8-3. The development of these radios led first to the radio telephone in cars and boats and then to cellular radios.

In the second half of the 20th century the telephone was extended to cars and boats. One difficulty was the loss of privacy since radio was used as the link to the wire telephone system. One day in 1974 a man on his boat going up the Hudson River called the marine telephone radio operator and gave her the telephone number of his home. He could be heard on weather radios in the area telling his wife that he was having engine trouble and would be coming home late. After his wife hung up the man called his mistress on the marine telephone to inform her that he was docking and would see her shortly.

A more serious problem was the limited number of frequencies available. As more and more cars and boats used the mobile telephone the system became jammed with people waiting to use it. The Bell Laboratories in the early 1970's started to develop the Advanced Mobile Phone Service (AMPS) to solve the problem. This became the cellular radio service.

This is basically a system to reuse a number of frequencies over and over again. A service area is divided into small cells, each with a low-powered base station. The cells are grouped into larger units called clusters. Within a cluster each cell is assigned a different frequency. Adjacent clusters repeat the frequencies. As a mobile goes from one cell to another, the mobile frequency is changed to correspond to the base frequency of the new cell.

The switching of the frequencies is accomplished by a Mobile Telecommunications Switching Office (MTSO). As the signal level decreases in one cell and increases in another, the MTSO signals the mobile to change frequency. Another cell site takes the hand-off and continues the phone calls without interruption as the mobile goes from one cell site to another.

There is one MTSO for each service area. The MTSO is connected by wire to the Public Switched Telephone Network (PSTN).

The mobile unit contains a combination radio transceiver and logic unit in the trunk. This is connected to the antenna and the mobile phone control unit near the driver.

At each cell site there is a designated set up channel. In this channel digital information is continuously transmitted between the cell site and the mobile unit. When the mobile unit is turned on, it selects and monitors a particular set up channel. The mobile unit listens for a paging signal containing the mobile phone number in binary form.

When the call originates from a fixed station, the person at the fixed station dials the mobile unit's number. This goes through the PSTN to the mobile's home MTSO in a specific service area. The MTSO converts the mobile's phone number to the mobile's identification number. The MTSO then instructs the cell sites that have paging channels to page the mobile service area. Since the mobile unit is continually observing its setup channel, it will detect its number and then seize control of the reverse set up channel. As soon as it does so, the mobile unit transmits its identification number to the cell site. The cell site in turn sends a message to the MTSO by way of a wire data link. This message tells the MTSO that the called unit has responded. The MTSO assigns an idle voice channel to the mobile unit by sending a message through the cell site and then over the forward set up radio channel to the mobile unit. The mobile unit moves to the assigned voice frequency. The MTSO then directs the cell site to transmit a data message over the radio voice channel. This data message rings the bell in the mobile unit. The mobile unit answers and the MTSO removes the audible ringing circuit and the conversation starts.

When the call originates from a mobile unit, the mobile subscriber punches the desired phone number while on-hook. This permits resending of the number without repunching if the number called is busy. It also allows for the mobile subscriber to correct errors before going off-hook. The mobile subscriber then depresses the "send" button on the handset unit. This causes the mobile unit to transmit a digital message over the reverse set up channel to a nearby cell site. The digital message contains the following: the unit's identification number, the called number, and a request for a voice channel. The cell site sends this to the MTSO over the wire line data link. In turn, the MTSO, through the cell site, tells the mobile unit which of the cell site's voice channels to use for the call. After the mobile unit is on the correct voice channel, the MTSO connects the call through the PSTN. When the called party picks up the phone, the connection is made.

One of the most important functions in AMPS is the hand-off as the mobile goes from one cell to another. The hand-off is the transfer of the mobile unit from one radio channel to another. The MTSO transfers the call to mobile radio voice channel at an adjacent cell site. At the same time the cell site blanks the voice signal for about 50 milliseconds. During the blanking of the voice signal, a burst of data on the voice channel instructs the mobile unit to switch to the new channel. The entire hand-off function takes about 200 milliseconds which allows the call to continue without interruption.

Roaming is the operation of a mobile telephone in a service area other than the service area initially assigned to a customer. In one method when a cellular customer roams away from the home service area, the out-of-town system receives the phone's data signal, recognizes that its not local and uses the existing Signaling System 7 (SS7) voice-messaging network to talk back to the home system. SS7 is a network switching protocol

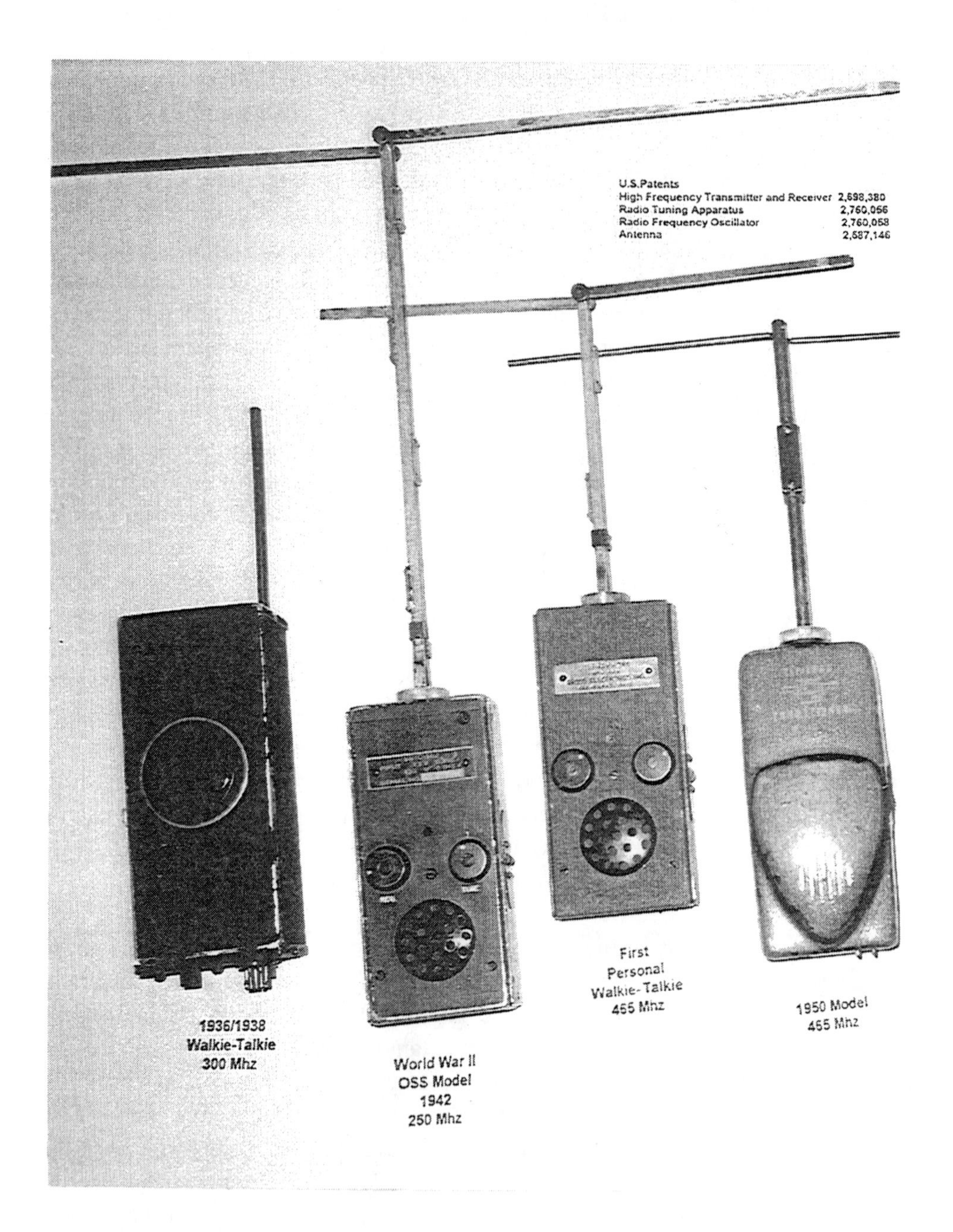

Figure 8-3 The Ancestors of the Portable Cellular Phone
Courtesy of Al Gross

capable of making high-speed connections. It can identify the caller. The out-of-town SS7 verifies the customer's identity and advises the home system to forward all calls.

It should be noted that the analog cellular phone conversations can be intercepted as politicians in the United States and the royal family in England found out.

The cellular phone installed in a vehicle was soon followed by the development of a small portable cellular phone which can be carried in a purse or attaché case, Figure 8-4. A demand was created for cheaper, smaller units. This was soon met by the advent of digital cellular systems.

A SECOND GENERATION CELLULAR SYSTEM

A second-generation cellular system is digital cellular. Digital cellular converts human voice into a digital format that is transmitted through the cellular network. There are a number of advantages for digital cellular.

First there would be lower cost since digital systems are generally lower in cost. Second, smaller portable units are possible since digital units require smaller batteries. Third, digital systems have increased capacity in the order of three to six times that of the conventional analog system. Fourth digital cellular is compatible to the Integrated Services Digital Network of the Public Switched Telephone Network. The compatibility extends many new services to the mobile cellular phone system. In addition, digitized voice ensures a degree of privacy.

In the late 1980's and early 1990's digital cellular was introduced around the world. The analog voice is converted into a digital format. Most of these digital cellular systems in the United States, Europe and Japan use a technology known as Time Division Multiple Access (TDMA). In this technology, time slots are used to increase the capacity of a cellular system by a factor of six.

Another cellular technology is Code Division Multiple Access (CDMA). In this technology different codes are used to increase the capacity of the cellular system. CDMA is used in a limited way in the United States and in other parts of the world.

HEALTH PROBLEMS WITH CELLULAR PHONES?

In January 1993 a Florida widower made the claim on a national television show that his wife died from a brain tumor caused by her cellular phone. His lawsuit against a cellular phone manufacturer and retailer caused cellular stocks to fall sharply.

In the ensuing furor representatives of the Food and Drug Administration (FDA), the FCC and the Environmental Protection Agency, and the National Cancer Institute testified at a briefing called by the House Telecommunications Subcommittee in the United States Congress. They all stated that there is no conclusive evidence linking cellular phones to

Figure 8-4a Wearable Cellular Phone, Closed Position
Courtesy of Motorola

brain cancer. The National Cancer Institute stated that among people under the age of 65 there had been no increase in brain cancer during the period when cellular phone use became popular.

The applicable standard is the ANSI/IEEE Standard C95.1-1992 level with respect to human exposure to radio frequency electromagnet fields, 3kHz to 300 GHz. Hand-held radio transmitters operating with 7 watts or less of RF input power at frequencies between 100 kHz and 1.5 GHz are excluded. Most cellular handsets in the 850 and 950 MHz bands operate below 7 watts and are thus excluded.

Figure 8-4b Wearable Cellular Phone, Operating Position
Courtesy of Motorola

Digital cellular phones have been shown to interfere with cardiac pacemakers, cardiac defibrillators, and even hearing aids. Cardiac pacemaker and defibrillator manufacturers are working on better shielding to prevent interference. In the meanwhile they are advising people using the cellular phone to hold it to the ear opposite the side of the body their cardiac instrumentation is on.

In Great Britain there have been some cases where operating digital cellular within a few feet of someone wearing a hearing aid produced a loud buzz in both devices even though the phone was on standby.

8.8 Personal Communication Services

Personal Communication Services (PCS) uses very small numerous microcells which will employ very low power transmitters and very small portable phones. PCS services are to include several advanced forms of voice and data, transmitted to and from pocket-sized telephones, wireless facsimile machines, and other portable devices. Later PCS is expected to include video and multimedia applications.

PCS in the United States of America

There are three categories of PCS in the United States of America, narrow band licensed PCS, broadband PCS, and unlicensed PCS. Narrow band licensed PCS for advanced paging and data messaging operate in the 900 MHz band. The receipt of a paging message will for the first time be acknowledged by radio transmission. Broadband licensed PCS will operate in the 2 GHz band. The wider bandwidth allows transmission of higher rates of information. This may be video, facsimile, or higher rates of data.

Unlicensed PCS operates in 40 MHz of the 2 GHz spectrum. These are low power, limited range devices owned and operated by end users on their own premises. Examples are wireless local area networks, wireless Private Branch Exchange (PBX) and personal data assistants. It includes high and low speed data links between computing devices in an office and some forms of cordless telephones.

Licensed PCS and Human Safety

The exclusions for hand-held cellular transceivers in the 850 and 950 MHz bands do not apply to licensed broadband licensed PCS which operate at higher frequencies. Here the ANSI/IEEE Standard C95.l-1992 states that the specific absorption rate at PCS frequencies shall not exceed 0.08 watts per kilogram when averaged over the whole body for any 6-minute to 30 minute period. The spatial peak specific absorption shall not exceed 1.6 watts per kilogram over any 1 gram of tissue or in the case of the extremities--hands, wrists, feet, and ankles--4 watts per kilograms over any 10 grams of tissue.

This standard was challenged by the Environmental Protection Agency (EPA) in Washington, DC on November 8, 1993. The EPA recommended against the 1992 ANSI/IEEE standard at 2 GHz frequencies. This question has not yet been resolved.

PCS in Other Areas of the World

In Europe there is the Digital European Cordless Telecommunications (DECT). There is also the cordless telephone - 2nd Generation (CT-2). In the United Kingdom there is the Digital Communication System at 1800 MHz (DCS 1800). In Canada there is an advanced form of CT-2 called CT-2+. In Japan there is a system called Personal Handy Phone (PHP).

8.9 FUTURE PUBLIC LAND MOBILE TELECOMMUNICATIONS SYSTEMS

Future Public Land Mobile Telecommunication Systems (FPLMTS), the official name designated by the UN International Telecommunications Union, is intended to become a worldwide personal communications network offering all of a telephone network's services including voice, facsimile, and data. FPLMTS includes cordless telephony, wireless pay phones, wireless private branch exchanges, and rural radio and telephone exchanges among terminals on land, on sea, and in the air. Calls within the mobile system would be routed to and from the existing telephone networks through satellites. FPLMTS will permit connection to the system operating in the mobile-satellite service band. Personal and mobile stations will be able to connect into the system by radio at any time and from any place in the world.

A number of different types of satellite systems can be used in FPLMTS, such as Little Leo (Low Earth Orbit), Big Leo and GSO.

Little Leo satellites provide low-cost, low-data-rate (up to 10 kb/s) two-way digital communications and position-location services to pocket-sized portable and mobile terminals at frequencies below 1 GHz. Such systems would employ 2 to 24 "lightsats," small inexpensive satellites launched into orbits 480-780 miles above the earth that are capable of serving the world.

A Big Leo satellite system consists of a large number of satellites in low earth orbits operating above 1 GHz. Big Leo systems use voice and a high-speed data service up to a few megabits/second. The Big Leo systems would also provide worldwide services in personal communication small cellular phones.

Motorola Satellite Communications is building a Big Leo system. Originally it was supposed to consist of 77 satellites in low earth orbit. The company named the system Iridium after the element whose atom has 77 orbiting electrons. Later Motorola changed the number of satellites to 66 but still retained the name.

Other companies have proposed three satellites in geostationary orbit. Hughes Aircraft Co. proposed such a system called Tritium, named after an isotope of hydrogen.

It is expected that FPLMTS will begin worldwide operation at the end of the 20th century.

BIBLIOGRAPHY

Dern, Daniel P. — *The New User's Guide to the Internet,* New York McGraw Hill, 1993

Garfinkel, Simon — *PGP.* O'Reilly & Associates, Sebastapol, CA 1995

Haddad, Alan David — *Personal Communications Network: Practical Applications,* Artech House, Boston, MA, 1995

Laquey, Tracy and Jeanne C. Ryrer — *The Internet Companion: A Beginner's Guide to Global Networking,* Addison-Wesley, 1992

Lebow, Irwin — *Information Highways & Byways from the Telegraph to the 21st Century,* IEEE Press, 1995

Logsdon, Tom — *Mobile Communications Satellites: Theory and Applications, New York,* McGraw Hill, 1995

Lynch, Daniel C. Marshal T. Rose — *Internet System Handbook,* Manning Publications Co., Greenwich, Conn., 1993

McConnell, Kenneth R. Dennis Bodson and Richard Schaphorst — *Fax: Digital Facsimile, Technology and Applications,* 2nd Edition, Artech House, Norwood, MA, 1992

Meurling, John and Richard Jeans — *The Mobile Telephone Book: The Invention of the Mobile Telephone Industry,* Communications Week International, 1994

Nellist, John G. — *Understanding Telecommunications and Lightwave Systems, An Entry-Level Guide,* Piscataway, NJ, IEEE Press, 1992

O'Neill, Judy E — "The Role of ARPA in the Development of the ARPANET, 1961-1972."
IEEE Annals of the History of Computing, Winter, 1995

Rappaport, Theodore S. — *Cellular Radio and Personal Communications,*
Piscataway, NJ, IEEE Press, 1995

Stallings, William — *Protect Your Privacy, A Guide to PC Users,*
Prentice Hall, 1995

CHAPTER 9

LEAVING THE HORSES BEHIND

The technology of the automobile has caused a number of deaths in the United States in the 20th century that was greater than the number of American service people killed during that time in wars. This led to different systems to reduce the number of deaths ranging from seat belts and air bags to the development of a future intelligent highway system which will use many of the technologies described in the previous chapters.

Although most of the automobile technology was developed in the 20th century, some of it goes back to the 18th century.

9.1 THE ROOTS OF THE AUTOMOBILE

THE STEAM CAR

The French monarch in 1769 needed a self-propelled artillery tractor to obtain military superiority over his neighbors. In response Nicholas Joseph Cugnot built the first full-scale vehicle powered by steam. It was a three-wheeled vehicle weighing about five tons. The machine had a boiler and a furnace mounted in front of the steering wheel which limited the driver's vision. The vehicle carried four people and traveled at a speed of two miles per hour. The first trial was in late 1769 but later the vehicle crashed into a wall because of the limited vision of the driver. In 1771, Cugnot built a second machine but it too crashed and the inventor was arrested. Cugnot proceeded with his steam car experiments until his death in 1804.

The development of the steam car continued in England from 1786 until 1840 when the steam railroad took over because it provided a safer and smoother ride.

The further development of the steam car took place mostly in the United States. John Fisher in 1851 formed the American Steam Carriage Company which produced a four-wheel steam carriage which could travel fifteen miles an hour over plank roads. Fisher was not successful in attracting investors. He decided in 1859 to build steam-powered fire engines. The Boston fire of 1872 made this market a profitable one. The horses that pulled the fire equipment came down with the flu. Buildings burned while the fire equipment remained in the firehouse. However, the steam passenger automobile did not become practical until the Stanley Steamer which was first demonstrated in 1898.

Francis and Frelan Stanley were originally schoolteachers in rural Maine. They became interested in photography and developed a process for making photographic dry plates. They sold it to Eastman Kodak for a large sum of money. They used the money to develop and sell steam-powered automobiles. The first Stanley Steamer weighed 600 pounds and was set on bicycle wheels. The vehicle could go twenty-five miles per hour. Two hundred orders came in at its first demonstration in 1898. An 1899 photograph of the Stanley Steamer is shown in Figure 9-l.

John B. Walker, the editor of Cosmopolitan Magazine, bought the Stanley Company in 1899 for $250,000. However, he sold the company which split and was finally brought back together with the Stanleys serving as consultants. They improved the steamer by moving the boiler to the front of the vehicle. By 1908 the Stanley Motor Carriage Company produced 650 cars per year.

The Stanley Steamer had two main disadvantages. It took thirty minutes to start and the range was limited to forty miles. Soon gasoline automobiles appeared that did not have these disadvantages. Francis Stanley died in 1917 at the wheel of one of his own vehicles. The last Stanley Steamer rolled out of the factory in 1924.

The First Electric Cars

In 1892, William Morrison drove his electric car at the Columbian Exposition in Chicago. The Columbia and Electric Vehicle Company of Hartford, Connecticut, produced the Columbia Daumon Victoria in 1899. It was an electric car designed by William Atwood along the lines of the popular horse-drawn carriage. The raised seat at the rear accommodated the driver. Behind the driver stood a footman as this was a vehicle for the wealthy. There were two electric motors which gave the vehicle a top speed of 30 miles per hour. When President Theodore Roosevelt visited Hartford, he was given a grand tour of the city in an electric carriage of this type.

C.E. Woods, another electric car manufacturer, in 1900 made electric cars with over thirty body types. The station wagon had hard rubber tires and wooden wheels, with a 3-l/2 horsepower motor at each rear wheel. A single lever to the left of the driver controlled the speed up to 12 miles an hour and also applied the brakes. It had an electric bell instead of a horn to alert traffic. The vehicle also had electric lights for night travel since it was advertised as a family vehicle for theater going which was mostly in the evening.

The advantages of the electric were low noise and no pollution. The great disadvantages at the beginning of the 20th century was the very limited range of about 20 miles between battery charges, the long charging time, and the expensive replacement of the batteries. In the last decade of the 20th century, there was renewed interest in the

Figure 9-1 Stanley Steamer 1899 Smithsonian Institution Photo No. 886499

electric car because of its zero tail pipe pollution and the development of improved electrical batteries. However, in the 20th century, the dominant automobile technology was the internal combustion engine.

The Development of the Internal Combustion Engine

An internal combustion engine is one in which a fuel is burned inside the cylinder of the engine itself. This contrasts with the steam engine where the fuel is burned outside the steam engine itself. The most common types are the gasoline engine and the diesel engine.

Gasoline, also called petrol, is a mixture of volatile, flammable liquid hydrocarbons derived from petroleum. Originally kerosene was the principal product from petroleum and gasoline was just a byproduct produced by distillation. Gasoline became the preferred fuel in early internal combustion engine vehicles because of its high energy of combustion and capacity to mix readily with air in a carburetor.

The Gasoline Engine

In the principal type of gasoline engine hot gases are obtained by burning a mixture of gasoline and air within the cylinder. The inflow of fuel gas and the exhausting of vent gas are controlled by valves. Ignition of the mixture of gasoline and air is caused at the proper instant by an electric spark. The rapid burning of the vapor pushes a piston, and a crankshaft converts the push to a rotary motion that drives a vehicle's wheels. A flywheel, a heavy steel wheel, attached to the rear end of the crankshaft, acts to smooth out the peaks and valleys of power flow. During the peaks, it stores power from the engine. During valleys, the flywheel delivers power to the engine. The crankcase is the housing of the crankshaft and associated parts.

This gasoline engine can be divided into two general types according to whether the operating cycle is completed in two or four strokes of the piston. In the four-stroke cycle the piston first descends on the intake stroke, during which the inlet valve is held open. Air mixed with gasoline in a carburetor is sent through a manifold (a manifold is essentially a tube) into the inlet valve. The fuel mixture is moved by the partial vacuum created by the descent of the piston. The piston then ascends on the compression stroke with both the inlet and exhaust valves closed, and the charge is ignited by an electric spark as the end of the stroke is approached. The power stroke follows, with both valves still closed and gas pressure acting on the piston because of the expansion of the burned charge. The exhaust stroke then completes the cycle, with the ascending piston forcing the spent products of combustion through the exhaust valve and into the exhaust manifold and out the tail pipe. Each cycle thus requires four strokes of the piston--intake, compression, power, and exhaust. The principle of the four-stroke cycle was established in 1862 by a French engineer, Alphonse Beau De Rochas. Most automobiles use the four-stroke cycle.

One of the important pioneers in the development of the internal combustion engine for vehicles was the German engineer Gottlieb Daimler. His portrait taken in 1881 is shown in Figure 9-2. The first Daimler internal combustion vehicle, Figure 9-3, was built around 1885. It weighed 110 pounds and could generate 1-1/2 horsepower.

Figure 9-2 Gottlieb Daimler, 1881
Smithsonian Institution Photo No. 88-6459

Figure 9-3 First Daimler Internal Combustion Vehicle, 1885
Smithsonian Institution Photo No. 38-555A

In 1895, Emile Constant Levassor, an early French automobile manufacturer, placed a Daimler engine in front of the chassis instead of under the seat or in the rear of the automobile. This made it possible for automobiles to accommodate larger engines with more powerful displacement. Levassor entered his Daimler equipped automobile in the 27 miles Paris-Bordeaux-Paris race. He won it with an average speed of fifteen miles per hour. In 1895, this was considered a fast speed for automobiles.

A racing enthusiast, Emil Jellinik, the Consul General in Nice for the Austro-Hungarian Empire, persuaded the Daimler Company to design a new racing model in 1900. Jellinik christened it the Mercedes in honor of his daughter, Mercedes Jellinik. The car won races and Mercedes became a popular name for Daimler cars. In 1901, the Secretary of the Automobile Club of France said, "We have entered the era of the Mercedes."

Another German car manufacturer was Karl Benz. He designed a lightweight internal gasoline combustion engine. In 1885, he built a three-wheeled vehicle incorporating his engine which lay horizontally in the rear to stabilize the auto. The vehicle had a differential gear system and it was the first gasoline-powered road vehicle designed as an integrated unit. In the fall of 1885, Benz publicly exhibited his vehicle. In 1888, his wife Berta, accompanied by her two teenage sons, borrowed her husband's latest model to drive 50 miles to visit her mother. However, on the return trip the road was mostly up hill and the vehicle could not make it. The two boys had to get out and push. The family returned tired but happy about their adventure.

In 1891 Benz built a four-wheeled vehicle that incorporated all the things that he had learned from his three-wheeled models. He sold two thousand vehicles by 1899 and eventually became Daimler's main rival. Later in 1926 Daimler and Benz merged to form the Mercedes-Benz Company.

The early internal gasoline combustion engine came to the United States in the late 19th century. George B. Selden, a New York patent attorney, in 1895 obtained U.S. Patent Number 549,160 for "an improved road engine." The Selden engine was adopted for use in many early American attempts at producing an automobile.

One of the first gasoline vehicles produced in the United States was the Lambert Gasoline Buggy of 1891 manufactured by John W. Lambert and test driven in Ohio City, Ohio. It had a single-cylinder, four cycle, water-cooled engine with the power transmitted by chain to a rear axle. There was no provision for reversing. The two passenger Buggy could be steered either with the hand lever or with a foot lever.

Other early American pioneers were the Duryea brothers. Figure 9-4 is a photograph of Charles Duryea driving his gasoline powered buggy in 1895. The most important American pioneer was Henry Ford who opened the way for the automobile in the 20th century.

Figure 9-4 Charles Duryea and his Gasoline Buggy, 1895
Smithsonian Institution Photo No. 11923

9.2 THE EARLY 20TH CENTURY AUTO DEVELOPMENT

In November 1900 the first auto show in America was held in Madison Square Garden, New York City. At the time there were not many outlets for selling autos. The show served as one of the few retail outlets for purchasing the vehicles. Five years later automobile dealers became the principal source of auto sales. The dealers helped spread the new technology in the early 20th century all over the United States.

One of the most prominent American auto manufacturers in the early 20th century was Ransom Olds. He became the first successful mass producer of gasoline-powered automobiles. In 1903, his company sold a record four thousand automobiles. In that same year, Olds made the trip between California and Detroit in sixty-two days. This encouraged sales.

In 1906, a great earthquake struck San Francisco. This demonstrated the advantage of the automobile over the horse-drawn carriage. Horses broke down from the heat of fires and the strain of continuously delivering needed supplies to disaster areas. Over two hundred private motor vehicles were used in the disaster relief effort. Caravans of motor trucks brought needed supplies to devastated San Francisco. After the great earthquake, many municipalities bought automobiles for police, fire, and other services. By 1907, there was widespread use of motorized fire trucks. The American Post Office was using motorized vehicles by 1909.

The use of motorized vehicles in the early 20th century grew very rapidly in America. In 1910, there were fifty thousand private automobiles and trucks in use in the United States. Motorized vehicles were owned mostly by upper income people. Auto manufacturers in the second decade of the 20th century wanted to expand the market to middle income people both in the cities and on farms.

Henry Ford was one of these people who made the auto available to the middle class. He was born into a farm family in Michigan on July 30, 1863. At the age of sixteen, he left home and went to Detroit where he worked for nine years at various mechanical jobs. He moved back to his family farm and in April 1888 Henry Ford married Clara Jane Bryant. He tinkered in a workshop attached to the sawmill on the farm. In September 1891, Henry Ford and his wife left the farm permanently and moved back to Detroit. He obtained an engineering job with the Edison Illuminating Company and within two years he was promoted to chief engineer. He worked on internal combustion gasoline motors at home after work.

Henry Ford designed and built his first engine on Christmas Eve 1893. The engine consisted of a one-inch diameter pipe, a homemade piston, and a lathe for a flywheel. Ignition was supplied by ordinary house current. Clara Ford poured gasoline into the intake valve and the engine ran successfully.

In 1896, Ford built his first automobile, called Quadricycle, in a shed behind his home. He did not realize that the door of the shed was too small for the car. He had to remove half of the side of the shed to get his vehicle out. The Quadricycle, Figure 9-5, had a two cylinder four stroke horizontal engine producing about four horsepower and it could go about 20 miles per hour. The drive was by leather belt and chain. On June 14, 1896, Henry Ford test-drove his first car which weighed slightly less than seven hundred pounds. He later sold the Quadricycle for $200 which was used to develop newer models.

In 1899, Henry Ford met William H. Murphy who owned most of downtown Detroit. Ford took Murphy for a ride from Detroit to Farmington, Michigan, in his second auto. On July 24, 1899, the Detroit Automobile Company was founded to manufacture and sell Ford cars. William H. Murphy supplied the funds for the new company. Ford resigned

95%

Figure 9-5 Quadricycle
From the Collections of Henry Ford Museum + Greenfield Village

Figure 9-6 Model T Fords, 1909. Leaving the Horses Behind
Smithsonian Institution Photo No. 60210

from the Edison Illuminating Company and devoted all his time to automobiles. At the end of 1899, the first vehicle was built by the Detroit Automobile Company. It was a 1600 pound racing car generating twenty-six horsepower. In early 1900, a second vehicle was finished by the Detroit Automobile Company. The 20th century and the Ford automobile were born almost simultaneously.

In early 1902, Henry Ford left the Detroit Automobile Company. Ford found new financiers and on June 16, 1903 he formed a new company, the Ford Motor Company. The first car out of the factory was the Model "A." In the first eighteen months, the Ford Motor Company sold over 1500 Model A cars. By the end of 1910, with the mass production of the Model "T," Ford Motor Company was recognized as the leader in building affordable automobiles. The Model T was manufactured from New Years Eve 1908 to May 1927. Fifteen million Model Ts were manufactured during this time. Figure 9-6 is a photograph of two Model Ts leaving the horses behind.

Most of this tremendous production was due to the adoption by Ford of the assembly line. In 1910, Ford borrowed the idea of a conveyor belt system from meatpacking plants. He installed one in his plant at Highland Park, Michigan. The various parts needed for assembly were moved from one workstation to another in a constant flow throughout the workday. The Ford assembly line reduced the time required for chassis assembly from twelve to less than two hours. In 1918, in one day Ford produced one thousand chassis.

Henry Ford had very strong views on a great many things. In 1918, he purchased the newspaper, "The Dearborn Independent." He called it "The Chronicler of Neglected Truth." In it he praised prohibition, conservation, and Wilson's plan for the League of Nations. It scorned the Gold Standard, and monopoly. In 1920, Ford launched a vicious 91-week campaign against the "International Jew." He used a forgery, the "Protocols of Zion" that the Russian Czar's secret police had made up in 1903. This forgery purported to be the secret minutes of a Jewish conspiracy to take over the world. Although it was shown to be fake, Ford used it to stir up hatred of Jews all over the world including Germany. Ford blamed the Jews for Bolshevism, Darwinism, liquor sales, gambling, short skirts, and cheap Hollywood movies. His venom was featured in German editions of the "Dearborn Independent." In the United States, Ford was sued over his statements but he chose to settle the suit out of court and published an apology to the Jews.

It is interesting to note that although Henry Ford was an avowed anti-Semite, he hired Albert Kahn, the son of a Rabbi, as the architect for his Model T factories. Kahn designed the Highland Park plant called the "Crystal Palace" which opened on New Year's 1910. In 1917, Ford built 700,000 Model Ts at the Highland Park plant. Kahn, a Jew, did not criticize Ford for his anti-Semitic tirades in the "Dearborn Independent." There were similarities between the two men that kept their relationship alive despite Ford's anti-Semitism. They both were self-educated and both disliked college graduates. Both loathed smokers and both could not stand subordinates who questioned their authority. Their relationship survived the storms and Kahn in 1917 started to work on the "Rouge," a 2,000-acre site in Dearborn, southwest of Detroit. This housed automated assembly lines flowing uninterrupted, all at one level, from raw materials to the finished car.

THE DIESEL

Another early internal combustion engine was the Diesel, invented by Rudolf Diesel (1858-1913) about 1890. He conceived the idea of injecting liquid fuel into air heated solely by compression. The result was a thicker, slower-flowing fuel source providing increased durability and efficiency. He obtained German patents in 1892 and 1893. In 1897, he demonstrated a 25 horsepower four-stroke, single cylinder internal combustion engine which used no ignition devices and relied on compression ignition. However, Diesel insisted that all engines manufactured under his license be made to operate at practically constant pressure, as described in his 1893 patent. This meant that the engines had to run at a very low speed. The only early application was large power generating plants including marine installations.

Diesel himself died in mysterious circumstances. He disappeared from the mail steamer "Dresden" in the English Channel en route to London on September 13, 1913. After his death, higher-speed diesel engines were developed that did not follow the constant pressure that Rudolf Diesel specified. Fuel is injected into the cylinder near the end of the compression stroke and burned rapidly, with sharply rising pressure while the piston is near its dead center.

The first diesel engine that was small and light enough for use in automobiles was built in 1922 in Germany. Diesel engines later became the main power source for trucks. Diesel fuel was cheaper and could be used for large vehicles going long distances. However, Diesel engines have high emissions of air pollutants including nitric oxide and soot.

9.3 TECHNICAL IMPROVEMENTS

In the early part of the 20th century, there were a series of technical improvements that transformed the horseless carriage of the turn of the century into the modern automobile.

PNEUMATIC TIRES

The earliest cars used solid rubber tires. An Irish veterinarian, J.B. Dunlop, developed pneumatic tires filled with air for bicycles in 1888. It was adapted later for use on automobiles by the Michelin brothers of France. In 1905, Harvey Firestone introduced air-filled tires on automobiles at the New York Auto Show. This meant that cars could go faster, safer, and give a more comfortable ride.

Glass Windshield and Windows

At the 1910 New York Auto Show, the glass windshield was introduced. This allowed drivers to go faster without the wind and dust blowing into their faces. In 1915, the "California Top" with side curtains and celluloid windows appeared. Soon glass windows replaced the curtains and celluloid. By 1920, one in every five new cars built came with completely enclosed bodies using glass windshields and windows.

The Self Starter

In 1912, Charles Kettering developed the self-starter, an electric motor that spared the chore of cranking and the risk of a broken arm if the engine's compression stroke kicked the crank handle back. Cranking took muscle and it was considered a man's job. The starter allowed women to become motorists.

Multiple Cylinders

Another technical development was the replacement of two cylinders by four, and then eight and twelve and even sixteen cylinders. More cylinders meant less vibration and faster acceleration but a large number of cylinders in two straight rows increased the size of the engine dramatically. For this reason, four and six cylinders remained standard until 1932, when the Ford V-8 was introduced. In the V-8 the eight cylinders were arranged in two rows angled toward each other in a V. This permitted the engine to be shorter and lighter. This would also enable the car to accelerate faster. Later gasoline shortage made the efficient four and six cylinder engines more popular than the eight-cylinder engine.

Hydraulic Brakes

The six-cylinder engine gave cars plenty of power. However, stopping the car with mechanical brakes on two wheels became a problem. Dusenberg came out with the first four-wheel hydraulic brakes in 1920.

The Development of an Electric Generating System

The function of the generating system is to restore the battery energy used up in starting up the engine and to supply all of the vehicles electrical energy requirements for lights, radio, etc.

The generating circuit is designed so that the battery "floats" on the line, that is, when the system is delivering sufficient current, the excess above vehicle requirements serves to keep the battery in a charged condition. When the system, due to low engine speed, is not delivering sufficient current, the battery furnishes the excess current needed.

The original type of generating systems in vehicles consisted of a DC generator, two regulators and the battery. The disadvantage is the DC generator produces voltage and current in relation to the speed at which the automobile operates. Later an AC alternator was used where the current output is independent of the speed of the vehicle. This type of system consists of an AC alternator, rectifier, a voltage regulator and the battery.

The Development of an Ignition System

The ignition system is part of the electric system of the automobile. Its purpose is to produce high-voltage surges (up to 20,000 volts) and to deliver them to the spark plugs in the combustion chambers of the engine. These high voltage surges then cause electric sparks across the spark gap of the spark plugs. The sparks ignite the air-fuel mixture in the combustion chambers so that it burns and causes the engine to operate.

The ignition system consists of three basic parts: the distributor, the ignition coil, and the spark plugs, together with the connecting wires. When the engine is running, the ignition coil is repeatedly connected to and disconnected from the battery by a rotating distributor. Every time the coil is connected, it becomes loaded with electrical energy. Then, when it is disconnected, the load of electrical energy is released in a high-voltage surge. This surge flows through the wiring to the spark plug in the engine cylinder that is ready to fire. This whole series of events happens in less than one three-hundredth of a second.

The Development of Lubricating Systems

It was realized early that automobile engines like any complex engine requires proper lubrication. There are a great many moving parts in the engine. These parts must be protected by lubricating oil so that there will be no actual metal to metal contact. The moving parts in effect float on films of oil. The lubricant also assists in cleaning the engine parts and it forms a seal between piston rings and cylinder walls to prevent blow-by of combustion gases.

The various types of lubrication system employed in internal combustion engines are:

a) Forced lubrication
b) Splash lubrication
c) Oil feed with fuel

The forced lubrication method soon became the dominant type of lubrication, especially in large high-speed engines. In this system of lubrication, the oil is drawn from a reservoir in the oil pan of the engine by a circulating oil pump. From the pump it is forced under pressure to the various bearings, cylinders, piston and piston rings. The oil is then returned to the pump where it reenters the pump and circulates again. In early vehicles, an oil pipe was connected to an oil pressure gauge mounted on the dashboard. In many later vehicles, the gauge was replaced with an idiot light. However, by the time the light became red, it was probably too late to save the engine. This was a case of going backwards with further development.

Some vehicles use a lubricating system combining forced lubrication with splash lubrication. The main bearing and the camshaft are lubricated by a pump (forced lubrication) and connecting rod bearings are lubricated by dippers which dip into oil filled troughs in the oil pan. The dippers also splash oil up into the cylinders and over the pistons and cylinder walls (splash lubrication).

Some motorcycles use two-stroke cycle engines. This type has no oil reservoir and no specific oiling system. Lubrication consists of mixing oil with fuel in certain specified amounts, and as the fuel circulates in the form of vapor in the engine crankcase, the heavier oil separates from the fuel and is carried to the working parts in the form of an oil mist (oil feed with fuel lubrication).

Oil purification is necessary in a lubrication system to keep out sand, dirt, and metal particles. One method is an oil strainer, a fine-mesh bronze screen located so that oil entering the pump is clean. A second type of oil purification is the oil filter which is placed in the oil line above the pump. It filters the oil and removes most of the impurities that escaped the oil strainer.

COOLING SYSTEMS

It was found in the early development of the automobile that a cooling system for the internal combustion engine was absolutely essential. The temperature in the combustion chamber during the burning of the fuel is estimated to range from 2700° F to 3200° F. If there was no cooling system, the intense heat generated within a gas cylinder would very quickly overheat the metal within the cylinder to such an extent that it would become red-hot, resulting in burned and warped valves, seized pistons, overheated bearings and a breakdown of the lubricating oil.

To avoid these conditions, means must be provided to carry off some of the heat, enough of it to prevent the temperature of the metal of the cylinder from rising above a predetermined point and low enough to permit satisfactory lubrication and operation. The excess heat is carried off by some form of cooling system.

There are four methods of cooling internal combustion engines:

a) Water circulation
b) Water and oil circulation
c) Air cooling
d) Air and oil circulation

Water-cooling is commonly obtained by means of a pump and associated piping, radiator, fan and system of jackets and passageways through the engine within which the water circulates.

Water cooling requires an anti-freeze solution to be added when the atmosphere temperature is below 32° F. The most popular anti-freeze solution is ethylene glycol which has an extremely high boiling point, does not evaporate in use, is noncorrosive and has no odor.

High performance and heavy-duty water-cooled engines may require additional cooling. The water and oil will usually receive extra cooling through added air-cooled systems using blades which act as a fan to circulate air over the fins cast integrally with the cylinders.

Air-cooled engines are used for small appliances such as lawn mowers and chainsaws and also for lightweight automobile engines. Air cooling of the oil to supplement the cylinder cooling is also used with many air-cooled engines.

The Development of Carburetion

Early in the history of the car, it was found that mixing gasoline with air produced a highly efficient fuel. This led to the gradual development of the carburetor and associated equipment forming a carburetion system.

A carburetor is essentially a device which mixes liquid gasoline with air. In this process it throws a fine spray of gasoline into air. The gasoline vaporizes and mixes with the air to form a highly combustible mixture. This mixture then enters the engine combustion chamber, where it is ignited by a spark from the spark plug. The burning fuel mixture causes the engine to produce power.

The carburetor gradually improved from the earliest models. It came to have a reservoir in which gasoline is stored and passages through which gasoline flows. The carburetor has larger passages through which air can flow.

It was soon found that the air going through the carburetor had to be filtered first to get rid of the dirt and dust which would otherwise get into the engine and damage it. An air filter was soon added to the system for this purpose.

Early on it was discovered that extra gasoline had to be delivered to the carburetor when going from a cold start. This was accomplished by a choke valve which chokes off the air in the carburetor producing a vacuum which causes an increased stream of gasoline to flow. At first, the choke was operated manually and the operator had to deactivate the choke after the engine warmed up. Later the operation was made automatic. Thermostats activated by temperature were added to control the choke valve, air valve, and other parts of the carburetion system affecting the mixture of air and fuel.

In some early cars, the flow of gasoline from the fuel tank to the carburetor depended on gravity. That is, on a level road the fuel tank was higher than the engine and gasoline naturally flowed down to the carburetor. However, when these early cars were driven up a hill, the fuel tank was lower than the engine and gasoline stopped flowing to the carburetor. The car could not make it up the hill. Some drivers found a solution by driving up the hill in reverse. Soon, however, a fuel pump was added to deliver gasoline to the carburetor even when the vehicle was going up a very steep hill.

9.4 Development in the Last Half of the 20th Century

A number of technical improvements again transformed the automobile, making it more efficient and much easier to drive and maintain. These included fuel injection, air conditioning, electronic ignition, automatic shifting, front wheel drive, four wheel drive, power steering, power brakes and computer control.

Fuel Injection

The fuel injection system uses, instead of a carburetor, a series of injection nozzles and a high-pressure fuel pump to spray the fuel into the air entering the engine cylinders. Mechanical fuel injection was first offered by Chevrolet and Pontiac in 1957 as a high-performance option.

The development of fuel injection led to a multi-programmed electronic system. One fuel injection valve is used for each cylinder. These are solenoid-operated valves that project into the intake manifold above the respective intake ports. Each valve is operated by a pulsed signal from the Electronic Control Unit (ECU) which is a solid state digital computer. The ECU is preprogrammed by the manufacturer to accept certain signals from the various sensors of the system and to translate these into a pulsed signal for operation of the fuel-injection valves.

One feature of at least some fuel injection vehicles is that the car will start as soon as it is in gear even though the operator does not put his foot on the accelerator. The driver must place his food on the brake before putting the car in gear. If this is not done, the automobile will move rapidly without the operation of the accelerator.

AIR CONDITIONING

Pontiac offered the first reliable unit that could be concealed in the dash around 1954. At first, this was an option but later it came with the car.

ELECTRONIC IGNITION

Before going into this development, the conventional electro-mechanical ignition system will be reviewed. This consists primarily of a storage battery and ignition coil, a capacitor, a distributor and spark plugs, one to each cylinder. The ignition coil raises the 12 volts of the storage battery in the primary to between 10,000 volts and 25,000 volts in the secondary winding of the ignition coil. This happens as the 12-volt supply is interrupted by breaker contacts. The opening of the 12-volt supply across the small number of turns in the primary induces a large voltage across the high number of turns in the secondary of the ignition coil. The breaker contacts, a breaker cam and a capacitor are in the distributor housing. The cam and breaker points serve to interrupt the primary circuit at certain definite intervals. The cam is located on the distributor shaft which is driven indirectly by the engine crankshaft. The capacitor prevents arcing across the breaker contacts. A rotating mechanical system in the distributor feeds the high voltage in turn to each spark plug at the proper time. A spark gap in the spark plug causes a spark which heats up the fuel in a cylinder of the internal combustion engine.

Electronic ignition system designs utilize electronic spark control. The complete control of spark timing is accomplished through a computer. A system of sensors signals the computer of the engine's operating conditions and the computer electronically adjusts the spark timing. The mechanically operated breaker points and cam are eliminated. Breaker-point pitting and cam wear are dispensed with, thus minimizing maintenance problems.

THE CHANGE FROM MANUAL TO AUTOMATIC TRANSMISSION

Before describing the change, it is necessary to review the elementary basics of what a transmission system accomplishes and why it is required.

A car starting from rest will require a greater torque or turning effort applied to the wheels than a car at cruising speed. The car starting from rest will have to overcome inertia to move while the car at cruising speed has to only overcome the relatively small friction from the wheels and the air.

A transmission system or gear changer provides a means of varying the torque ratio between the engine and the wheels which meets the varying requirements of the automobile under different conditions.

In stick-shift vehicles, this is done by changing the gear ratios between the engine and the driven wheels. Gear size ratios determine the relative speed and torque between the engine crankshaft and the driven wheels.

Two meshing gears of different sizes will have different speeds and torque. For example, in Figure 9-7 gear B is twice the size of gear A. Gear B will have a speed of 1/2 that of gear A. However, the torque applied to gear B's shaft will be twice that of A. Torque on shafts or gears is measured as a straight line force multiplied by the distance from the center of the shaft or gear. For example, suppose the tooth of gear A is pushing against the tooth of gear B with a 25-pound force. This force at a distance of 1 foot (the distance from the center of gear A)means a torque or turning effort of 25 pound-feet.

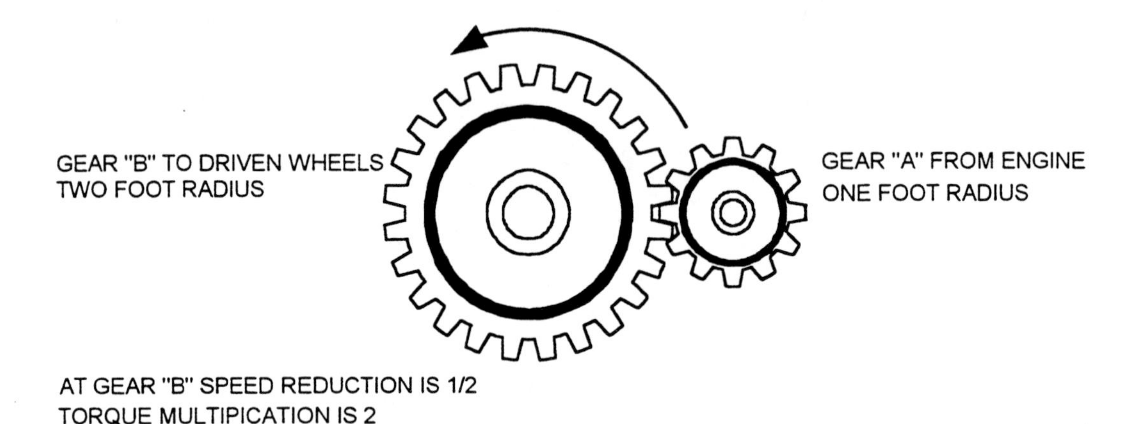

Figure 9-7 Principle of the Gear

The 25-pound push from the gear teeth of the smaller gear is applied to the gear teeth of the larger gear but it is applied at a distance of 2 feet from the center. The same force is acting on the teeth of the larger gear, but it is acting at twice the distance. The large gear is going at half the speed of the smaller gear but it is delivering twice the torque or turning effort.

In a particular actual vehicle when the transmission is in low gear there is a speed reduction of 12 to 1 from engine to wheels. This means that the crankshaft turns 12 times to turn the wheels once. It also means that the torque increases 12 times. If the torque delivered by the engine is 100 pound-feet, then 1200 pound-feet of torque is delivered to the wheels. Assume the wheel radius is 1 foot. Then the push on the wheel axis, and thus the car, is 1200 pounds which is sufficient to move the car from a stationary position.

As the car moves slowly from a stationary position, it does not need as much torque but it needs more speed. This is done by changing the gear ratio in second shift to apply more speed and less torque. As the car approaches cruising speed, the gear ratio is changed once more, this time to a ratio of one. These are the basic three shifts in a car.

Additional ones are added for going up a steep hill, down a steep hill, and going in reverse.

A clutch is required to disengage the gears while the shifting of gear takes place. A certain amount of coordination is needed to shift smoothly from one gear position to the next. If this is not done, then the clutch will wear out prematurely. To prevent the clashing of gears in more modern vehicles, synchronizing devices are added. These devices assure that gears which are about to mesh will be rotating at the same speed and thus will engage smoothly. However, it still takes some skill and coordination to shift gears in a car with manual transmission.

Automatic transmissions in cars vary with the automobile manufacturer but they all have a torque converter and planetary gear sets which are controlled by brakes and clutches to provide two or three forward gear ratios. The torque converter is a type of fluid coupling that uses a fluid to transmit the turning force from one shaft to another. Figure 9-8 shows a simplified version of two members of a fluid coupling. The torque converter, Figure 9-9, is encased in a special transmission fluid. The one on the right is the driving member from the crankshaft of the engine and it is called the pump. The one on the left is the driven member and is called the turbine and it rotates the transmission

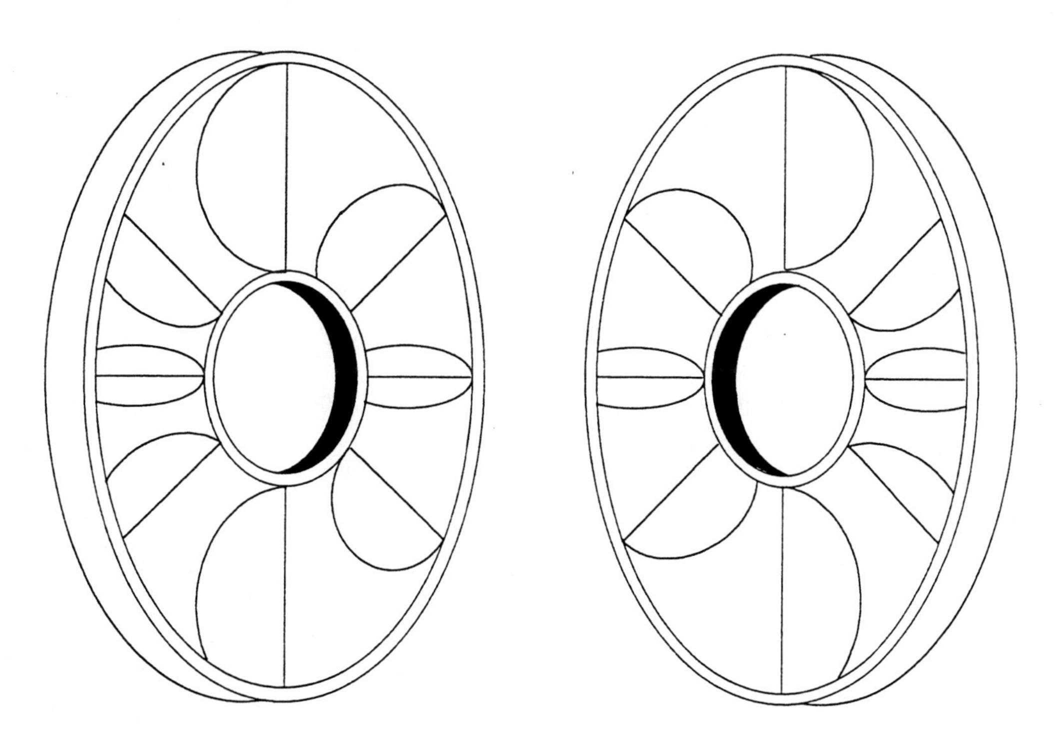

Figure 9-8 Hydraulic Fluid Coupler

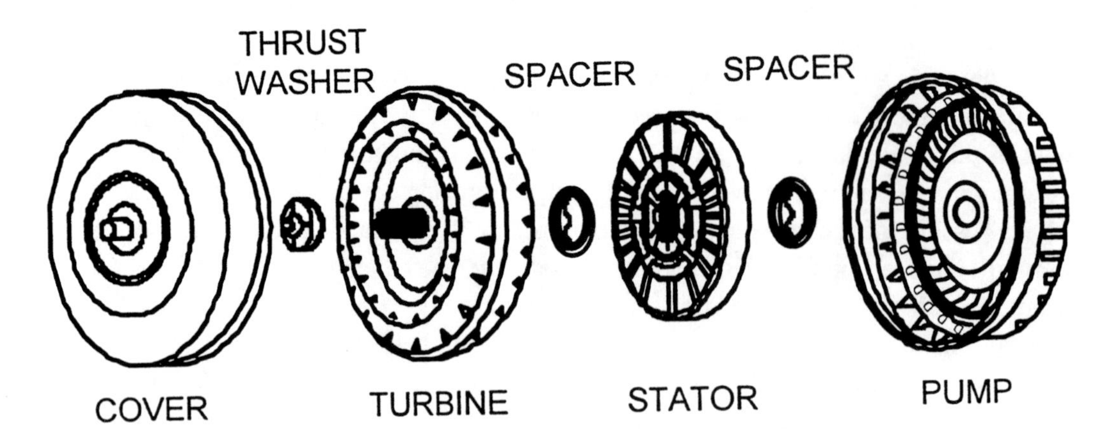

Figure 9-9 Hydraulic Torque Converter

shaft. Oil from the rotating pump causes the turbine to rotate thus coupling the crankshaft of the engine to the transmission shaft. A third member, the stator, is added to increase the efficiency of the fluid coupling. Without the stator, the fluid leaving the inner part of the turbine would be thrown into the pump in a hindering direction, opposing the direction of the pump. To prevent this, the curved vanes of the added stator change the direction coming out of the turbine into a helping direction. The effect of the stator is to cause the fluid to exert a harder push as it enters the turbine. This is called torque multiplication. In many torque converters, the torque is more than doubled. This can be compared with two gears of different sizes. If the small gear drives the larger gear, the larger gear will turn more slowly but will have a greater torque.

In the same way, when the pump turns faster than the turbine, the turbine will have a greater torque or turning force.

The stator causes the torque converter to multiply torque, when the pump is turning faster than the turbine. This speed difference and increase in torque is equivalent to what the manual transmission does in low gear. It allows the engine to turn fast while the car wheels are turning slowly so that high torque can be applied and the car can accelerate.

As the car comes up to speed, the turbine speed begins to catch up with the pump. When this happens, the fluid leaving the trailing edges of the turbine vanes is moving at about the same speed as the pump. Therefore, it could pass directly into the pump in a helping direction without being given an assist by the stator. In fact, under these conditions, the stator vanes get in the way.

To get the stator vanes out of the way as the turbine catches up with the pump, the stator is mounted on a freewheeling mechanism. As long as the fluid from the turbine strikes the faces of the stator vanes, the stator is stationary. When the fluid strikes the back of the stator vanes, the stator starts to freewheel. It begins to rotate so that the stator vanes, in effect, get out of the way.

Actually, the turbine can never quite catch up with the pumps. The pump must always be turning a little faster than the turbine in order for the fluid to have an effect on the turbine blades. A typical ratio is 9:10. That is, the pump turns 10 times while the turbine turns 9 times. This would be the ratio during cruising at a steady speed. During acceleration, the ratio increases. This produces a torque multiplication for good acceleration.

The above is a simplified description of a torque converter used in automatic transmission systems. The transmission shift going from one forward speed to another is automatic and is accomplished by clutches and brakes within the transmission. These devices release or lock up various components of the planetary-gear system.

The first successful automatic transmission car, the Hydra-matic Oldsmobile, came on the market in 1940. However, it was not until after World War II that hydraulic fluid systems were introduced generally as options. Soon the automatic transmission system became very popular and many new drivers were unfamiliar with the manual transmission system. This was illustrated around 1970 when a jewel thief in New York City ran to his get-away car with over a million dollars worth of jewels. The car had been stolen and left across the street by an accomplice who disappeared when he saw a policeman around the corner. The jewel thief hurled the jewels into the car and tried to drive away. However, he had learned to drive on an automatic shift and the car was a stick shift type with mechanical gears. When the policeman investigated the noise of clashing gears, the crook asked for instructions. The policeman spotted the jewels in the car and the thief had a long time in jail to think about it.

THE CHANGE FROM REAR WHEEL DRIVE

In the 1950's, front wheel drive was introduced commercially. The transmission drove the front wheels instead of the rear wheels. This gave better control especially in bad weather. Four wheel drive systems were originally introduced in the 1970's to enhance mobility for light trucks. In the mid-1970's, some automobile manufacturers made improvements in the four wheel-drive system for regular automobiles. In the 1980's every car manufacturer in the United States had at least one four-wheel-drive car.

THE COMPUTER AND THE CAR

The microprocessor, the heart of a small computer, was invented in 1971. By 1985 microprocessors were installed in vehicles sometimes with disastrous results. However, as designers and auto mechanics became more familiar with small computers, cars used an increasingly number of microprocessors. As time went on into the 1990's microprocessors fine-tuned the operation of the vehicle. This included the transmission system, the electron ignition, and other systems. By 1995 practically every new car had a number of microprocessors as an integral part of the operation of the vehicle.

9.5 SAFETY PROBLEMS AND SOLUTIONS

In 1955 in England a "Keep Death Off the Roads" campaign was launched to address the growing problem of auto safety on the highways.

In the United States in 1966, Congress passed the National Traffic and Motor Vehicle Safety Act mandating recall notices for safety defects found in automobiles.

Automobiles in the United States killed more U.S. citizens than all the wars put together. Table 9-1 shows the motor vehicle traffic fatalities from 1899 to 1993.

MOTOR VEHICLE TRAFFIC FATALITIES

YEAR	TOTAL FATALITIES	YEAR	TOTAL FATALITIES	YEAR	TOTAL FATALITIES	YEAR	TOTAL FATALITIES
1899	26	1923	17,870	1947	31,193	1971	52,542
1900	36	1924	18,400	1948	30,775	1972	54,589
1901	54	1925	20,771	1949	30,246	1973	54,052
1902	79	1926	22,194	1950	33,186	1974	45,196
1903	117	1927	24,470	1951	35,309	1975	44,525
1904	172	1928	26,557	1952	36,088	1976	45,523
1905	252	1929	29,592	1953	36,190	1977	47,878
1906	338	1930	31,204	1954	33,890	1978	50,331
1907	581	1931	31,963	1955	36,688	1979	51,093
1908	751	1932	27,979	1956	37,965	1980	51,091
1909	1,174	1933	29,746	1957	36,932	1981	49,301
1910	1,599	1934	34,240	1958	35,331	1982	43,945
1911	2,043	1935	34,494	1959	36,223	1983	42,589
1912	2,968	1936	36,126	1960	36,399	1984	44,257
1913	4,079	1937	37,819	1961	36,285	1985	43,825
1914	4,468	1938	31,083	1962	38,980	1986	46,087
1915	6,779	1939	30,895	1963	41,723	1987	46,390
1916	7,766	1940	32,914	1964	45,645	1988	47,087
1917	9,630	1941	38,142	1965	47,089	1989	45,582
1918	10,390	1942	27,007	1966	50,894	1990	44,599
1919	10,896	1943	22,727	1967	50,724	1991	41,508
1920	12,155	1944	23,165	1968	52,725	1992	39,250
1921	13,253	1945	26,785	1969	53,543	1993	40,115
1922	14,859	1946	31,874	1970	52,627		

Table 9-1 Deaths Due to Automobile Accidents
Source: U.S. Department of Transportation

In 1899 there were 26 deaths from automobile accidents. A peak death of 54,589 occurred in 1972. The total deaths of Americans in the Vietnam War was about 53,000 over an eight-year war period. These statistics were obtained from the U.S. Department of Transportation, National Highway Traffic Safety Administration, Washington, DC.

One of the first safety additions to the automobile was the mandatory seatbelt use laws (MUL). The U.S. Department of Transportation July 1984 rulemaking on automatic occupant protection began a wave of legislative action that resulted in the enactment of safety belt use laws in many states. The goal of those laws was to promote belt use and thereby reduce deaths and injuries in motor vehicle crashes.

The first mandatory belt use law in the United States was enacted in the State of New York in 1984. By 1985 nine states and the District of Columbia had passed MULs. A study of the effectiveness of MULs in 1985 was made by the U.S. Department of Transportation, National Highway Traffic Safety Administration, Office of Research and Development. It was estimated that because of MULs 285 lives were saved in 1985. That Department predicted the potential life savings of universal seatbelt usage would be about 40 percent of the 26,138 front seat occupant fatalities that had occurred in 1985.

As of December 1992, 42 states and the District of Columbia had belting use laws in effect. These laws differ from state to state according to the type and age of the vehicle, occupant seating position, etc. The first mandatory child restraint use law was implemented in the State of Tennessee in 1978. Since 1985, all 50 states and the District of Columbia have had child restraint use laws in effect.

Research* has found that lap/shoulder safety belts, when used, reduce the risk of fatal injury to front-seat passenger car occupants by 45 percent and the risk of moderate-to-critical injury by 50 percent. For light truck occupants, safety belts reduce the risk of fatal injury by 60 percent and moderate-to-critical injury by 65 percent.

In 1992,** 26,412 front-seat occupants of passenger vehicles (cars, light trucks, vans, and utility vehicles) were killed in motor vehicle traffic crashes, 67 percent of the 39,250 traffic fatalities reported for the year. Among front-seat passenger vehicle occupants over 4 years old, safety belt saved an estimated 5,226 lives in 1992 and prevented approximately 136,100 moderate-to-critical injuries.

At the high use rates achieved in other countries (85 percent), safety belts could have saved the lives of 10,837 front-seat occupants (that is an additional 5,611) for the nation as a whole in 1992. If all front-seat occupants wore safety belts, 14,138 lives (that is, an additional 8,912) could have been saved in 1992. Ejection from the vehicle is one of the most injurious events that can happen to a person in a crash. In fatal crashes, three-quarters of the occupants who were ejected from passenger cars were killed. Safety belts provided the greatest protection against occupant ejection: in fatal crashes in 1992, only 1 percent of restraint occupants were ejected, compared to 18 percent of unrestrained occupants.

---

*Reference, "Effects of Mandatory Seatbelt Use Laws of Highway Fatalities in 1985," by Paul Hoxie and David Skinner, Transportation Systems Center, Cambridge, MA 02142, Final Report April 1987.

** From Traffic Safety Facts 1992, Occupant Protection, National Center for Statistics and Analysis, Washington, DC 20590

According to observational surveys conducted by the states and reported to National Highway Traffic Safety Administration, 62 percent of passenger vehicle occupants used their safety belts in 1992. The reported restraint use rate among all occupants of passenger cars involved in fatal crashes was 47 percent in 1992. The use rate for drivers was higher (51 percent), and the highest use rate was reported for children age 4 and under (62 percent).

Another passenger constraint used in conjunction with the shoulder/lap seatbelt was the air bag introduced in 1972. The air bag is a very effective safety device that is built into the steering wheel or dashboard of a car. In a frontal crash, it inflates rapidly to cushion the occupant from violent impact with the hard interior surfaces of the car.

In a serious frontal crash--equivalent to hitting a brick wall at a speed greater than 12 miles per hour--a crash sensor activates the air bag. Within 1/24 of a second after impact, the bag is inflated to create a protective cushion between the occupant and the vehicle interior, such as the steering wheel, dashboard and windshield. The air bag inflates and then deflates rapidly.

The sensors are switches which are activated by a crash. They discriminate between impacts intended to inflate the bag (those severe enough to cause injury) and events not intended to inflate the bag (i.e., “fender-benders” in parking lots, or panic stops). As of December 1992 it was estimated that over 14 million passenger vehicles on the road were equipped with air bags. Between 1988 and 1992 there were an estimated 132,000 air bag deployments in crashes. It is estimated that 558 lives were saved by air bags and/or safety belts equipped with air bags. In addition the presence of air bags prevented an estimated 40,000 moderate-to-critical injuries.

Beginning with September 1997 all new passenger cars in the United States will be required to have passenger and driver air bags, along with manual lap/shoulder safety belts. The same requirement applies to light trucks beginning in September 1998.

Another automobile safety technology, anti-lock brakes were introduced around 1984. When a driver tries to stop suddenly on a slippery road with ordinary brakes the car will skid and spin around and the driver will not be able to steer. With anti-lock brakes a wheel-speed sensor signals a microprocessor that one or both wheels are slipping and that corrective action is necessary. The microprocessor orders the brakes applied to slow down by pumping the brakes four to six times a second. The car will not skid and spin around and the driver can steer the car around obstacles.

On January 4, 1994, Congress passed a law initiating proposed rule making on anti-lock brakes. The National Highway Traffic Safety Administration was given 36 months to examine whether anti-lock brakes should be required in new cars. The study was to quantify benefits, make a cost benefit analysis and to test different systems.

An adjunct to antilock braking systems is traction control. This includes the addition of cutting engine power automatically. The microprocessor controls both the brakes and the engine. Traction control is replacing four-wheel drive for passenger car applications because of the economics involved. Some traction control systems use only engine control.

Another factor contributing to auto safety is the 55-mph speed limit in the United States. This was enacted during the Arab oil embargo to conserve fuel. The 55-mph speed limit became effective by March 1974 in all states. Motorists slowed down on all major highway systems. Total travel declined in 1974 for the first time since 1946. Accompanying these reductions in speed and travel, 9,100 fewer persons died in motor vehicle accidents. The number of highway fatalities declined from 54,052 in 1973 to 45,196 in 1974. Such a sudden annual decline in highway fatalities was unprecedented outside of wartime, and as the fuel shortage receded, safety soon became the paramount issue surrounding the 55-mph speed limit.

A study by the Transportation Research Board, Special Report 204, "55:A Decade of Experience" was issued in 1984.

In 1982 Congress expressed concern that eroding compliance with the 55-mph national maximum speed limit threatened the safety of America's highways. Several states had weakened their penalties for violations of the 55-mph speed limit, and some members of Congress became concerned that this would encourage faster driving and would impair highway safety. In the Surface Transportation Assistance Act of 1982, Congress requested that the National Academy of Sciences "investigate (1) the benefits, both human and economic, of lowered speeds due to the enactment of the 55 mph National Maximum Speed Limit, with particular attention to savings to the taxpayers, and (2) whether the laws of each state constitute a substantial deterrent to violations of the maximum speed limit on public highways within such state."

One of the conclusions of the study was that the 55-mph speed limit saves 2,000 to 4,000 lives per year. Because of the substantial benefits to safety, the preponderant view of the committee's study was that the 55-mph speed limit should be retained on almost all of the nation's highways. However, despite all of this evidence, the U.S. Congress passed a law effective December 7, 1995 allowing the states to set maximum speed laws in their jurisdictions. Montana abolished the speed limit on some of its highways and some other states increased the maximum speed to 75 miles per hour.

9.6 Pollution and Solutions

In 1948 a chemist at the California Institute of Technology, Arie Jan Haagen, had documented the contribution of automobile emissions on smog (a combination of smoke and fog) in the Los Angeles area. Haagen stated that the automobile was the primary culprit of the increasing incidence of smog which he declared was detrimental to public health. According to Haagen the nitrogen oxides and unburned hydrocarbons coming out of the tail pipes of cars mixed to form the clouds of thick brown smog. The smog smelled bad, irritated the eyes and caused trees and plants to take on a brownish color.

California passed the first emission control law in 1959. This was made mandatory nationwide by the Vehicle Air Pollution Act of 1965. In 1970 the Clean Air Act was passed mandating that auto emission pollutants be reduced nationwide by 90 percent over a six-year period. Improvement to the engine, unleaded gasoline, and the development of the catalytic converter followed.

POLLUTANTS FROM AUTOMOBILES

There are three basic pollutants from internal combustion automobile engines. They are unburned hydrocarbons, nitrous oxides, and carbon monoxide. In addition there was lead from additives to gasoline. The unburned hydrocarbons mix with nitrous oxides to form ozone. Table 9-2 shows the health effect of automobile pollutants.

Table 9-2
Auto Pollutants and Health Affects

AUTO POLLUTANT	*HEALTH EFFECT*
Nitrous Oxides	Acid Rain
Ozone	Smog, Lung Damage
Lead	Child Growth and Development
Particulates	Smog, Lung Cancer
Carbon Monoxide	Deadly in Enclosed Spaces,

SOLUTIONS USING THE GASOLINE INTERNAL COMBUSTION ENGINE

The catalytic converter in use since the 1975 model year in the United States reduced carbon monoxide and unburned hydrocarbons and oxides of nitrogen.

A catalyst is an agent which promotes the rates at which chemical reactions occur but which itself remains unchanged. For automotive exhaust applications, the pollutant removal reactions are the oxidation of carbon monoxide and hydrocarbons and the reduction of nitrogen oxide. Metals such as platinum, palladium and rhodium are used as the catalysts.

In addition to the active metal, the converter contains a support component whose functions include adding structural integrity to the device, providing a large surface area for metal dispersion, and promoting intimate contact between the exhaust gas and the catalyst. Two types of supports are used: pellets and monolith. The pelleted converter consists of a packed bed of small, porous ceramic spheres whose outer shell is impregnated with the active metal. The monolith is a honeycomb structure consisting of a large number of channels parallel to the direction of exhaust gas flow. The active metal resides in a thin layer of high-surface-area ceramic placed on the walls of the honeycomb. In either system the support is contained in a stainless steel can installed in the exhaust system ahead of the muffler.

Two types of catalyst systems, oxidation and three way are found in automotive applications. Oxidation catalysts remove only carbon monoxide and hydrocarbons, leaving nitrous oxides unchanged. An air pump is often used to add air to the engine exhaust upstream of the catalyst, thus ensuring an oxidizing atmosphere. Platinum and palladium are generally used as the active metals in oxidation catalysts. Three-way catalysts are capable of removing all three pollutants simultaneously, provided that the catalyst is maintained in an environment that is neither overly oxidizing or reducing. Platinum, palladium, and rhodium are the metals must often used in three-way catalysts. In addition, base metals are frequently added to improve the ability of the catalyst to withstand small, transient perturbations in air-fuel ratio. In both oxidation and three-way catalyst systems, the production of undesirable reaction products, such as sulfates and ammonia, must be avoided.

Maintaining effective catalytic function over long periods of vehicle operation is often a major problem. Catalytic activity will deteriorate due to two causes, poisoning of the active sites by contaminants, such as lead and phosphorus, and exposure to excessively high temperatures. Catalyst excessive temperature is often associated with engine malfunctions such as excessively rich operation or a large amount of cylinder misfire. To achieve efficient emission control, it is very important that catalyst-equipped vehicles be operated only with lead-free fuel and that proper engine maintenance procedures are followed. In such cases catalytic converters have proved to be a very effective means for reducing emissions without sacrificing fuel economy.

Changes in Fuel to Reduce Pollution in Internal Combustion Engines

The U.S. Government in the 1990 Clean Air Act set a cap on carbon monoxide levels at nine parts per million of air. In 1992 New York City had levels that averaged 12.7 parts per million. This violation was also true in many other cities. In many states reformulated fuel increasing the amount of oxygen thus converting carbon monoxide to carbon dioxide was required during cold-weather months starting in the fall of 1992. Other reformulation of gasoline to meet the Clean Air Act of 1990 were planned to reduce the amount of hydrocarbons in 1995.

Another attempt to reduce fuel pollution is the use of ethanol, a mixture of a fuel derived from corn and gasoline. This did not spread over the United States.

The use of natural gas instead of gasoline is another attempt to reduce pollution. The advantage of natural gas is the elimination of particulate emissions. The U.S. Postal Service, United Parcel Service and some bus lines adopted natural gas vehicles.

The Use of Electric Cars to Eliminate Pollution

The California state legislature in 1990 passed a law that stated if a company wants to sell at all in California, then 2 percent of its sales must consist of vehicles with no tail-pipe emission by 1998. By 2003 at least 10 percent of the autos sold in California must meet that standard. This meant electric cars because they were the only automobiles that would have no tail-pipe emission.

In March 1996 the California Air Resource Board amended the rules on zero tailpipe emission suspending them until 2003. By that year at least 10 percent of the autos sold in California must have no tail-pipe emission.

One of the arguments raised against the electric vehicle (EV) is that the electric utilities that would charge EV batteries emit as many harmful pollutants as the internal combustion vehicles that EVs would replace. The Environmental Protection Agency (EPA) of the U.S. Government issued a Preliminary Electric Vehicle Assessment in November 1993 which stated the above. However, critics, such as the Edison Electric Institute, the Electric Power Research Institute, and the Electric Transport, charged that the EPA did not take into account the latest EV technology and used old electric utility emissions data. The EPA agreed that the report was premature and should not have been released to the public. Electric utilities in the last decade of the 20th century drastically reduced harmful emissions from electric power plants.

The electric car consists of a powerful battery, an inverter and two induction motors. The inverter changes the current from DC to AC and sends it to the two induction motors, each of which turns one of the front wheels. The limitation of electric cars is the weight and expense of the battery, the range limitation between charges and the time it takes to charge. The conventional lead acid battery has many limitations.

General Motors Corp. in 1996 began leasing its GM EV1 Electric Car, Figure 9-10, using a lead acid battery. It is expected to have a range of about 70 miles and a top speed of 80 miles per hour.

Honda is building an electric two-door, four-passenger car using nickel-metal hydride batteries. It is expected to have a range of about 125 miles and a top speed of 80 miles per hour.

Another version of the electric automobile is the hybrid vehicle. This consists of both electric motors and an internal combustion motor. Hybrid electric vehicles can operate close to a zero-emissions mode, but do not suffer the range limitations of pure EVs. Hybrid EVs do have a fuel-burning heat engine but they are nowhere nearly as polluting as conventional vehicles even when their heat engines are operating. There are two reasons. First, the heat engines can be quite small since the vehicles rely on their electric motors for accelerating and climbing hills. Second, their heat engines operate at a fixed or slowly changing speed which makes it easy to keep emissions under tight control.

Figure 9-10 General Motors EV_1 Electric Vehicle
Copyright 1996 GM Corp. Used with Permission GM Media Archives

There are two types of hybrid vehicles, parallel and series. The parallel hybrid vehicle has an internal combustion and transmission for one set of wheels along with an electric motor drive for the other set of wheels. It has a high rate energy storage flywheel. The chemical battery is used for low-rate energy storage.

Figure 9-11 The Dodge Intrepid ESX; A Hybrid Concept Car
Courtesy Chrysler Corporation

The series hybrid vehicle contains an internal combustion engine-alternator combination. This provides power to the electric motor which drives the vehicle. A chemical battery is used for slow charging and discharging. A flywheel is used for fast charge and discharge. A special flywheel will be described later.

Chrysler, in 1996, introduced the Dodge Intrepid ESX, a series hybrid electric concept car using two electric motors with lead acid batteries and a three-cylinder turbo-diesel driving an alternator. Figure 9-11 is a photograph of the Dodge Intrepid ESX Hybrid Car.

Another version of the electric car is the fuel cell which combines hydrogen and oxygen to generate electricity. Fuel cells produce only water and in some cases carbon dioxide as byproducts. In 1995 Daimler-Benz AG of Stuttgart, Germany, developed a small research van powered by fuel cells, Figure 9-12. It has a range of 150 miles with a

Figure 9-12 Daimler-Benz Fuel Cell Electric Car
Courtesy Daimler-Benz

top speed of 66 miles per hour. Daimler-Benz engineers believe that the next fuel cell vehicles will make hydrogen on-board from methanol. The vehicles would refuel at modified filling stations.

Another possible future development that departs completely from the familiar chemical battery is the flywheel battery. This innovation is a refinement of the flywheel best known for its use in automobile engines. The flywheel attached to a rotating shaft, modulates and smoothes out the delivery of power from motor to machine. In theory, because of its angular momentum, any spinning wheel will rotate indefinitely, storing for future use all energy that set it spinning in the first place. In practice, however, friction can quickly dissipate the energy. One major source of friction in the traditional flywheel design is the need for a shaft that transmits energy into or out of the wheel. The frictiongenerated by ball bearings at the point of attachment between support frame and shaft robs the spinning disk of its pent-up power.

However, new technology has changed the concept of the flywheel. The new flywheel is built of high-strength carbon composites which enable the wheel to spin at extremely high speeds without breaking apart. The next step is to eliminate the shaft. The flywheel is placed in a vacuum to eliminate air friction and then is suspended in space by a magnet attached to the bottom of the flywheel. The magnet acts in concert with a superconducting magnet just beneath the wheel to levitate the wheel above the superconductor and holds it in place. Virtually all friction in the system is eliminated because there is no physical contact with the flywheel. The magnetic bearing is several thousand times more efficient than the best ball bearings.

To get energy into and out of the system, a permanent magnet is built into the flywheel. As it spins, the magnet passes stationary electric coils surrounding the wheel. To draw energy out, the flywheel's permanent magnet is made to generate an electric current in the coils, which slows down the flywheel. To recharge the flywheel, magnetic fields are generated in the coils, pushing the flywheel faster and faster as its permanent magnet whizzes by.

Companies such as United Technologies of Hartford, Connecticut, have focused on making small flywheel batteries for use in electric cars, achieving rim speeds of nearly 2,400 feet per second. Flywheel batteries can quickly absorb or release enormous amounts of energy, enabling quick recharges as well as fast acceleration on the highway. A company, American Flywheel System of Medina, Washington, designed a car in 1994 powered entirely by flywheels.

9.7 PREVENTING AUTOMOBILE ACCIDENTS BY USING TECHNOLOGY

The seat belt and the air bag technology helped reduce fatalities due to automobile accidents. However, they did not reduce the number of the accidents themselves. New technologies employing radio, radar, computers, microchips, lasers, fiber optics and satellite systems can be used to help prevent accidents. Much of the planning for this is being done in the last five years of the 20th century.

One example is the use of radar in a vehicle to eliminate blind spots as the operator decides to change lanes. The radar can also be used to alert drivers to problems on their front and rear. In the United States in 1995 General Motors, Chrysler Corporation, and Vorad Systems filed comments with the FCC outlining plans to use millimeter radar to prevent automobile accidents. The frequencies to be used are located above 40 gigahertz. At these frequencies the transmissions are extremely short in range. This made this type of radar suitable for avoiding vehicular accidents.

THE INTELLIGENT VEHICLE HIGHWAY SYSTEM (IVHS)

In the last decade of the 20th century plans were drawn up by the U.S. Government for an Intelligent Highway System that would incorporate many of the 20th century technologies described in previous chapters. The purpose of IVHS is to reduce automobile highway accidents and substantially increase the efficiency of highways.

The principle will be demonstrated in a 7.6 miles stretch of highway near San Diego, California, in 1997 using cars and buses. This will be a prototype Automated Highway System (AHS). In order to develop the concept a National Automated Highway System Consortium (NAHSC) was formed. This is part of IVHS.

AHS is a fully automated vehicle operation (hand-off, feet-off) on dedicated lanes. Dual mode vehicles will be able to operate on both AHS and non-AHS lanes. It will include automobiles, buses, and trucks.

In the AHS the driver will drive the vehicle into a special lane. The system will assume control of the vehicle. This includes the brake, throttle, headway and lane keeping. When the driver chooses to exit, the vehicle is moved to a transition lane and the driver resumes control. Proponents of AHS forecast a 2 to 3 fold in safety and efficiency.

IVHS will include technologies like the GPS satellite navigation system, computer control of vehicles, wide area wireless such as cellular digital, lasers, and optical fibers, radar and many other 20th century technologies.

While IVHS including AHS is scheduled for the early 21st century, the planning and development occurred in the 20th century. One of the possible problems is the governmental financing in an era of tight budget restrictions.

BIBLIOGRAPHY

Duffy, James E. and Chris Johanson — *Auto Drive Trains Technology,* Goodheart, 1995

Duffy, James E. — *Auto Electricity and Electronics Technology* Goodheart, 1995

Gartman, David — *Auto Slavery: The Labor Process in the American Automobile Industry,* Rutgers University Press, 1986

---------------------- — *Auto Opium: A Social History of American Design,* Rutledge, 1995

Graham, John D. — *Auto Safety: Assessing America's Performance,* Greenwood, 1989

Mackenzie, James and Michael P. Walsh — Driving Forces*: Motor Vehicle Trends and Their Implications for Global Warming, Energy Strategies and Transportation Planning,* World Resources Institute, 1990

Reyes, Gary — *The Automobile: Horseless Carriages to Cars of the Future,* Mallard Press, 1990

Schiffer, Michael Brian — *Taking Charge: The Electric Automobile in America,* Washington, DC, Smithsonian Institution Press, 1994

PUBLICATION WITHOUT LISTED AUTHORS

Review of the Research Program of thePartnership for a New Generation of Vehicles. By National Research Council. Published by National Academy Press, Washington, DC 20418, 1994

CHAPTER 10

ON THE WINGS OF EAGLES

10.1 TRYING TO GET OFF THE GROUND WITH FLAPPING WINGS

Humanity has envied the freedom of birds to fly above the earth for thousands of years. The Greeks had the ancient legend of Daedalus and his son Icarus. Daedalus was a mythical builder and inventor on the Island of Crete in the Eastern Mediterranean. He constructed the labyrinth for King Minos of Crete. Minos later imprisoned the father and son. Daedalus planned an escape from Crete by building wings of feathers attached to the body by wax. Flapping the wings like a bird would enable them to fly away. Daedalus made his escape but Icarus flew too high and the sun melted the wax and he fell into the sea.

Leonardo da Vinci (1452-1519) studied the flight of birds and bats. He built mechanical wings copied from those of the bat which he regarded as the essential basis of any flying machine. He drew a sketch of a flapping wing controlled by a spring. It was one of a number of drawings that were clearly intended for the study of the flapping motions.

In the mid-20th century, James L.G. Fitzpatrick built and flew an "Ornithopter" based on the drawings of Leonardo da Vinci. The machine, powered by a one-horse power air compressor engine, relied on flapping wings. It is stored at Princeton University.

Leonardo da Vinci also designed an aircraft powered by a pedal operated by a human. Unfortunately the materials available at that time were too heavy for human powered flight. In 1990 employing modern materials an Olympic class bicycle rider using only human leg muscle power moving pedals to drive a propeller flew 90 miles following the legendary flight of Daedalus from Crete to the mainland of Greece. This flight connected the ancient myth to the technological revolutions of the 20th century including the use of the fixed wing.

10.2 The Beginning of Flight with A Fixed Wing

Heavier than air modern airplanes with their fixed wings depend for their lift on Bernoulli's principle. Daniel Bernoulli (1700-1782) was the most distinguished of a family of mathematicians. His father was Johann Bernoulli who later became jealous and tried to obtain priority for himself for his son's discoveries.

In 1738 Daniel Bernoulli published *Hydrodynamica* in which he studied the basic principles of fluids. The Bernoulli principle stated that the pressure in a fluid decreases as its velocity increases. His father then published *Hydraulica* in a vain attempt to get credit for his son's achievements.

It took approximately 150 years before engineers applied Bernoulli's principle to flying machines. They curved the top surfaces of fixed wings leaving the bottom surface flat. The airflow on the curved upper surface has a longer distance to go than the air under the flat lower surface. Consequently the air on the upper surface must travel faster than the air along the lower surface. In accordance with Bernoulli's principle, there is less pressure on the upper surface and, therefore, there will be a force of lift pushing up on the lower wing surface.

There is another component of lift in an airplane. If the wing is at a shallow angle to the horizon and moving, the resultant air pressure on the bottom of the wing will push upwards. This is normally a small component of lift on the runway and not enough to take off. However, the lift is proportional to speed and after the plane is in flight this component of lift becomes much greater. It is what permits an aircraft to fly upside down even though the Bernoulli effect then does not contribute to lift.

In 1890 Clement Ader (1841-1926), a French self-taught engineer, built a fixed wing monoplane which he named the Eole (after the God of wind, Aeolus). Ader fitted his plane with a furnace, boiler, steam engine and a condenser so that the water could be circulated. On October 9, 1890 he flew it a distance of 160 feet. Although the steam engine was unsuitable for sustained and controlled flight, Ader's short hop was the first demonstration that a manned heavier than air machine could take off from level ground under its own power. In 1990 France celebrated the centenary of the "First Pilot of the World."

In the years from 1891 to 1896 a German Engineer, Otto Lilienthal, built gliders with fixed wings and flew them on exploratory flights using Bernoulli's principle. He constructed over 2,000 gliders and flew a maximum distance of nearly two miles. In 1896 he crashed from a height of 50 feet and died. He left tables to show the shape of the wing for maximum lift.

Samuel Pierpont Langley (1834-1906) was an American aeronautical pioneer. In 1896 he built unmanned model aircraft powered by steam engines. His No. 5 unmanned model was powered by a small steam engine driving twin pusher propellers. It flew a number of flights of almost a mile.

In 1898 Langley was asked by the U.S. government to build a full size flying machine operated by a pilot. Langley built the aircraft and he asked Stephen M. Balzer to

Orville and Wilbur Wright designed their own engine and it was manufactured in house mainly by their mechanic Charlie Taylor. It was a four-stroke engine with four in-line water-cooled cylinders. The weight was drastically reduced by casting the largest part, the crankcase in aluminum. The engine was bolted by its four feet to two ribs on the 1903 flyer to the right of the centerline. The weight of the engine was balanced by the prone pilot on the left. The two pusher propellers were driven in opposite directions to cancel out any twisting motion. With the pilot settled in place, the fuel cock was turned on and the engine was started by the other brother turning a propeller. The pilot then adjusted the ignition until the revolutions per minute were in the order of 1,100 to 1,200.

The brothers first ran the engine in Dayton, Ohio, in February 1903. It operated smoothly and they started the long preparation for the first manned flight.

On the morning of December 17, 1903 the Flyer I, Figure 10-l, made history's first powered, sustained, and controllable airplane flights without any assistance at Kill Devil Hills, North Carolina. The first flight lasted 12 seconds. The last flight that day lasted 59 seconds and covered 852 feet of ground.

An improved Flyer II equipped with a new engine was flown at Huffman Prairie near Dayton, Ohio, in 1904. This was followed in 1905 with Flyer III, Figure 10-2, the world's first practical airplane. It could turn, bank, circle, do figure eights with ease, and stay airborne for more than half an hour. In October 1905 this machine flew 25 miles in approximately 40 minutes.

The Wright brothers dominated world aviation until the end of 1909, building their machines in both Europe and the United States. Improved Wright machines made excellent flights in 1910 and 1911. Wilbur Wright died of typhoid in 1912. Orville lived on until 1948 and continued to make contributions to aeronautics until his death.

Another early pioneer was the American Glen Curtiss (1878-1930). He began by building engines for bicycles. In 1904 he designed and built a motor for the dirigible, "California Arrow." He then joined the Aerial Experiment Association founded by Alexander Graham Bell.

Glenn Curtiss fought Wright in the courts for the patent on the heavier than air manned plane, but he lost. Curtiss then got permission from the Smithsonian to rebuild Langley's Aerdrome A aircraft. He made some modifications and flew the aircraft 150 feet. Curtiss removed the modifications and installed the plane in the Smithsonian where it was billed as the first airplane.

Orville Wright became angry and shipped the Wright 1903 Flyer to the Science Museum in London. During World War II it was stored in a subway for safety. After the war the Smithsonian decided that the Wright 1903 Flyer was the first one. Negotiations were started with Orville Wright. He died in 1948 but gave permission to move the aircraft before his death. On December 17, 1948, forty-five years after the first flight,

Figure 10-2 Wright 1905 Flyer III
National Air and Space Museum, Smithsonian Institution Negative A317

the 1903 flyer was installed in the Smithsonian. The Wright brothers finally got the recognition they deserved.

In the years just before World War I a number of aircraft designers appeared in Europe who made advances in the building of planes. Among them were Charles and Gabriel Voisin and Louis Bleriot of France and Henry Farman (Anglo-French), Rumpler of Germany and many others.

Bleriot first worked with Gabriel Voisin and then worked on his own. His Bleriot V of 1906 was his first plane after he left Voisin. It made a few hopping flights and then crashed and was scrapped. Bleriot continued his series of monoplanes culminating in the Bleriot XI with a 25 horsepower engine. He flew the Bleriot XI across the English Channel from Calais to Dover on July 25, 1909. This plane had a maximum speed of 75 mph. The Bleriot XI-2 and -3 were also developed for the military as a reconnaissance machine with two and three seats.

10.4 FLYING IN WORLD WAR I

World War I (1914-1918) saw the rapid development of heavier than air flying machines as both sides converted planes into instruments of war. On October 5, 1914 the French fitted a Voisin aircraft with a machine gun. Sergeant-Pilot Joseph Frantz with his observer Lois Quenault shot down a German Aviatik. The French kept improving their fighters and developed the Morane-Saulnier Type N. It had a top speed of over 100 miles per hour and it could climb to 9,840 feet. It was fitted with a machine gun which fired between the blades of a propeller. Another famous French fighter was the Spad. The Spad XIII with a 220-hp water-cooled engine was capable of a maximum speed of 137 mph and altitudes of over 21,300 feet. It had two 0.303-inch machine guns.

In Britain, the Royal Aircraft designed a number of combat aircraft in World War I. The best, designed by H.P. Folland, was the S.E.5a and used a 200 hp water cooled engine. It was a biplane and had a maximum speed of 138 mph at sea level. It carried two 0.303-inch machine guns and 100 lbs. of bombs. Another British fighter was the Bristol F-2 biplane. It was considered the best two seat combat aircraft of World War I. It used a 275-hp Rolls Royce Falcon water-cooled engine. It had a maximum speed of 125 mph at sea level. It used two or three 0.303 inch machine guns and up to 120 lbs. of bombs. A later British fighter aircraft was the Sopwith F1 Camel. It was a biplane with a 130 hp air-cooled engine with a maximum speed of 104.5 mph at 10,000 feet. Another Sopwith plane was the Sopwith triplane which entered the war in the spring of 1917.

In Germany the triplane concept was taken up by a number of manufacturers including Fokker, a native of the Netherlands. The Fokker Dr I had an air-cooled rotary engine with 110 hp. The maximum speed was 115 mph at sea level. Another German fighter was the Albatros D III, a biplane, with a maximum speed of 109 mph at 3,280 feet. It had a 175-hp Mercedes water-cooled engine. The most famous German pilot was the "Red" Baron Manfred von Richthofen who used the two machine guns of his planes to great effect, destroying a large number of allied aircraft until he was shot down.

In the United States in March 1915 Congress authorized the formation of the National Advisory Committee for Aeronautics (NACA). This was a prestigious body with representatives from the National Bureau of Standards, the civilian scientific community, the military services, the Weather Bureau and the Smithsonian Institution. The National Advisory Committee for Aeronautics had a mandate "to supervise and direct scientific study of the problems of flight, with a view to their practical solution...." This organization was to advance the cause of aviation in the United States by research including wind tunnel development in the period between the World Wars and later. In 1916 it recommended the establishment of an air mail service which was to be a profound step in aviation development between the two World Wars.

10.5 Between the World Wars

World War I had seen a tremendous improvement in the development of the airplane. Fighters reached heights of 23,000 feet and speeds up to 145 mph. Bombers flew up to 800 miles to deliver up to 4 tons of bombs. Most of the aircraft was made of wood and canvass. Many were biplanes, some were triplanes and a few were monoplanes. The end of the war saw the evolution of commercial passenger flight from the fighters and bombers of World War I.

The great advances in aviation in World War I carried over into commercial aircraft right after the war. In Germany the Junkers J10 saw limited service as an escort fighter in the war. Then Junkers developed a commercial plane, the F13, from the J10. The Junkers F13 was one of the first commercial planes to fly after World War I. It had two pilots plus four passengers. It was the most important air transport of the 1920s in Europe. It was operated by Junkers Luftverkehr which later became Deutsche Lufthansa. Production continued to 1932. It had one 185 hp water-cooled engine and could cruise at 75.5 mph at sea level. The Junkers F13 was a monoplane using the metal construction patented by Dr. Hugo Junkers in 1910 for thick-section cantilever monoplane wings.

In France in 1920 the former Spad fighter was changed to transport six passengers up to 250 miles. The plane was named the Bleriot Spad 33. It was followed by the Spad 46 and in 1923 by the Spad 56. Bleriot then designed a four-engine biplane, the 135, which was operated by Air Union on the Paris to London route from 1924 to 1926. It carried 10 passengers at 85 mph with a maximum range of 375 miles.

After World War I Dr. Fokker left Germany for his native Holland. There in 1924-25 he built the F.VII commercial aircraft powered by a 360-hp Rolls-Royce Eagle engine. From that Fokker developed the commercial eight-passenger F.VIIA that first flew in March 1925 with a 400 hp Packard Liberty 12 engine. This was followed by the F.VIIB. The Fokker F.VIIB 3m had a crew of two plus up to eight passengers. It was a monoplane with three 240-hp engines. It used a welded steel tube fuselage and a high-set cantilever wing of thick section and wooden construction. Fokker made a Ford-reliability tour in the United States with the F.VIIA-3 m. The power plant consisted of three 240 hp Wright Whirlwind engines.

The United States Air Mail Service

In the United States in World War I Curtiss' best-known plane was the JN-4, known as the Jenny, Figure 10-3, a trainer for prospective pilots. It was a biplane that had a speed of 60 mph at sea level and could fly to over 9,800 feet. The Jenny had a 90 hp water-cooled engine and was a two seater for an instructor and one student. In the 1920s most American pilots learned to fly in the Jenny. The plane was used by barnstormers and became famous for such exploits as the first Canadian Rocky Mountain air mail flights.

Figure 10-3 The Jenny
National Air and Space Museum, Smithsonian Institution Negative A52773C

Between 1918 and 1927 the U.S. Government ran an AirMail service which delivered the mail by airplanes. The Government owned the aircraft, the continental radio navigation system, and hired the pilots. The AirMail Service in September 8, 1920 opened a transcontinental service across the U.S. In February of the following year the U.S. Post Office demonstrated day-and-night transcontinental service.

In 1926 things began to change as the U.S. Government started to contract out services for local runs between cities. One of these contracts was awarded to Robertson Aircraft Corporation for the run between St. Louis and Chicago. They hired as their chief pilot a young ex-army and stunt flyer, Charles Lindbergh. On September 1, 1927 the Post Office turned the entire AirMail Service including the transcontinental run to commercial operators.

Charles Lindbergh (1902-1974)

Charles Lindbergh was the international star of aviation, especially between the two wars. In his second year at the University of Wisconsin his interest in aviation led him to quit school and enroll in a flying school in Lincoln, Nebraska. He purchased a World War I Curtiss Jenny with which he made stunt-flying tours through Southern and Midwestern states. He spent a year at the Army flying school in Texas (1924-1925) and became a second lieutenant in the Army Reserves.

Charles Lindbergh was then hired at $400 per month by the Mil-Hi Airways and Flying Circus, which operated out of Denver, Colorado. He not only took up passengers in an old biplane but in the evening Roman candles and streamers were attached to his wings to give a spectacular fireworks display in the darkness.

Charles Lindbergh was now 23 years old and the Flying Circus almost killed him. The fireworks displays took place at night in small towns without nearby airfields. At one time he was forced to land after the fireworks at a makeshift farmer's field without any lights on the ground except for a flashlight held by his employer.

After that experience he was happy to accept a job offer from the Robertson Aircraft Corporation in St. Louis. They had obtained the Air Mail contract between St. Louis and Chicago as the Post Office was getting out of the air mail business.

Charles Lindbergh was given full charge of the program and he was authorized to hire two other pilots to serve under and share the run with him. He plotted the route between St. Louis and Chicago, picked his own landing fields and ordered the equipment for them. He chose two Army cadets from his class at San Antonio to fly with him. Together they picked out nine landing grounds between St. Louis and Chicago and arranged for gas dumps and some night lighting equipment.

Lindbergh inaugurated the service on April 15, 1926 taking off just before dawn from Chicago. In the afternoon he loaded mail at St. Louis and began the return journey to Chicago at 4 PM, landing at two fields along the way to pick up more mail. He landed in Chicago at 7 PM with a full load of mail.

Lindbergh had some close calls flying the mail in all kinds of weather. On September 15, 1926 he was flying to Maywood Field, Chicago, when fog rolling in blanketed that area to a height of nearly a thousand feet. Searchlights at the field failed to pierce the thousand-foot cover and Lindbergh cruised the area and then decided to fly south in search of a clear area where he could land. He ran out of gas and jumped out. His parachute worked perfectly but the falling plane followed him down and he was afraid that he would be hit. He finally landed in a cornfield. He found that the plane had crashed some three miles away. The mailbags were intact and Lindbergh had them put aboard the 3:30 AM train for Chicago.

Bad weather forced him to make three more emergency parachute jumps within a year. He was the first pilot in the United States to make four emergency parachute jumps. The 24-year-old pilot even made the pages of The New York Times. He was beginning to earn a reputation as the AirMail pilot who flew in all kinds of weather and miraculously escaped with his life.

In Chicago, a day after he had made an emergency parachute jump, Lindbergh went to see a movie to relax. In the newsreel that preceded the main feature, there was a picture of a Sikorsky biplane being readied for a flight across the Atlantic to Paris. The newsreel caption stated that Captain Rene Fonck was the pilot and that the four-man crew was competing for the Orteig Prize. Lindbergh hurried out of the theater and bought newspapers which described the Orteig Prize.

It was named for the man who originally proposed it, Raymond Orteig, a fat Frenchman who operated two hotels in New York City. The original offer in 1919 was $25,000 to any flier or group of fliers who crossed the Atlantic from Paris to New York or from New York to Paris non-stop. This offer stipulated that the flight must be made within five years. Two British fliers, John Alcock and Arthur Whiten in 1919 did fly across the Atlantic from Newfoundland to a crash landing in an Irish bog. The distance was less than 2,000 miles while the New York to Paris run was 3,400. In 1926 Orteig renewed his offer and a French pilot, Captain Rene Fonck, took up the challenge in a Sikorsky plane. It crashed on take-off and killed two of the four-man crew. Fonck escaped but the contest was now open to others.

In his spare time Lindbergh had given flying instructions to a number of rich businessmen in St. Louis. Through these contacts and others in the city he collected $15,000 to buy a plane for the solo flight across the Atlantic. He circulated a large number of aircraft firms with the specifications of the craft he was seeking. He received only one affirmative reply. It was from Ryan Aircraft Company of San Diego, California.

Lindbergh gave up his mail-flying job and went out to San Diego to look over the Ryan plant. He was convinced that they could do the job for $10,580. His backers closed the deal on February 24, 1927, and by mid-April the plane, a single engine monoplane was ready to roll out of the workshop. It was christened the Spirit of St. Louis.

The plane had a top speed of 130 mph and could carry 400 gallons of gas with a range of 4,000 miles. On April 8, 1927 Lindbergh flew the plane for the first time. After some further flight tests he was ready.

In the meantime a number of pilots were attempting to win the Orteig Prize. While Lindbergh was waiting in San Diego for a storm area to clear, he saw a story on the front page of a newspaper that interested him. French Captains Nungesser and Coli had taken off from Paris to New York in a biplane named L'Oiseau Blanc (White Bird). Nungesser was a decorated World War I hero. Several sources reported that the White Bird had passed over Portland, Maine, and other places on the western coast of the Atlantic Ocean. However, as time passed it became evident that the plane had come down in the Atlantic.

Lindbergh took off from San Diego on May 10, 1927 and made a stopover in St. Louis where he briefly met his financial backers. He took off and landed in Curtiss Field, Mineola, Long Island. A few days later when the weather across the Atlantic was clearing, he had the plane towed to nearby Roosevelt Field.

At 7:54 Eastern Daylight time, May 20, 1927, Lindbergh in the Spirit of St. Louis took off bound for Paris. His only instruments were a compass and a sextant. On his knees he carried a chart showing the Great Circle route which would save mileage. He had rejected a radio to save weight. Another instrument that Lindbergh employed was a homemade periscope. He used it for looking ahead when he was airborne.

Lindbergh flew over St. John's in Newfoundland and headed over the Atlantic. After 27 hours of flying from Roosevelt Field he spotted a fishing fleet. He turned off the engine and glided to within fifty feet of one of the vessels. He shouted, "Which way is

Ireland?" No answer came and he continued on his way. An hour later he flew over the southwest coast of Ireland, 16 hours after leaving Newfoundland. At 9:52 PM local time he sighted the Eiffel Tower and circled over it once. He landed at Le Bourget Field, Paris, at 10:24 PM, Saturday, May 21, 1927. This was a milestone in the history of aviation between the two World Wars.

Lindbergh became a world hero and went on a series of goodwill flights. On one of these flights to Mexico, he met Anne Morrow, daughter of the U.S. Ambassador to Mexico. They were married on May 27, 1929 and together they flew all over the world. During this period Lindbergh acted as technical adviser to Transcontinental Air Transport and Pan American Airways, personally pioneering many of their routes.

The Lindbergh's first child Charles August Lindbergh III was born on June 22, 1930. On March 1, 1932 the baby was kidnapped and murdered. The subsequent arrest, trial and execution of the accused kidnapper, Bruno Hauptman, made headlines all over the world. Afterwards, the Lindberghs left the country and went to live overseas.

After 1936 Lindbergh visited Nazi Germany. He was fascinated by the growing air power of a country preparing for war. He seemed to admire some aspects of Nazi Germany. In 1938 he was decorated by Hitler's government which led to a great deal of criticism. He refused to return the decoration afterwards. Lindbergh went back to the United States and in 1940-41 he made speeches that were deemed by many people to be anti-Semitic. Criticism of his public statements by President Franklin D. Roosevelt led Lindbergh to resign his Air Corps Reserve commission on April 27, 1941.

When the United States entered the war, Lindbergh offered his services to the War Department but he was turned down because he was seen by people in the U.S. Government to have been pro-Nazi. He attempted to work for Pan America, United Aircraft and Curtiss-Wright but they all turned him down under Government pressure. Lindbergh finally went to Henry Ford who loathed Roosevelt, despised democracy and was anti-Semitic. Ford hired Lindbergh as a technical consultant at the Willow Run Bomber plant at Detroit on April 3, 1942. Later, he flew 50 combat missions during a tour of duty in the Pacific. After the war, in 1954 he was appointed Brigadier General in the Air Force Reserve by President Dwight D. Eisenhower. He wrote a book, "The Spirit of St. Louis," describing the flight to Paris. It was published in 1953 and won the Pulitzer Prize. He also wrote a number of other books. He died on August 25, 1974 in Hawaii.

THE DEVELOPMENT OF U.S. COMMERCIAL AIRCRAFT

In the United States, Ford developed a Ford 4-AT-E-Tri-Motor which first flew in June 1926. It was a high-wing monoplane. The Ford plane had a corrugated all-metal construction which resulted in the nickname, "Tin Goose." The 4-AT, shown in Figure 10-4, had two pilots and eight passengers. This was increased to thirteen passengers with the 5-AT in 1928. The cruise speed was 107 mph at sea level.

Figure 10-4 Ford 4-AT-E Tri-Motor Plane
National Air and Space Museum, Smithsonian Institution Negative 90-16830

Another U.S. commercial company between the World Wars was Boeing. Their Model 200 was a mailplane with limited passenger capacity. Boeing claimed that its Model 247 was the first "modern" air transport. It was initially flown in February 1933. The plane featured all-metal construction, cantilever wings, pneumatic de-icing of the flying surfaces, a semi mono-coque* fuselage, retractable landing gear and fully enclosed accommodation for ten passengers. The 247D had drag-reducing engine cowlings and controllable-pitch propellers.

The Lockheed L10-A Electra was a passenger transport that first flew in February 1934 with a pair of 450 hp Pratt & Whitney air-cooled engines. It was a monoplane with a crew of two plus up to ten passengers. The Lockheed L10-A Electra had a maximum speed of 190 mph at 5,000 feet. Its operational ceiling was 19,400 feet with a maximum range of 810 miles. The Electra had all metal construction with retractable landing gears.

The Sikorsky S-42B was an intermediate/short range passenger flying boat designed for Pan American Airways for their routes across the Caribbean and South America. It

*Mono-coque is a type of construction in which the outer skin carries a major part of the stresses.

was a parasol-winged flying boat with a conventional boat hull. There were four radial engines on the leading edge of the wing. The first S-42 flew in service in August 1934 between Miami and Rio de Jainero. It was later used on transpacific flights including some pioneering ones across the South Pacific to New Zealand. The power plant consisted of four 800 hp Pratt & Whitney air-cooled engines. The S-42B had a cruise speed of 140 mph at 2,000 feet. It had a crew of four and two cabin attendants plus up to 32 passengers.

The Boeing 314a is considered by many as the greatest flying boat ever built for the Civil Air Transport Service. It was designed for Pan American Airways and first flew in June 1938. In May 1939 the 314 entered service as a mailplane and then in June of the same year it carried its first passengers. The flying boat had accommodations for three crew members plus seven flight attendants and up to 74 passengers. It had a cruise speed of 183 mph at sea level and was powered by four Wright engines with fuel for a range of 3,500 miles.

Transition from Commercial to World War II Service

The Douglas DC-3 passenger transport is said to have opened the era of modern air travel in the mid-1930s and later became the mainstay in World War II of the Allies' transport effort. Production of 10,349 aircraft was completed in the United States; at least another 2,000 were produced in the Soviet Union as the Lisunov Li-2, and 485 were built in Japan as the Showa L2D. The DC-3 has crew accommodations for three plus two cabin crew and up to 24 passengers. It was a cantilever low wing monoplane with features such as retractable landing gear and trailing edge flaps. The plane was powered by 1,200 hp air-cooled engines. It had a maximum speed of 230 mph at 8,500 feet with an operational ceiling of 23,200 feet and a range of 2,125 miles. Later in World War II the DC-3 was ordered for the U.S. Army as the C-47 Skytrain, and in the U.S. Navy it was called the R4D. The U.S. supplied the plane to the British under the Lend-Lease Act in the late 1930s and early 1940s. They were called Dakotas. After the war large quantities of these aircraft were sold cheaply to civil operators all over the world. It contributed to the development of air transport in most of the world's remote regions. At the end of the 20th century there were still a number of DC-3 aircraft in service.

10.6 World War II Propeller Aircraft

World War II saw the rapid development of aircraft culminating in jet aircraft which opened up the jet aviation age after the war.

In Germany the best German fighter of World War II was the Focke Wulf FW 190 whose engine developed more than 2,000 hp. It reached speeds of 440 mph and climbed to over 32,800 feet. Armed with both machine guns and cannon it was a formidable fighter.

The Focke Wulf FW 200 Condor in Germany was another example of a plane originally designed for commercial use just before World War II and then adapted to military use. The plane was developed as a transatlantic passenger and mail aircraft for Deutsche Lufthansa and flew during July 1937 with room for a maximum of 26 passengers in two cabins. Later the FW 200C series became Germany's most important maritime reconnaissance bomber of World War II. It played a major part for Germany in the Atlantic and Arctic convoy campaigns. The aircraft had four 1,200 hp air-cooled engines. It had a cruise speed of 208 mph at 13,124 feet. The FW 200C-3 Condor had one cannon and two machine guns plus up to 4,630 lbs. of bombs. Later models also used anti-ship missiles.

British Fighter Aircraft

The chief winner of the Battle of Britain in 1940 was the Hurricane. This was a monoplane powered by a 1,000-hp engine and armed with eight machine guns. It could reach speeds of 312 mph and ascend to altitudes of over 32,800 feet. The Hurricane's partner in the Battle of Britain was the Spitfire. It received much more publicity even though the Hurricane destroyed more German aircraft. The Spitfire was first flown in March 1936 but it was developed in 24 versions throughout the war. The output of its 12 cylinder Rolls-Royce engine was increased from 1,030 to 2,050 hp, its speed from 357 mph to 450 mph and its ceiling from 33,780 feet to 44,600 feet. Originally equipped with eight machine guns, the 1944 version had two cannons and four machine guns. An attack plane, the Typhoon was also developed by the British. It was widely used in 1944, supporting the Normandy landings with bombs and rockets. Its 2,200 hp 24 cylinder engine gave a maximum speed of 406 mph.

American Fighter Planes

The U.S. Air Force's most popular fighter was the P-51 North American Mustang. It flew at up to 437 mph and had a ceiling of 41,000 feet. It had a range of 2,300 miles. Its armaments consisted of six heavy machine guns and could deliver a bomb load of 2,000 lbs. Another fighter, the Republic P-47 Thunderbolt, saw service on D-Day and later as an attack aircraft against ground troops. It had a 2,000-hp 18-cylinder engine with a top speed of 437 mph. The P-38, a twin tail boom fighter, had two engines which gave it a top speed of 412 mph and a ceiling of 42,600 feet.

Carrier-based fighter aircraft played an important part in the war in the Pacific. The most formidable fighter was the Vought F4U Corsair. It flew at over 425 mph and had a ceiling of over 36,000 feet.

Japanese Fighters

The most famous Japanese fighter in World War II was the Zero or Mitsubishi A6M Reisen. It was powered by a 950-hp engine and it flew at a maximum speed of over 312-mph and a ceiling of 32,800 feet. It was armed with two 20-mm cannon and two 7.7-mm machine guns. The Zeros accompanied by the dive-bomber Aichi D3Al were the main aerial force attacking Pearl Harbor on December 7, 1941. The dive-bomber was powered by a 1,000-hp engine and could fly at about 245 mph and had a ceiling of over 29,500 feet. It carried three 7.7-mm machine guns and heavy bombs which destroyed a large part of the American battleships at Pearl Harbor.

Soviet Fighters

The best Soviet fighters in World War II were the Yak single seater fighter series designed by Aleksandr Yakovlev. More than 35,000 Yak fighters were produced in World War II. In 1944 the Yak-3 joined the fighting. It flew at around 437 mph and had a ceiling of 32,800 feet. The Yak-3 had two 20-mm cannons and two 12.7-mm machine guns.

American Bombers

The most important long distance American bombers in World War II were the Boeing B-17 Flying Fortress, the Consolidated B-24 Liberator, and the Boeing Superfortress B-29. The B-17 flew at 220 mph carrying a bomb load of four tons to targets more than 1,000 miles from base. The B-24 flew 1,250 miles transporting four tons of bombs at nearly 300 mph. The B-29 had a maximum speed of 357 mph and dropped its 10,000-lb. bomb load at distances of over 1,875 miles. The B-29 Enola Gay dropped an atomic bomb on Hiroshima, Japan, that contributed to the end of Japanese hostilities in World War II.

German Bombers

In the early part of World War II German bombers rained tons of bombs on London in an attempt to bring Britain to its knees. The main German bomber in the early stages was the Heinkel He 111. Its top speed was 250 mph and had a bombing radius of 375 miles. Another German bomber that attacked London was the Dornier Do 17 which flew at over 220 mph to deliver its 2,200 lbs. A superior German bomber was the Junkers Ju 88. It had a maximum speed of 280 mph and a bombing radius of 625 miles. Another German bomber was the Dornier Do 217 which flew at 315 mph with an effective range of 625 miles. Later the Heinkel He 177 had an improved bombing radius of 1,250 miles.

Soviet Bombers

In the 1930s Tupolev developed the TB-3 four-engine bomber. It was a monoplane of all metal construction. It carried a payload of two tons with a radius of 690 miles at a speed of 125 mph. Later in World War II he built the Tu 2, a twin engine bomber with a radius of 810 miles at 340 mph.

10.7 Turbojet Aircraft

Turbojet propulsion involves the issuing of a fluid from a nozzle that produces a force on the aircraft in the opposite direction. This is in accordance with Newton's Third Law of Motion which states that, "To every action there is always opposed an equal reaction." The turbojet is an air-breathing propulsion engine used in military aircraft and in commercial airlines.

The turbojet is a heat engine in which heat is added by burning fuel at constant pressure in a combustor. The hot gas turns a turbine whose shaft is connected to a compressor which rotates and compresses the incoming air. Further expansion through the narrow jet nozzle converts the gas into a stream of high velocity, producing thrust for propulsion power.

There are two general types of jet fuel fractionated from crude petroleum. One is a naphtha-kerosene blend used by the U.S. Air Force as JP-4. The other jet fuel is a kerosene used by the world's airlines as Jet A or Jet A-1.

Introduction of Jet Aircraft in World War II

Jet aircraft were introduced at the end of World War II. They came too late to alter the course of the war. However, the development of jet fighters and bombers led to commercial jet aircraft.

In England, the jet engine pioneer was Frank Whittle, a Royal Air Force Officer. In 1937, he formed a company, Power Jet Limited, to build jet engines. He had conceived of the idea while a student at the Royal Air Force College. In April 1937, he built his first jet engine which accelerated out of control. An improved engine led to his being allowed to continue by the RAF. Whittle in April 1941 made the first flight with his jet engine. However, the war made it difficult to continue and the British government turned the Whittle engine over to the United States government for future development.

The General Electric Company, which had turbine experience, was given the job of

Figure 10-5 The XP 59a, The First American Jet Aircraft, 1942
National Air and Space Museum, Smithsonian Institution Negative A37216a-

building an improved Whittle jet engine. Bell Aircraft was selected to build the aircraft. Bell built the X-59a, Figure 10-5, a two-engine jet aircraft, and it was tested at Edwards Air Force Base in Muroc Dry Lake, California. This was the first jet aircraft in the U.S. In October 1942 testing began. It flew to 46,000 feet which propeller aircraft could not reach. Fifty XP 59s were placed in service for the U.S. Army Air Corps but none were in combat. They were used mostly for training pilots in jet aircraft.

In early 1943, another company in the United States, Lockheed, started the design of a jet fighter, the P-80 Shooting Star, Figure 10-6. It made its first flight on January 8, 1944. At the end of the war there was one P-80 in England and one in Italy. It flew at 593 mph with a range of 1,250 miles and a ceiling of 45,930 feet.

In England, the development of jet fighters continued and culminated in the Meteor, Figure 10-7. It was a twin-engine aircraft which first flew in 1943. A more advanced version in 1944 reached speeds of almost 600 mph. At this time, Germany was devastating London with the V-1 rocket which flew faster than any British fighter except the Meteor. The pilot could not shoot down the V-1 because the exploding rocket would destroy his own aircraft. The jet fighter pilot placed a wing tip under the V-1 and pushed it off balance, tumbling it to the ground.

Another British jet aircraft fighter was the single engine Vampire Fighter made by de Havilland. It made its debut in 1943 and flew at 543 mph. The de Havilland was the only British company that at the end of the war had built both a jet engine and a jet plane. This would have consequences for commercial jet planes after the war in Britain.

In Germany, Hans von Ohain was the pioneer in jet engine development. In 1939, a test jet aircraft, the He 178, was built by Heinkel to test the concept. Messershmitt, another German aircraft company, built the Me 262 which flew at 500 mph. This was a deadly fighter which came too late in the war to make a difference. Although 1200 planes were built, only 200 were actually in service. By this time, there were very large numbers of Allied propeller planes. In 1945, a group of P-47s attacked one Me 262, seriously wounding the pilot, Luftwaffe General Adolf Galland.

10.8 The Development of Commercial Jets

The first passenger flight using a turbojet aircraft was made in 1952 by the de Havilland D.H. 106 Comet, Figure 10-8, flying between London and Johannesburg. The plane flew at 35,000 feet with a pressurized cabin. It had problems of metal fatigue and explosive decompression of pressurization that caused a number of fatal flights.

The first great commercial success of the jet age was the four engine Boeing 707 introduced in 1958. It flew at close to 600 mph with a range of 4,375 miles. The plane had titanium tear stoppers to prevent a repeat of the Comet's problems.

Another early American jet was the Douglas DC-8 which made its debut in 1959. An improved model DC 8-50 introduced in 1960 was a four-engine swept-wing jet with a

Figure 10-6 P-80 Shooting Star, 1944
National Air and Space Museum, Smithsonian Negative 1B-14832

cruising speed of 580 mph and a range of 5,600 miles. Its carrying capacity was 189 passengers, increasing to 259 in 1965.

In France in 1959 the Caravelle was a medium range two-engine jet that started airline service. The Caravelle flew at 500 mph carrying 100 passengers 1,430 miles.

The British in 1964 introduced a four-engine jet, the BAC vc-10, carrying 135 passengers. It cruised at 560 mph and had a range of 5,000 miles.

The Soviets began producing civil jets in 1956 with the medium range Tupolev Tu-104. A twin engine aircraft, the Tu-104 carried from 50 to 100 passengers, cruising at 500 mph and covering 1,875 miles. Its successor, the Tu-204 carried 200 passengers at a cruising speed of 530 mph. The three engine Yakovlev YAK-42 started in service in 1975. It carried between 100 and 120 passengers and flew at 500 mph covering up to 1,250 miles.

Figure 10-7 Meteor Jet, 1944
National Space Air and Space Museum, Smithsonian Institution Negative 90-G686

THE TURBOPROP

The turboprop is a modification of the turbojet. The turboprop uses additional turbines to make the jet exhaust crank a propeller that provides thrust. The rapid spinning of the turbines and their shaft must be geared down to run the propeller at a usable speed. It was employed for short runs where it was the most economical. The turboprop was more fuel efficient at low altitudes and moderate speeds. It could take off and land on short runways like a conventional piston-engine plane. Moreover, its vastly greater power enabled it to carry larger loads at higher speeds than comparable conventionally powered aircraft.

The first turboprop, the British Vickers Viscount, went into commercial service in 1953 and became one of the most widely used of all short-haul planes. Soon other turboprops were manufactured. One of them was the Lockheed Electra, the only United States turboprop in regular commercial service but soon lost out to faster turbojets.

Figure 10-8 D.H. 106 Comet, The First Commercial Jet, 1952
National Air and Space Museum, Smithsonian Institution Negative 95-1794

Another was the Fokker F-27, which became Europe's largest selling airliner. A few turboprops were still flying in the last decade of the 20th century wherever distances are relatively short, traffic light and runways limited. However, they were mostly superseded by turbojets and later by turbofans.

THE TURBOFAN

The turbofan is another modification of the turbojet. The fan consists of a large number of rotating blades in the front of the plane which act as a huge propeller supplying an enormous amount of thrust. The blades of the turbofan rotating at very high speeds compress air bypassing the core. This cool air channeled by the fan's shrouding forms a surrounding buffer for the rapidly moving, noisy exhaust and reduces engine noise.

One of the first turbofan jets was Boeing 727 which rolled out of the factory on November 27, 1962. Forty percent of its thrust came from a large fan in front*. This fan made it much more economical to run than any other engine of the day and much quieter as well. It had three engines, each developing 15,000 pounds of thrust.

Another feature of this aircraft was the change in the structure of flaps which are used to slow the plane in landing. In the space of just a few seconds the wing's normal blade-like shape, ideal for high speed flight could be modified into a kind of parasol which permitted the plane to float to the ground. This allowed the 727 to land on short runways.

The 727 is considered to be one of the most successful commercial transports in history. It could seat up to 131 and had a range of 2,500 miles. It was quieter and more efficient than most of its competitors.

THE JUMBO JET

Passenger traffic increased dramatically in the early 1960s. This led to the concept of the jumbo jet carrying large numbers of passengers. The Boeing 747 made its first commercial flight in 1970 flying across the Atlantic. One version of this plane was capable of carrying 498 passengers at over 580 mph and a range of 7,000 miles. It had four turbofan engines, each with a thrust of 41,000 pounds. The 747 had an upper level lounge reached by a spiral staircase.

Because of the large number of passengers Boeing appointed a safety committee to consider every aspect of the aircraft. Three separate back-up hydraulic systems were installed to insure against catastrophic failure. Four main landing gears were installed so that the plane could land safely if two landing gears failed. Wings were built to support a load 50 percent greater than normal. Flame-resistant, nontoxic materials were used for the cabin wall lining.

---

* In more modern turbofan jets 80% of the thrust comes from the fan.

Other jumbo jets were the Lockheed three engine L-1011 Tristar with 345 passengers and the McDonnell Douglas DC-10 (1971) a three-engine turbofan jet carrying up to 380 passengers. A later entry (1983) was the Boeing 757 with two turbofan engines carrying up to 218 passengers.

In Europe the twin turbofan engine Airbus A-300 was first put in service by Air France in May 1974. It was constructed by Airbus Industrie, a European consortium. It carried 331 passengers and it was more economical to operate on short haul routes.

10.9 Supersonic Flight

Supersonic flight is the passage through air at greater than the local velocity of sound. Velocity of supersonic flight is referenced to the velocity of sound in terms of Mach 1, 2, etc. where Mach 1 is the velocity of sound. The first aircraft to fly at supersonic speeds was a Bell X-1 rocket-powered research plane, on October 14, 1947. This was followed by many military aircraft capable of supersonic flight which were built. Their speed was generally limited to Mach 2.5 because of frictional heating of the skin of the plane.

In September 1962 Dr. William Strange of the British Aircraft Corporation and Lucien Servanty of the French Sud Aviation started the design of a supersonic airliner. The cost seemed prohibitive for any private company. On November 29, 1962 the Anglo-French Supersonic Aircraft agreement was signed, committing both governments for the entire cost of the development. Work began in May 1963 on the prototypes, one in France and one in England. The aircraft was named the Concorde in honor of the agreement.

The planes were designed to carry 118 passengers at a speed of Mach 2.2 or 1,450 miles per hour at a cruising altitude of 50,000 feet. The wings were Delta shaped to obtain efficiency for supersonic flight and would also permit the plane to land at 177 mph. However, the wings would lower the plane gently to earth only if the pilot pointed the Concorde steeply into the air. This meant that the pilot would not be able to see the runway. This was solved by giving the plane a nose that at the touch of a button could be tilted down, thus giving the pilot a view of the runway as the plane landed at 177 mph. The nose is also tilted down during the takeoff and then straightened out for the flight.

It was decided that the Concorde should use four turbojet engines instead of turbofans since the large air intakes of the latter would cause excessive drag at supersonic speeds. This meant that the Concorde with four giant turbojet engines would be noisy. On top of this, there was the problem of the shock wave or sonic boom from any plane flying faster than sound. The shock wave could shatter windows, crack walls and shake plaster from ceilings. All of this made public acceptance of many regular supersonic airlines taking off and landing near a city doubtful.

However, work went ahead on the two prototypes and the French plane was completed in August 1968. Tests showed problems with the brakes and landing gear. Solving these took time and the Soviets on December 31, 1968 rushed their supersonic airliner, the Tupolev 144, into the air for a test flight. It looked so much like the Concorde that the British suspecting industrial espionage called it the "Concordski." The French had the first flight test of their Concorde on March 2, 1969 and the British flew theirs on April 9, 1969. Both Concorde flights were at subsonic speeds.

At the Paris Airshow in 1973 the Soviet Tu-144 supersonic plane crashed killing the jets crew of six, and seven villagers, mostly children. In November 1977 the Tu-144 went into weekly service between Moscow and Alma Ata in Soviet Central Asia. Ten months later the aircraft was grounded for good without an explanation.

The Anglo-French Concorde was put into commercial service on January 21, 1976. There was bitter opposition to the plane in the United States because of the sonic boom. After the United States government approved a 16-month trial to Washington, DC and New York, the Environmental Defense Fund sued the government to try to halt the flights.

The British and French governments lost money on the Concorde and they could not sell any planes to the airlines. The American government declined to finance a supersonic airliner and the Concorde had the field to itself. The price of the ticket between New York and Europe was so high that passengers were limited to the wealthy or business executives.

10.10 Death from the Skies

There are many causes of fatal aircraft accidents and there are a number of technologies to alleviate the problems. A number of the problems and partial solutions will be examined. In some cases a combination of causes will result in accidents but here in each case the dominant problem will be examined as determined by a pertinent investigation board.

In several countries there are investigative bodies which try to determine the cause of an aircraft accident and make recommendations to avoid it in the future. In the United States there is the National Transportation Safety Board (NTSB). In Canada there is the Canadian Aviation Safety Board (CASB), and in the United Kingdom there is the Air Accident Investigation Branch (AAIB). The Japanese government has the AAIC or Air Accident Investigation Commission, Ministry of Transport. There is also an International Civil Aviation Organization, the ICAO with headquarters in Montreal, Canada.

There are three general tools in investigating aircraft accidents. The CVR is a Cockpit Voice Recorder and the DFDR is a Digital Flight Data Recorder which can record as many as 100 special items. After every accident an effort is made to recover and examine the two recorders in a laboratory. A third tool is the collection of the parts of the plane and its reconstruction to pinpoint the exact cause. These three tools together with eyewitness accounts of survivors in the air and from people on the ground can often give the cause or causes of the accidents. Some examples of specific cases and prevention methods and technologies will now be examined.

STRUCTURE FAILURE

On January 10, 1954 a BOAC Comet, Yoke Peter, departed from Rome bound for London. The plane radioed that it broke through the overcast at 26,000 feet headed for the cruising altitude of 36,000 feet. A second radio transmission was suddenly stopped and the plane plunged into the sea. Six crewmembers and 29 passengers perished. All Comets were grounded.

An investigation recommended some 50 modifications to reduce the possibility of an explosion. These included armor-plated shields between the engines and fuel tanks, reinforcement of fuel lines and safeguards against the accumulation of explosive hydrogen gas from the batteries. The recommendations were accepted by BOAC on February 4, 1954, and 47 days later the Comets began flying again. However, on April 8, 1954 only two weeks and two days after the resumption flights a Comet, Yoke Yoke on a flight from Rome to Cairo disappeared at night. The next day five bodies and some equipment were found in the sea off the Island of Stromboli in the Mediterranean. The Prime Minister, Winston Churchill, ordered an investigation by the Royal Aircraft Establishment headed by the Director, Sir Arnold Hall.

Hall suspected metal fatigue which could cause a crack in the fuselage. This in turn would cause the pressurized cabin to explode. To test this theory he placed a grounded plane in a water tank. Water pressure in the cabin was raised to 8-1/4 pounds per square inch to simulate flying above 35,000 feet. The pressure was held there for three minutes and then lowered while the wings were moved up and down by hydraulic jacks. Sir Arnold Hall calculated that each pressurization cycle in the tank was the equivalent of a three-hour flight. When carried on 24 hours non-stop the testing would age the Comet 40 times faster than actual service.

After 9,000 simulated flying hours, the fuselage split. This would have caused the cabin in an actual flight to explode with bomb like force. The split began with a fracture in a corner of a window and extended for eight feet. Royal Air Force Establishment metallurgists found the telltale discoloration and crystallization of metal. This method of testing was to be used by other airlines to test their new jetliners, which replaced the ill-fated Comet.

DC-10 EQUIPMENT FAILURE

An equipment failure occurred with the DC-10 on June 12, 1972. An American Airline DC-10 had just taken off from Detroit when a cargo door blew off as the jetliner was being pressurized. The sudden decompression buckled the cabin floor, rupturing some of the hydraulic control lines that ran underneath. Luckily the plane had the 15 rows of seats empty in the rear of the plane. If the seats had been filled the additional weight might have severed all the control lines. The pilot managed to land the plane in one piece. An investigation showed that the culprit was the latch on the cargo door. The design of the latch was changed and new latches were installed in all DC-10s. However, two DC-10s came off the assembly line without the latch modifications. One of them was found and modified before an accident could occur. However, a special model with seats for more than 300 passengers was delivered to Turkish Airlines without the modifications.

On March 3, 1973 a Turkish plane with a full load of passengers had the latch give way and the cabin floor collapsed completely. The plane plunged to earth killing all 346 people aboard.

Crew Failure

In Nairobi on November 20, 1974 at 4:42 in the morning, the crew of Lufthansa 747 started the engines. As the aircraft was lifting off the runway, the plane experienced severe buffeting. At about a hundred feet up the rate of climb fell rapidly to zero. A passenger who was a former airline pilot looked out from this window seat and saw that the wing leading edge flaps were not extended. The aircraft crashed and exploded in a fierce fire. Fifty-nine people were killed and the rest survived.

An investigation revealed that the leading edge flaps were in a retracted position which caused the plane to crash at takeoff. Lufthansa dismissed the pilot and the engineer. The latter was prosecuted for homicide for gross negligence but a court acquitted him.

It turned out later that this type of event had happened before with 747s but without any accident happening. In March 1975 the FAA issued an AD (Airworthiness Directive) pointing out the problem and adding an additional warning to alert the crew by means of an air horn if the leading edge flaps were in a retracted position at take off.

Personnel

People failure includes maintenance personnel as well as crewmembers. On May 25, 1979 an American airline left Chicago bound for Los Angeles. As it took off the left wing tore off. The plane climbed to about 300 feet and then rolled to the left until the wings were almost vertical. The plane crashed and there were no survivors. An investigation showed that the maintenance crew had cracked the engine mount while removing the turbofan for maintenance.

Icing on the Wings and Crew Failure

On January 13, 1982 a 737 from Miami landed at Washington National Airport. It was scheduled to take off for Florida as Air Florida Flight 90 with seventy-one passengers and three infants. It was snowing heavily and ice built up all over the aircraft and it had to be de-iced.

The de-icing was performed by American Airlines providing the service under contract. The American maintenance manual had specific instructions for de-icing 707s, 727s and DC-10s but nothing for 737s. The Boeing manual for 737s cautioned against de-icing solutions as their dilution with melted snow can result in the mixture refreezing and becoming more difficult to remove. This became a problem if the plane was delayed in taking off while it was still snowing. This was the case with Air Florida Flight 90 and fresh snow, which could be blown off during takeoff, melted into a slushy mixture which would freeze on the leading edges and the engine inlet nose cone during the wait for takeoff.

The pilot never switched on the de-icing equipment in the plane. The plane took off and at 16:01 it struck the 14th Street Bridge over the Potomac River. The plane killed the drivers of four cars and then plunged into the waters below. A handful of passengers were saved by a helicopter and one survivor was saved by a civilian who dived into the ice cold water.

Fire

On June 2, 1983 an Air Canada DC 9 was on route from Dallas, Texas, to Montreal, Canada, when three circuit breakers, associated with the aft lavatory flush motor, tripped in quick succession. This might have been due to motor malfunction. Nine minutes after this event, smoke was noticed in the cabin. Flight Attendants found smoke in the lavatory, which they saturated with CO_2 from an extinguisher while one of them walked to the cockpit to inform the Captain. Since the aircraft was less than half full, they moved all passengers to the front section.

The first officer went back to investigate but he did not have portable breathing apparatus. Firefighting by the crew was inhibited by the lack of a full-face smoke mask with self-contained breathing apparatus. Black acrid smoke began to fill the passenger cabin from ceiling down to knee level.

The pilot made a successful emergency landing in Cincinnati thirty minutes after the first signs of trouble. The pilots tried to enter the passenger cabin but were driven back by thick black smoke. Only the crewmembers and eighteen of the passengers managed to get out in sixty to ninety seconds, before flash fire engulfed the cabin and killed twenty-three passengers. They never made it though two doors and three overwing exits were fully open. The inhaled smoke made some of the passengers unable to negotiate the open overwing exits.

The Cincinnati fire led to the installation of smoke detectors in aircraft. It also led after a long fight to the mandatory requirement in the United States for compulsory oxygen masks for cabin crews after 1990. Over the years, preventive design and construction, better detection and extinguishers for engines as well as cargo holds helped to reduce the frequency of in flight fires.

The fight against post-crash fires was not as successful although there was one triumph in the use of less fire active fuel. JP was more inflammable than kerosene, but JP had a weight and price advantage. Lord Barbazon of the Air Safety Group in the United Kingdom proposed a fuel-duel. He stated that he would stand in a pool of kerosene and his opponents in a pool of JP. Both duelists would then light a match. No one took up the challenge. America followed Britain's lead and by 1990 most passenger lines switched to kerosene fuel.

Terrorism

On December 21, 1988 at 19:03 a bomb exploded in a Pan Am 747 enroute from London to New York. The plane came down in the village of Lockerbie, Scotland, killing everybody on board plus eleven people in their homes and cars, for a total loss of 270 people.

The disaster was traced to a suitcase housing a radio-cassette plastic bomb. After a long investigation two Libyan agents were named as the culprits but Libya would not surrender them for trial.

The Lockerbie explosion alerted security at airports all over the world. Detailed studies of various steps that could be implemented to avoid the repetition of the Lockerbie tragedy were made. At one extreme is the El Al system where all passengers and all baggage are systematically and thoroughly searched. This prevented bombs going off in the air. However, it resulted in long hours of check-in time, which most other airlines objected to. At the other extreme were airports where security was so lax that people could walk through a door bypassing the metal detector gate without being challenged. In between there were great variations in security depending on the quality of personnel and equipment.

The Semtex plastic bomb in the Lockerbie disaster could not be detected by the conventional X-ray equipment used in airports. By 1994 two sophisticated expensive systems that could detect plastic bombs were developed in the United States but used primarily in Europe and the Middle East. The first was a variation of the CAT scanner used in hospitals. The second was a portable chemical analyzer. The reason that these systems were not used in the United States was that the considerable expense would be borne by the airlines whereas in other countries the governments picked up the tab.

On July 17, 1996, TWA Flight 800, a Boeing 747, left Kennedy Airport bound for Paris. It blew up off the coast of Long Island, killing 230 people. This prompted an overhaul of airport security and a resolve by the U.S. government to use the latest equipment and replace the antiquated X-ray equipment at airports.

Semetex chemicals were found in the wreckage of TWA Flight 800 but it was later disclosed that these were the remains of a previous test with bomb sniffing dogs.

Wind Shear (Microbursts)

Wind shear is a very localized downdraft which hits the ground and fans out like an inverted mushroom creating horizontal winds in all directions. It is referred to as a microburst because of its very restricted area. It can slam an aircraft to the ground without touching two others nearby.

On approaching a microburst a pilot during takeoff might fly into a headwind of eighty knots that rapidly increases his indicated airspeed. The pilot pulls up to try to maintain climb out airspeed but barely a mile away, the aircraft is slammed down by the vertical wind shaft forcing it to descend rapidly. Then the wind changes abruptly to the opposite direction and the plane crashes to the ground as the pilot is helpless. During a landing approach the problem is similar.

Microbursts were named and studied by T.T. Fujita, Professor of Meteorology at Chicago University. He studied an accident in 1975, Eastern Airlines Flight 66, which crashed at JFK during a landing. He explained the phenomenon and then many other unexplained accidents were now clarified.

To avoid wind shear ground based Doppler radar is being developed to warn pilots. A direct data-link will alert all pilots directly without going first through an air controller who may be busy dealing with heavy traffic. In addition, air borne wind shear detectors are being developed.

Collisions

The skies in and around airports have become crowded with both large and small aircraft trying to takeoff and land. In this situation many collisions are apt to happen. Some eight hundred accidents of the type occur every year in American airspace alone.

One example of a fatal collision is that of a DC-9 Air Mexico Flight 498 with a Piper Cub light plane over Cerritos, California, on August 31, 1986. The DC-9 took off from Tijuana in Mexico at 11:20 AM Pacific Daylight time for Los Angeles with fifty-eight passengers and a crew of six. At 11:40 a Piper Cub was cleared for take-off at Torrance, California. The DC-9 began its descent towards Los Angeles at 11:44. The weather was clear. At 11:50 the arrival radar controller sent a message to the Captain. "Traffic ten o'clock, one mile, northbound, altitude unknown." The DC-9 acknowledged the message.

The unknown craft was a Grumman Tiger who had contacted Los Angeles for guidance, and that is how he was spotted on radar. The Grumman Tiger pilot had no idea that he had strayed into the Terminal Control Area without permission. The radar controller told off the Grumman Tiger pilot. Just as he finished his lecture the controller noticed that the DC-9, Flight 498, had disappeared from his radar screen. It was 11:52 and the Piper Cub and DC 9 wreckage was showering down on Cerritos. Everyone on board was killed plus fifteen on the ground. Five houses were destroyed and seven others damaged.

One technology which can help avoid tragedies like the Cerritos incident is the Traffic Alert and Collision Avoidance System or TACAS. The investigation of the Cerritos collision concluded that if the DC-9 had been equipped with TACAS then the probability of avoiding the Piper would have increased to 95 percent. The U.S. Federal Aviation Agency ruled that beyond December 1991, every domestic and foreign passenger aircraft must have TACAS mounted in the cockpit if it flies in United States airspace.

Aircraft Deaths in Perspective

The number of deaths from commercial aircraft is relatively very small compared with automobiles. On December 19, 1994, the FAA announced that there were 0.3 deaths per 100,000 departures for large planes and 0.4 for small planes. In 1996 there were 1,187 plane deaths in commercial flights worldwide, the highest in a number of years. This compares to about 40,000 U.S. fatalities from cars. However, an airplane crash is much more spectacular than most automobile accidents.

10.11 Noise from Above

Noise from aircraft especially jets has grown to the point where some citizens living near airports have a serious problem. When jets were introduced at an airfield, the people in the adjacent city painted a sign on their roof "JETS GO HOME." To counter this and many other protests, early jets were equipped with a cluster of tubes at the rear of each engine. These decreased the jet noise to some extent. Unfortunately they reduced thrust by 10 percent and added a large amount of fuel costs.

The reason for the huge roar of a turbojet is primarily the high speed of the exhaust from the combustion chambers. An improvement was the use of turbofan jet aircraft. The exhaust is diluted with slower air passed around the engine by the fan. This makes the turbofan aircraft considerably quieter than the turbojets. Modern turbofan engines have the capability in the future of reducing the noise further.

Supersonic aircraft have two noise problems. The first is that the less noisy turbofan has large air intakes that causes excessive drag. The turbojet which is very noisy has been the engine choice for supersonic aircraft.

An even more serious problem produced by supersonic planes is the sonic boom. The shock wave radiates outward from a supersonic aircraft. This shock wave or sonic boom can shatter windows, crack walls, and shake plaster from ceilings.

In 1964 the Federal Aviation Administration sponsored an experiment called Operation Bongo Mark II which was a study of the effect of supersonic flight on people below. Air Force B-58 supersonic bombers flew over Oklahoma for a five-month period. The flights produced 1,254 sonic booms which shattered windows and shook buildings. The U.S. Government had to pay $200,000 in damages to the people of Oklahoma City. Because of the problem of the sonic boom, supersonic commercial aircraft have been inhibited from flying across populated areas.

10.12 FLYING INTO THE 21ST CENTURY

A new method of designing and constructing commercial aircraft was developed in the 1990s for the Boeing 777. It is the first U.S. built commercial transport to rely solely on fly-by-wire controls rather than cables. In fly-by-wire the control surfaces are activated by electronic controls. The Boeing 777 is the biggest twin engine commercial airplane of the 20th century. It uses the largest most powerful jet engines ever developed in this century. The concept basically started on October 15, 1990 with a hand-written agreement between Boeing and its first 777 customer, United Airlines.

The aircraft was the first to be designed completely by computer. Some 2,200-work stations were networked to a cluster of 8 main frame computers. Using three-dimensional digital software, designers could manipulate parts as solid images on computer screens. The simulation of the assembly of parts on computer screens enabled the designers to make sure all the parts fit together correctly. Any design error between parts (called interference) could be detected and corrected without building mock-ups. It also eliminated hundreds of thousands of drawings. The result was Boeing's first "paperless plane."

Testing of the aircraft systems was accomplished in three big testing laboratories by 43 parallel processor computers supplemented by others. One of the three areas was the systems integration laboratory which tested avionics with real time simulations of the 777 in flight. The avionics includes the Airplane Information Management System (AIMS). This consists of dual cabinets that contain the electronics needed to handle flight management, flat panel cockpit displays, digital communication management, and engine data interface. The aircraft can still be flown if one of the two AIMS cabinets fail.

The second test area was the flight control test rig used to evaluate the fly-by-wire systems. The controlled aircraft surfaces are the aileron, a roll control area on the wings, the rudder on the rear vertical stabilizer, an elevator on the tail surface for pitch control, a set of spoiler* plates on top of the trailing edge of the wing used to slow the airplane down when the airplane is landing, and flaps on the trailing edge of the wing to provide more lift for the wing at low landing speeds.

*The spoiler plates lie flat on the wings and pop up when landing.

The fly-by-wire control system itself had effectively nine computers, anyone of which could fly the plane. This built in redundancy was also tested in the flight control test rig.

The third test area was the cockpit simulator. This enabled the crew to work out flight problems long before the aircraft was built. The cockpit simulator had a long distinguished history of many years in the aircraft industry but this facility used the latest computers available toward the end of the 20th century.

The Boeing 777-200 took off on its first flight on June 12, 1994, Figure 10-9. It weighed 506,000 pounds (229,250 kilograms), had a length of 209 feet, 11 inches (60.9 meters). Some 10,000** people worked on the aircraft in the Boeing plant in Seattle. Some of the component parts were built as far away as Australia and Japan. The 777 after very vigorous flight and ground testing was turned over to United Airlines. The first commercial flight took place on June 7, 1995. Boeing announced that the plane would be in use for 50 years, well into the 21st century.

The 777-200 initial model had 375 passengers with a range of 4,350 statute miles (7,000 kilometers). There are a number of variations of the Boeing 777 with different passenger loads and different ranges. For example, the Boeing 777-300 has a maximum of 550 passengers and a range of 6,200 statute miles.

10.13 The Beginning and the End of the 20th Century

The Wright Flyer I first flew on December 17, 1903 at the beginning of the century. The Boeing 777 first flew commercially on June 7, 1995 towards the end of the century. It is interesting to make a comparison of the two aircraft in Table 10-1 to show the explosion of aircraft technology between the beginning of the 20th century in 1903 and towards the closing in 1995.

** Source, Sabbogh in Bibliography

Figure 10-9 Boeing 777-200, 1995
Courtesy of the Boeing Company

	WRIGHT FLYER I	*BOEING 777-200*
Date	December 17, 1903	June 7, 1995, First Commercial Flight
Power	One 12 Horsepower Gasoline Engine	Two Turbofan Jet Engines, Each with 80,000 Pounds of Thrust
Crew	One	Seventeen
Passengers	None	375
Range	852 Feet	4,350 Statute Miles
Speed	30 Miles Per Hour	0.84 Mach
Number of People, Design and Construction	3	10,000
Length	21 Feet	209 Feet, One Inch
Width	40 Feet	199 Feet, One Inch
Weight	564 Pounds	506,000 Pounds
Flight Control	Manual	Fly-By-Wire Computer Control

Table 10-1 Aircraft at Beginning and End of 20th Century

BIBLIOGRAPHY

Barlay, Stephen — *The Final Call: Why Airline Disasters Continue to Happen,* New York, Pantheon Books, 1990

Boyne, Walter J. — *Clash of Wings: World War II in the Air,* New York, Simon & Schuster, 1994

Chant, Christopher — *The World's Greatest Aircraft,* Crescent Books, 1991

Gunston, Bill — *The Development of Piston Aero Engines,* Somerset, Patrick Stephens Limited, 1991

Hamlin, George W. — *Skyliners, Mainliners, Falcons, and Flagships,* Miami, World Transport Press, 1991

Krause, Shari Stamford — *Aircraft Safety, Accident Investigations, Analyses, and Applications,* McGraw Hill, 1996

Petzinger, Thomas, Jr. — *Hard Landing: The Epic Contest for Power and Profits That Plunged The Airlines Into Chaos,* Times Books, 1995

Sabbagh, Karl — *Twenty First Century Jet: The Making and Marketing of the Boeing 777,* Scribner, 1996

Scott, A. — *The Shoulders of Giants: The History of Flight Until 1919,* Addison Wesley, 1995

Taylor, Richard L. — *Understanding Flying,* New York, Delacorte Press, 1977

CHAPTER 11

MARRYING TECHNOLOGY TO MEDICINE

11.1 SEEING INSIDE THE HUMAN BODY

The 20th century saw the development of different methods of looking into the body. The first of these developments was the X-ray. While the X-ray was discovered at the end of 1895 the device itself was developed in the early 20th century for medical use.

X-RAYS

In 1895 Wilhelm Konrad Roentgen (1845-1923) accidentally discovered the X-ray while studying the action of a Crookes tube. This was a glass enclosure which had been partially evacuated down to a pressure of about 0.01 millimeters of mercury. Enclosed in the Crookes tube are two electrodes, one at each end. The negative electrode is called a cathode and the positive one is designated the anode. A high electric voltage is applied across the electrodes. Crookes tubes come in various versions with different modifications.

One type is shown in Figure 11-1. The cathode was a cup shaped piece of metal. A piece of thin metal foil was placed at the center of curvature of the cup. When a high electric potential is applied across the electrodes, the metal foil will soon become white hot. Rapidly moving charges later called electrons caused the metal to glow.

On November 8, 1895 Roentgen discovered that a Crookes tube emitted a kind of ray which passed through the walls of the tube and produced a peculiar greenish, glowing light on a paper screen containing barium platino-cyanide. This kind of glowing light was called fluorescence. Roentgen placed a black card between the Crookes tube and the screen which continued to glow.

Roentgen spent the next six weeks repeating and extending his observation on the properties of the new rays which he called X-rays. He soon learned that they affected photographic plates and that they penetrated some materials. He produced X-rays of balance-weights in a closed box and other objects. He noticed the outline of the bones of his fingers when he held the object being X-rayed. On December 22, 1895 he brought his wife into the laboratory and took an X-ray photograph of her hand.

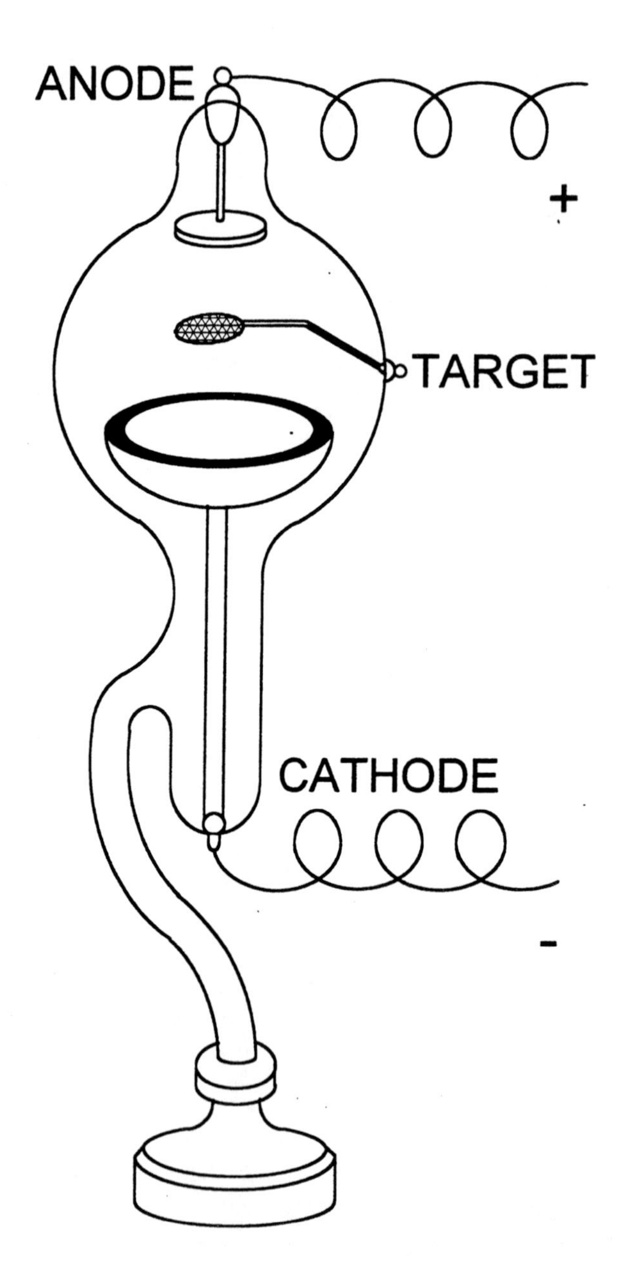

Figure 11-1 Crookes Tube

Other scientists had come very close to discovering the X-rays long before Roentgen's demonstrations. Crookes, the inventor of the Crookes tube, had in 1879 complained of fogged photographic plates that happened to be stored near his tubes. Two scientists in Philadelphia in 1890 noticed a peculiar blackening of photographic plates after having demonstrated a Crookes tube but they failed to follow up on this.

Roentgen performed a number of experiments on the nature of the new ray. He demonstrated X-rays were not reflected like light. Later other scientists showed that X-rays were electromagnetic waves whose wavelength was just below ultraviolet rays as shown in Figure 11-2.

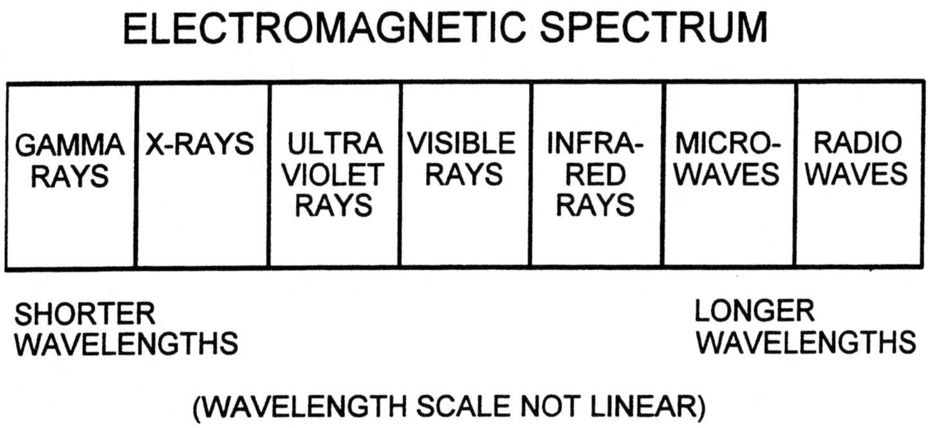

ELECTROMAGNETIC SPECTRUM

GAMMA RAYS	X-RAYS	ULTRA VIOLET RAYS	VISIBLE RAYS	INFRA-RED RAYS	MICRO-WAVES	RADIO WAVES

SHORTER WAVELENGTHS

LONGER WAVELENGTHS

(WAVELENGTH SCALE NOT LINEAR)

Figure 11-2 Relative Wavelength Position of X-Rays and Gamma Rays

In the early 20th century the modern X-ray machine evolved from Roentgen's equipment. Figure 11-3 shows the basic X-ray machine used in medicine. X-rays are generated by the bombardment of a dense target such as tungsten by electrons. The electrons are extracted by heating the cathode by means of a filament. The X-ray tube is exhausted to the highest possible vacuum and very high voltages are applied to the cathode and anode.

In the early 20th century X-rays became an indispensable part of medicine and dentistry. X-rays enabled doctors and dentists for the first time to look inside the body without surgery. They were also used inappropriately in view of the dangers of excessive X-rays which were discovered later.

Fluoroscopy

Roentgen's fluoroscopic unit, developed in 1896, provided the basis for the

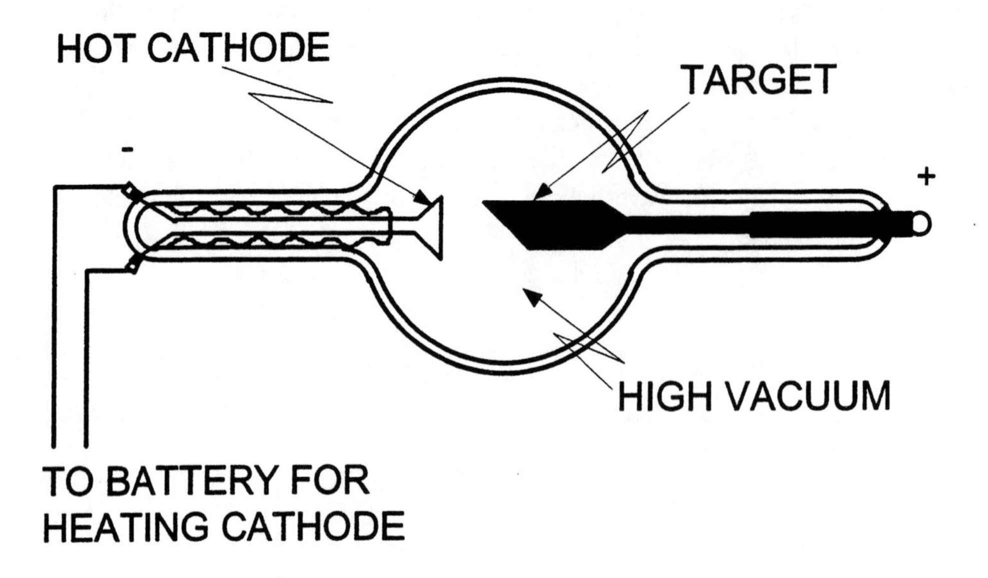

Figure 11-3 Basic X-Ray Machine

fluoroscope developed in the later part of the 20th century. This included an X-ray tube and a fluoroscopic screen incorporated into an image intensifier. This is an electronic instrument that brightens and intensifies the fluoroscopic image many thousands of times. This has resulted in a dramatic improvement in image quality while markedly reducing X-ray exposure. Fluoroscopy is a dynamic X-ray imaging technique that produces a moving image over time. It is very useful for evaluating organ movement such as the beating of the heart or movement of the diaphragm.

The gastrointestinal (G.I.) series and the barium enema are the most common fluoroscopic studies. These begin with the administration of a barium mixture either by ingestion or by an enema that fills the stomach or large intestine. The barium mixture, like dense tissues, blocks the X-ray beam. Fluoroscopy then reveals the location of the barium coated lining of the stomach and intestine and enables the radiologist to observe as they contract and distend. Abnormalities such as ulcerations and tumor of the intestine can be seen by fluoroscopy. It is also used in angiography.

Angiography

Angiography is the X-ray study of blood vessels. Arteries and veins are not normally rendered visible by using X-rays. An iodinated compound which is opaque to the X-ray is injected to make the examination of blood vessels possible. This technique is used to show the extent that arteries have become clogged and narrowed by arteriosclerosis, which can lead to strokes and heart attacks.

A radiologist performs angiography by inserting a long narrow tube, called a catheter, through a needle puncture in an artery, usually at a point in the groin. The catheter, guided by fluoroscopy is threaded through the branching arteries of the body to the location to be examined. The appearance of the arteries is then recorded on a series of rapidly exposed X-ray film images, two to six pictures per second. The result is an X-ray moving picture called a cine-radiograph. Cine-radiograph is often used to study the coronary arteries, which supply the heart muscle.

Digital subtraction angiography is a computerized technique for recording X-ray images. Two images are made of the same artery, one before and one after injection of contrast medium. This produces a third image, which shows an unobstructed image of the blood vessels. This procedure is safer because the amount of contrast solution required is less than that needed for conventional angiography.

Computerized Axial Tomography (CAT)

CAT is a scanning technique combining computers and an X-ray imaging system which uses an array of detectors to collect information from a narrow beam through a human body. The information collected is then used by a computer to reconstruct the image. CAT was the work of two men, Allan Macleod Cormack, a South African born U.S. physicist, and Sir Godfrey Newbold Hounsfield, an English electrical engineer. They shared the Nobel Prize for Physiology or Medicine in 1979.

Hounsfield led the design team that built the first all transistor computer in Great Britain. Later, while investigating the problem of pattern recognition, he developed the basic idea of CAT. In 1972 the first clinical test of CAT scanning was performed successfully in Great Britain. It rapidly became a diagnostic tool in the hospitals of that country.

Allan Macleod Cormack held a part-time position as a physicist for an American hospital radiology department. This experience aroused his interest in the problem of X-ray imaging of soft tissues. By the early 1970's he had established the mathematical and physical foundation of computerized scanning. CAT rapidly became a standard piece of equipment in many American hospitals. Initially a technician was assigned to each CAT by the manufacturers to keep the complex equipment operating.

Computerized axial tomography is an imaging technique which uses an array of detectors to collect information from an X-ray beam that has passed through the human body. To produce the scans, the patient is positioned on a narrow table called a gantry. The gantry slides into the scanner which is a tube-like structure. An X-ray tube built into the scanner emits a very narrow X-ray beam as it rotates around the body. The radiation that passes through the body is recorded by detectors that lie opposite the X-ray tube. Each detector emits tiny flashes of light in proportion to the amount of radiation it receives. Because X-rays are differentially absorbed by body tissues, the brightness of the light is also a measure of tissue density. A computer measures the flashes of light, magnifies them a thousandfold, and then stores their location and brightness in its memory. From this information, the computer constructs a two-dimensional anatomic image that represents a cross-sectional slice through the body.

CAT can produce images of internal slices of organs, such as the liver and kidney, that are more precise than film X-rays. It is also useful in visualizing the brain and abdomen to detect trauma damage or the size and location of some tumors. Three-dimensional images can be generated by using special computer software. They are especially useful in planning reconstructive orthopedic or plastic surgery.

Early CAT scanners used linear motion of both a pencil shaped X-ray and a single detector. Later models used a wide fan-shaped X-ray beam with an array of as many as 300 detectors. These machines can complete a CAT scan in 7 to 10 seconds. In order to obtain enough information to calculate one image, the scanners can take as many as 90,000 readings. The computer must solve many thousands of equations to generate images.

The same technology developed for X-ray tomography was later used for different types of medical imaging. These include SPECT, PET, MRI, and ultrasonic imaging which will now be described.

SINGLE PHOTON EMISSION COMPUTERIZED TOMOGRAPHY (SPECT)

SPECT is another tomography technology used to obtain two-dimensional images that are thin slices of internal organs such as the heart, brain, and liver. A computer can put together three-dimensional views of organs and also produce cine displays used to evaluate heart muscle contraction. SPECT uses radioisotopes.

This technology began after the discovery by Enrico Fermi in 1935 that chemical elements can be made radioactive by bombarding them with neutrons. Neutrons consist of an electron and a proton combined. It has zero electric charge and the mass of a proton. Fermi found that atoms of elements so bombarded capture those neutrons assuming an additional nuclear mass while remaining the same elements. These radioisotopes are unstable and dissipate excess energy by spontaneously emitting radiation in the form of gamma rays. Gamma rays are an extremely energetic form of electromagnetic radiation just below X-rays in wavelength as shown in Figure 11-2.

Different radioisotopes tend to concentrate in particular organs. Iodine-131, for example, settles in the thyroid gland and is useful for thyroid studies. Carbon-14 is useful in studying abnormalities that underlie diabetes, gout, and anemia. For most studies the radioisotopes are injected into the body and gamma rays are emitted from the organ or tissue under examination. The radioisotopes are also called radionuclides.

A gamma camera suspended above the patient converts the gamma rays into an image that shows the distribution of the radioisotopes. The image is recorded on film and is called a scan. Scans of the heart and bone are the most common nuclear medical examinations. A bone scan may detect the presence of cancer or other problems. A cardiac scan is used to evaluate heart functions. In addition, scans are taken of other organs such as the lungs, kidneys, brain, liver, and thyroid gland. Computerized tomography similar to that used in CAT scans are employed with radioisotopes to obtain images.

POSITRON EMISSION TOMOGRAPHY (PET)

This is a radiological technique that is used to study the metabolic activity inside an organ. It was developed by Professor Michel Ter-Pogossian of Washington University in the 1970's. The patient is injected with a short-lived radiotracer mixed with a compound that the organ uses such as glucose and then placed inside a doughnut-shaped array of sensors. Before the injected substance loses its radioactivity, the scan reveals how the organ is using the compound.

A CAT scan or MRI can reveal anatomical changes but PET can detect problems before such changes result. In the case of a tumor, sometimes even before the mass is visible on another scan, PET can show whether it's an active malignancy. This is a great advantage but the disadvantages are the need for a nearby cyclotron to make the short-lived radiotracers and the need for a nearby chemistry laboratory.

The Yale/VA PET Center built in 1991 at West Haven cost $5.8 million, and $800,000 a year to run. The Center charges about $1,900 per scan. PET has been used in the study of brain-related disorders such as Alzheimer's disease and epilepsy. It accurately tracks brain function.

IONIZING RADIATION IN RADIOLOGIC IMAGING

There is a problem in the imaging technology just described from X-rays through PET. Before discussing the bad effects of ionizing radiation on the human body, terms should be defined. An ion is an atom or group of atoms that have acquired an electric charge by gaining or losing electrons. Ionization radiation is the formation of ions by the addition or removal of electrons by radiation from X-rays, gamma rays and particles such as alpha and beta rays. Alpha particles are the positively charged helium nuclei and beta particles are negatively charged electrons emitted by the decay of radioactive atoms.

Ionizing radiation can cause mutations in the sex cells of the gonads. This can produce mutations in offspring with dire consequences. Lead shields around the gonad region can prevent this but many medical practitioners neglect to do this.

In the late 1960's a professor of radiology gave a lecture at a well-established medical school in the United States. He emphasized the use of lead shields around the gonads when using X-rays. The students then proceeded to the X-ray section of the hospital wing of the medical school. One medical student, observing that no shields were being used, went back to the professor and asked, "Why?" The teacher was surprised and said that in 16 years of lecturing, no student had ever brought this up before.

Another problem with ionizing radiation is that it can affect cells in such a way as to cause cancers. The effect of ionizing radiation is cumulative over the lifetime of a patient. In the past, X-rays were given for trivial reasons. In the 1940's children were routinely fluoroscoped for shoe fittings. Sometimes the use was not trivial but became part of an annual physical exam using chest X-rays as a routine procedure.

New York State in its document, "Ionizing Radiation," Part 16, stresses the ALARA Principle (as low as reasonably achieved). In the "Guide for Radiation Safety/Quality Assurance Program," the New York State Department of Health states:

> "The Regulations in Part 16 and this guide have been established on the ALARA Principle to assure that the benefits of the use of ionizing radiation exceed the risks to the individual and the public health and safety."

New imaging technology has been developed which does not have ionizing radiation. One of these is MRI which came into use gradually beginning in the early 1980's.

Magnetic Resonance Imaging (MRI)

The foundation of the MRI technology can be traced to the discovery of a phenomenon called nuclear magnetic resonance. This was discovered simultaneously and independently by two physicists in 1946. They were Edward M. Purcell of Harvard and Felix Bloch of Stanford University. The two were awarded the Nobel Prize in Physics in 1952.

Purcell and Bloch started with the knowledge from Quantum Mechanics that the atomic nuclear particles, such as protons, spin like little tops and behave as miniature magnets. These spinning protons are normally randomly oriented. When an external magnetic field is applied, the magnetic field causes them to wobble like a gyrating top and also to line up in two energy bands. They divide up into a lower, more populous energy group whose magnetic field lines up with the direction of the magnetic field and a higher energy group in which the nuclear magnets oppose the external magnetic field.

Purcell and Bloch discovered that if a radio wave is sent across the external magnetic field, the wobbling protons in the lower group absorb the radio energy and jump to the upper energy band. However, for this to happen, the radio frequency must match the frequency of the proton's natural spin. This is similar to a resonant circuit in a radio receiver which is tuned to the frequency of a radio transmitter. This gives the name, nuclear magnetic resonance to NMR.

Once the radio pulse is switched off, the protons in the higher energy level will fall back to the lower energy level emitting the energy they absorbed in the form of a radio signal. The magnitude of the energy is directly proportional to the density of the protons in a region. The resonant frequency of a particular nucleus in one environment differs from the frequency of the same nucleus in a different environment. The nuclear signal can be used to report what the atom's neighbors are. An antenna wrapped around the sample is connected to a radio receiver whose faint output can be monitored on an oscilloscope. The sample was usually a test tube of chemicals in solution.

This technique was used by organic chemists as a spectrometer for the analysis and identification of chemical compounds. The NMR furnished information on the structure of molecules, chemical bonding, and internal motions in solids and liquids. Felix Bloch poked his finger into his spectrometer and obtained an NMR signal. Later in the 1960's the NMR machines were built commercially by a small manufacturer in New Kensington, Pennsylvania, called NMR Specialties. The device was a cylinder about 18 inches in diameter and four and a half feet tall. The center of the cylinder was a hollow vertical bore into which the operators of the spectrometer placed their samples.

In 1964 one of the visitors to NMR Specialties was Dr. Raymond Damadian, a medical doctor in the Biophysical Laboratory of the Department of Medicine of the Downstate Medical Center in Brooklyn, New York. He conceived of the idea of a giant NMR device which could detect cancer in a human being. He knew that the proton in the nucleus of the hydrogen atom of water had a strong NMR signal. He reasoned that cancer cells had a different water structure than normal cells and therefore the protons of water ought to give a different NMR signal for cancer cells.

In June 1970 Damadian went to NMR Specialties in New Kensington with some rats with cancerous tumors. He compared the tumor tissue with normal rat tissue and found that he could distinguish the two kinds of tissue by the NMR signals. He repeated his experiments in July 1970 with the same result. He soon realized that NMR could be used to detect many diseases, not only cancer.

On March 19, 1971 Dr. Damadian published an article entitled, "Tumor Detection by Nuclear Magnetic Resonance" in the magazine *Science*. In it he pointed out that many tumors can't be detected by X-rays which go right through them. He wrote about the possible use of NMR to detect internal cancers. The idea of putting people in a large NMR machine provoked disbelief because up to that time NMR was used by chemists on test tube samples.

Dr. Damadian envisioned a powerful movable magnetic field and a large coil to emit radio waves. The coil would be wrapped around the patient's chest while the magnet passed back and forth across the body. A detector would pick up NMR emissions for analysis.

Damadian thought up a method of focusing the magnetic field on a particular part of the body. He called this method FONAR, an acronym for "Field Focused Nuclear Magnetic Resonance." It worked by shaping the magnetic field in the form of a saddle. NMR signals would be used only from the nadir or saddle point of the magnetic fields. He envisioned that a new NMR machine would replace the X-ray.

The invention of NMR imaging in medicine is like most inventions--the subject of much controversy. The other main contender for the invention was Paul Lauterbur, a chemist. While in the Army in 1953 he was assigned to an Army NMR laboratory to study the complex chemistry of nerve gas. After he was discharged from the Army, he finished his doctoral studies, and in the 1960's he went to the State University of New York at Stony Brook to do NMR studies in biological areas.

In 1971 the financially ailing NMR Specialties in New Kensington, Pennsylvania, asked Lauterbur to take over the company. At New Kensington, Lauterbur thought about using the NMR on a human body to detect cancer. He decided on a novel method of focusing the magnetic field. His idea was to superimpose on the uniform magnetic a smaller adjustable field that varied across the patient. This adjustable field would give each part of the patient a different magnetic history. The corresponding slight differences in the NMR resonance frequencies could be used by a computer to construct an image.

He set down this idea in a notebook suggesting that it could allow NMR images to be done on the body. He had a witness sign the notebook as a safeguard for his ideas. He resigned from NMR Specialties and returned to Stony Brook to work out the details of his idea. His system differed from the FONAR method of Damadian in that Lauterbur's system collected data a plane at a time instead of a point at a time. Lauterbur's method thus was faster and clearer for imaging. On the other hand, Damadian's FONAR was superior for gathering chemical information to help diagnose suspicious tissue seen on an image. Lauterbur's technique was published in the March 16, 1973 issue of *Nature* magazine. Included in the article was the first NMR image ever made. It was two tiny tubes of water immersed in a larger tube. In an addition to the original submission he wrote that this technique could be used in the study of malignant tumors in the body.

Lauterbur did not reference Damadian's paper. This made Damadian furious because he thought Lauterbur was trying to steal his idea. This ignited a conflict between the two men. To this day there is a controversy about who invented what is known now as MRI, Magnetic Resonance Imaging.

Raymond Damadian was awarded U.S. Patent No. 3,789,832, "Apparatus and Method for Detecting Cancer in Tissue" on February 5, 1974. He together with his associates, Michael Goldsmith and Larry Minkoff, worked very hard to develop the use of the NMR technology to scan the human body.

Paul Lauterbur was also working to be the first to scan the human body with an NMR machine. By 1976 he had images of a rat showing a tumor growing in it. He ordered a magnet large enough to put a human in it. The magnet was supposed to have a circular bore of 24 inches in diameter, inside of which would be an antenna and then inside that was supposed to go a human being. However, when the magnet arrived, it had a 16-inch bore, too small for a human being. Lauterbur had lost the race.

Damadian and his team worked long hours, sometimes all night, to be the first to make a scan with an NMR machine. After some failed heroic efforts to get money, including a stay in Plains, Georgia, to ask President Carter for funds, Damadian finally got $40,000 from four rich men in March 1977. It was enough to finish his machine.

Minkoff built a skinny wooden rail about 6 inches wide and about 20 feet long. It was designed so that it could move backward and forward as well as sideways. The patient sat on this board as it was moved allowing different parts of the body to be scanned. Someone would have to shove the patient around.

A coil was wrapped around the subject to allow the radio waves to enter the body. To make the antenna, Goldsmith used tough garbage-bin cardboard wound with copper foil tape which worked after much trial and error. It was the last part of the machine to be built. The scanner was completed in June 1977 and Damadian christened it Indomitable.

The Indomitable shown in Figure 11-4 had a storage chamber at the very top for liquid helium used to cool the superconductive magnet at the periphery of the machine. In the center was Minkoff's narrow wooden plank where the patient sat. Goldsmith's cardboard coil antenna was attached to it.

As one test before he put a human being in the Indomitable, Damadian wanted to try a dead fresh turkey. The only place to get a dead fresh turkey in Brooklyn was from a kosher butcher shop. A phone call was made to a nearby shop and Damadian was informed that the butcher was closing the store. Damadian said that he needed it for a medical experiment and after some discussion the butcher agreed to stay open for a few minutes longer. Damadian ordered a graduate student to rush down to the butcher and bring back a turkey. This was placed in the Indomitable and the interior of the turkey was outlined in detail.

The Indomitable was ready for a human trial. It had never been done before and several doctors came over with a defibrillator and an EKG as a precaution. Damadian was to be the first human to try the machine on May 11, 1977. The cardboard coil antenna was slipped over Damadian bare chest but it was a very tight fit. He sat down on the hard wooden rail and the machine was turned on. There was no NMR signal at all. However, his vital signs were good and the electrocardiogram showed no significant change.

After looking at all the possibilities, the team concluded that Damadian was just too fat for the very tight antenna. Minkoff was thin and finally on July 3, 1977 he got into the machine almost at midnight. Graduate students moved him inch by inch inside the machine. After two hours both Minkoff and the students were very tired and they all had to take a break. Minkoff stayed in the machine for four hours and forty-five minutes while his entire chest was scanned.

As data was received Goldsmith sketched out an image by hand with colored pencils on a sheet of graph paper. The picture Goldsmith drew showed a cross section through the chest, revealing the body wall, the right and left lung, the heart and a cut through the descending aorta. Later a computer was fed the data to reconstruct a better image.

After the first success many improvements were made. Then Damadian sister's father-in-law, Sou Chan, was scanned on February 5, 1978. It was known that he had a cancer and it showed up. This was the first MRI image of cancer in a human being.

Figure 11-4 First Human MRI Machine, 1977
Smithsonian Institution NMAH/Medical Science, Negative 86-10890-23

In March 1978, Damadian decided to use the name FONAR for a commercial company which would manufacture the machine. He and his team moved to Plainview, Long Island, to start the manufacturing of the new FONAR scanner. Damadian and his team designed a whole new machine which was much more practical for medical work than the Indomitable. By March 1980 the new machine was ready; it was called the QED 80. It used a permanent magnet instead of the superconductive magnet.

In the QED 80 the patient lay on what looked like a stretcher and entered the tunnel of the machine by computer control. Radio energy was fired at the patient through an antenna embedded in a plastic hood. The patient heard a noise from the pulsing magnetic field. When the radio energy was turned off, the antenna picked up the radio signals emitted by the hydrogen protons in the tissues. These signals were sent to a computer which depicted the result in pictorial form.

Soon competitors appeared with new and better machines. FONAR answered with an improved Beta 300. Diseases, including cancers, that could not be detected by CAT scans were observed by the new MRI machines. The *Wall Street Journal* printed a story of a young man who had terrible seizures. A CAT scan revealed nothing but an NMR image showed a mass of fibrous tissue on the left lobe of his brain. Surgeons removed it and the seizures ceased.

Some of the new manufacturers called the process Magnetic Resonance Imaging to avoid the term Nuclear in Nuclear Magnetic Resonance (NMR). By the 1980's nuclear had bad connotations and most companies used MRI, FONAR and a number of other MRI companies received U.S.F.D.A. approval in 1985.

One of the problems with the first generation of MRI machines like those made by FONAR was claustrophobia which some patients suffered inside the tunnel. An open MRI was developed to solve this problem. Later Larry Minkoff founded his own company, MAGNA-LAB Inc., which manufactured another kind of open MRI. (See Figure 11-5). This is used for appendages and can show an image of a weight-bearing knee which was not possible with previous MRI machines.

MRI looks at the plentiful protons in water molecules in living tissues. There are a lot of water molecules in the body but there are places where there is no water. Examples are the empty spaces in the lungs and fatty tissues in breast cancer. Research is being performed on the use of xenon to enhance MRI imaging. Magnetized xenon was blown into a mouse's lungs and the MRI picture improved dramatically.

MEDICAL ULTRASOUND

Ultrasound is sound transmitted at frequencies above the highest frequency that humans can hear, 20,000 Hertz. In medical ultrasound the frequencies from 2 to 10 megahertz are used most often. Medical ultrasound depends on the piezoelectric effect first discovered in 1880. Piezo comes from a Greek word meaning to press. The piezoelectric principle is that certain materials, quartz, ceramics, and others will produce an electric voltage when deformed by an applied pressure. Conversely an electric voltage

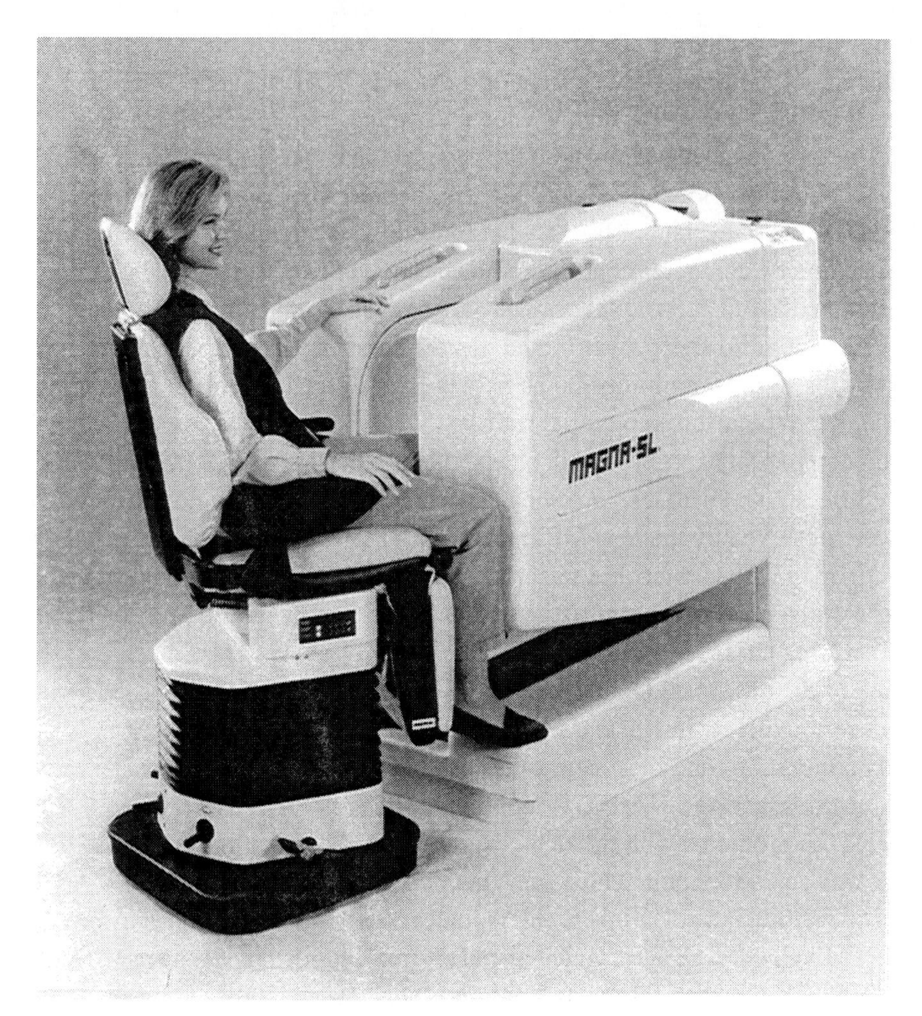

Figure 11-5 Appendage MRI
Courtesy of Magna-Lab Inc.

will produce a varying pressure or sound. In medical ultrasound imaging the piezoelectric material or transducer is a ceramic, lead zirconate titanate (PZT). Medical ultrasound for medical imaging came into use in the late 1950's. Ceramics such as PZT are not naturally piezoelectric as quartz is. PZT is made piezoelectric by placing it in a strong electric field while it is at high temperature.

In medical ultrasound the transducer is held against the body of a patient. A short pulse, typically two or three cycles of electricity at the desired frequency is applied to the PZT transducer which then emits a burst of ultrasound. Some of the ultrasound is reflected back from the first tissue but some continues till it is also reflected back by farther away tissues. The reflected sound energy is converted by the transducers to electric voltages which are displayed on a screen. This gives a picture of the inside of the body which can be used for medical diagnosis without the ionizing effect of X-rays.

There are two general types of scanning used in medical ultrasound. A single linear rectangular image is made up of many parallel scan lines. Each scan line represents a series of echoes returning from a pulse traveling through the tissues. One scan line is produced from one pulse. The pulse is then transmitted from a different position to produce a second scanning line parallel to the first. About 100 parallel scan lines will produce a complete linear rectangular frame.

The second type of scanning is a sector image shaped like a pie slice. Here the scan lines all originate from the same location and travel out in different directions to give a pie shaped sector image.

Doppler Ultrasonography

This is useful in studying the speed of blood flow in the body. The Doppler effect is a change in the frequency of a wave as a result of motion. For example, the sound of a train whistle will increase in pitch or frequency as heard by an observer at a train station in the path of an oncoming train. When the train is going away from the observer the frequency of the whistle is heard as a lower pitch. In blood flow measurement the moving red blood cells in the blood vessel reflect the sound from an ultrasound transducer. The frequency of the reflected sound changes due to the motion of the blood cells. This change of frequency is a measure of the velocity of the blood and can be displayed.

Ultrasound is used to image many parts of the body but is used most often in obstetrics. It is used frequently to observe the fetus in the womb. This raises the question of safety in the use of medical ultrasound in general and for fetal studies in particular.

There is an absence of known risks but recognizing that bioeffects can be subtle or low incidence, a conservative approach to the medical use of ultrasound is recommended by authorities in the field. Prudent caution (P.C.) means that ultrasound imaging should be used for imaging only when medically indicated. Exposure is minimized by minimizing instrument output intensity and exposure time. Doppler instrument outputs are significantly higher than those for imaging. Fetuses are presumably more sensitive to possible bioeffects. Fetal Doppler studies appear to be the most liable to cause problems although no special risk has been identified even in this extreme case.

11.2 Using Body Electric Currents and Magnetic Fields

Electroencephalogram (EEG)

Electroencephalogram is the recording of the minute electric currents produced by the brains of human beings. Hans Berger developed this technology starting in the late

1920's. It was soon found that EEG has important clinical significance for the diagnosis of brain disease.

The electroencephalograph is the machine that records brain waves for medical study. The device produces a 16 channel ink-written record of brain waves. To produce the record, 20 equally spaced electrodes are pasted to the surface of the scalp in accordance with the standard positions of the International Federation of EEG. This organization has a standard called the 10/20 system. In addition to the position of the 20 electrodes, it specifies the 10 patterns or combinations of electrode pairs that are used for transforming the spatial location from the scalp to the channels which are traced on the EEG.

Normal EEG waves are defined by both form and frequency. There are a number of normal waves such as alpha, beta, theta, and delta rhythms. Alpha consists of 8-12 Hertz sinusoidal waves. They are normally present during the waking hours and enhanced by closing the eyes. Beta are faster waves around 14-30 Hertz. They are prevalent with the use of some sedatives and tranquilizers. Theta waves of 4-7 Hz replace the alpha rhythm during drowsiness and light sleep. Delta waves of 0.5-4 Hertz are present during deep sleep in normal people of all ages. Delta waves are the primary waves present in the records of normal infants. They are almost always pathological in the waking records of adults.

The EEG technology is used in the study of functional abnormalities such as epilepsy. During an epilepsy episode, spikes become repetitive and synchronized over the whole surface of the brain. The EEG is also used in studying comatose states. The delta waves predominate and the whole record is slowed down. If the EEG becomes flat for several hours the coma is considered terminal. An EEG recorded at the highest gain with widely spaced electrode positions and the absence of cerebral reflexes and spontaneous respiration is considered an indication of "brain death."

ELECTROCARDIOGRAM (EKG)

The electrocardiogram is a method of graphic tracing of the electric current generated by the heart muscle during a heartbeat. Electrocardiograms are made by applying electrodes to various parts of the body and recording the tiny heart currents. The electrodes are placed on the ankles, wrists and the chest wall. The normal electrocardiogram shows the contraction of the atria, the two upper heart chambers, and the following contraction of the two lower heart chambers, the ventricles. Any deviation from the normal is indicative of a possible heart disorder such as an irregular heartbeat.

Superconducting Quantum Interference Device (SQUID)

SQUID is essentially a magnetometer which picks up and accurately locates very low magnetic fields generated in the brain. It gives a measurement of the activity taking place in different parts of the brain. It is useful in diagnosing local problems in the brain.

11.3 Changing Surgical Procedures

Surgery in the 20th century changed radically incorporating new anaesthetic techniques, and blood transfusions from the discovery of blood groups to blood substitutes, laser surgery and laparoscopic surgery.

Anesthetics

Various methods have historically been used to alleviate the pain of surgical operations. These included the clubbing of the patient into unconsciousness in ancient Egypt to the use of opium by the Ancient Greeks and then by Arab physicians in the Middle Ages. In the early 19th century British sailors imbibed freely of rum just before emergency amputations aboard ship right after a battle.

In 1799, Sir Humphry Davy, a British chemist inhaled nitrous oxide and discovered its anesthetic qualities but its implications for surgery were ignored. In the early 1840's nitrous oxide parties became fashionable in certain quarters in Britain and the United States. Nitrous oxide was passed around in bladders and inhaled for its soporific effect. Soon it was discovered that ether had the same effect but it had the advantage that it could be carried in small bottles.

Dr. William Morton, a U.S. dentist, demonstrated the use of ether as a general anesthetic in a surgical operation in October 1846 at the Massachusetts General Hospital in Boston. This was followed by other operations using ether in British and United States hospitals. In 1853 British physicians administered chloroform to Queen Victoria when she gave birth to her eighth child, Prince Leopold.

Early anesthetics in the second half of the 19th century were hazardous since the dose needed to produce unconsciousness was fairly close to the amount that would paralyze the breathing center of the brain. No precise control of the dosage was given because the anesthetics were administered by sponges soaked in ether or chloroform.

In the early 20th century methods of precisely controlling the dosage and new and safer gases were introduced in the use of inhalation anesthetics. These included trichloroethyl and halothene which are mixed with oxygen and nitrous oxide in an anesthetic machine. The anesthetists can control the flow and composition of the gas mixture precisely. The patient can be maintained on respiration by mechanical means, if necessary, while the gas mixture is delivered directly to the lungs by a close-fitting endotracheal tube. This prevents the accidental inhalation of mucus, saliva, and vomit.

In the 20th century a specialty of medical anesthesiology was gradually developed as anesthetics became an essential part of surgery. In the United States in the mid-1930's the specialty was officially recognized with the establishment of the American Board of Anesthesiology for certifying physician anesthetists. Anesthesiology was originally concerned only with the administration of general inhalation anesthetics. The anesthetists' activities were later broadened with the introduction of local anesthetics injected into the fluid surrounding the spinal cord. These spinal anesthetics were used for major operations in the lower half of the body.

Also the anesthetists began to inject an agent such as the barbiturate thiopental sodium into a vein to put the patient to sleep for short operations. Soon the anesthetists used this agent also to put the patient to sleep before the administration of an inhalation anesthetic is begun.

Although the safety of anesthetics has improved dramatically in the 20th century, there is still a small but significant risk in major surgery. However, the benefits to risk ratio is very high and anesthetics make modern surgery possible.

BLOOD TRANSFUSION

The Inca Indians in Peru had long before Columbus practiced blood transfusion successfully. However, nearly all South American Indians are of the same blood type O-Rh positive. Europeans tried blood transfusion around 1628 in Italy. So many patients died that blood transfusion was banned in Italy, England, and France after the late 17th century.

In 1900 the ABO classification of blood groups were found. The ABO and Rh blood groups are among those most commonly considered. The blood groups of donor and recipients are determined before transfusions. The discoveries of the ABO and Rh blood groups in the 20th century have made major surgery practical. There are four major ABO blood groups: A, B, O, and AB.

An African-American surgeon, Dr. Charles R. Drew, at Howard University, discovered the life-saving properties of plasma, the liquid portion of the blood minus the cells. Dr. Drew showed that plasma, unlike whole blood, could be stored for relatively long periods of time which led to the beginning of blood banking. Plasma could be freeze-dried and stored as a powder for two or three years. However, the plasma powder in the field during the early part of World War II had to be mixed carefully in the field with sterile water. This took as long as ten minutes which was a long time during battle conditions.

Another major advance in blood transfusion occurred during World War II. Dr. Edwin Cohen, a Harvard biochemist broke plasma down into different proteins. In 1940 Dr. Cohen and his team at Harvard separated out the various proteins of plasma by a process called fractionation. He poured plasma into a test tube, added ethyl alcohol and spun it in a centrifuge for a half-hour. A pellet of protein settled on the bottom and was set aside. The remaining liquid would be fractionated at a slightly different temperature and salt content. Each step isolated a different fraction of the plasma protein. Every fraction of the plasma serves a different function. The fifth fraction was albumin which had strong osmotic properties. Albumin remains stable at room temperature for years. It could be packaged in small bottles and became very useful in World War II. The Red Cross did the collection of blood and the distribution of plasma.

After the war in the late 1940's, the Red Cross began collecting blood again. It became the world's largest handler of blood products. Cohen's fractionating process became big business producing 20 products. One of these, Factor 20 was the blood-clotting factor that helped hemophiliacs to clot their blood.

Dr. Cohen held a number of patents to the fractioning process he developed. Right after the end of World War II he gave away all his patent rights. Cohen thought that blood should be used for the good of all humanity. He never made a cent out of his discoveries. Dr. Edwin Cohen died of a stroke and a cerebral hemorrhage in 1953 at the age of 61.

Major problems in blood transfusion arose in the 1980's. Diseases such as AIDS and hepatitis B were transmitted through blood transfusions. AIDS was not known as a disease before thousands of people were infected in a number of countries by transfusions. Later precautions were taken to avoid the problem. However, it is very difficult to guarantee that AIDS will not be spread by human blood transfusion.

BLOOD SUBSTITUTES

Blood consists of four basic components, Platelets (blood clotting), White Cells (infection fighting), Plasma, and Red Cells (hemoglobin, *oxygen carrying*).

Baxter Healthcare Corporation has developed an oxygen-carrying solution to overcome the storage, blood typing and safety issues inherent to blood. This product is called Diaspirin Cross-Linked Hemoglobin (DCLHb™).

It is expected to temporarily augment the oxygen-carrying capacity of the patient's red blood cells. Its life-saving potential may become apparent first in emergency clinical use where typed and cross-matched blood is not available, and an oxygen carrier could serve as a vital bridge to a blood transfusion.

DCLHb may also be useful in strokes where a lack of oxygen delivery to part of the brain is the cause of cell death or ischemia. Ischemia is the underlying defect in heart attacks as well.

In 1996 the Baxter hemoglobin therapeutic was under a continuing clinical investigation for surgery and trauma applications, and for the treatment of ischemic disease.

Figure 11-6 shows a model illustrating the DCLHb molecule. It consists of two Alpha and two Beta subunits and four hemoglobin groups. The hemoglobin groups each contain an iron atom (Fe) where oxygen binds to the molecule. The patented cross-link

85%

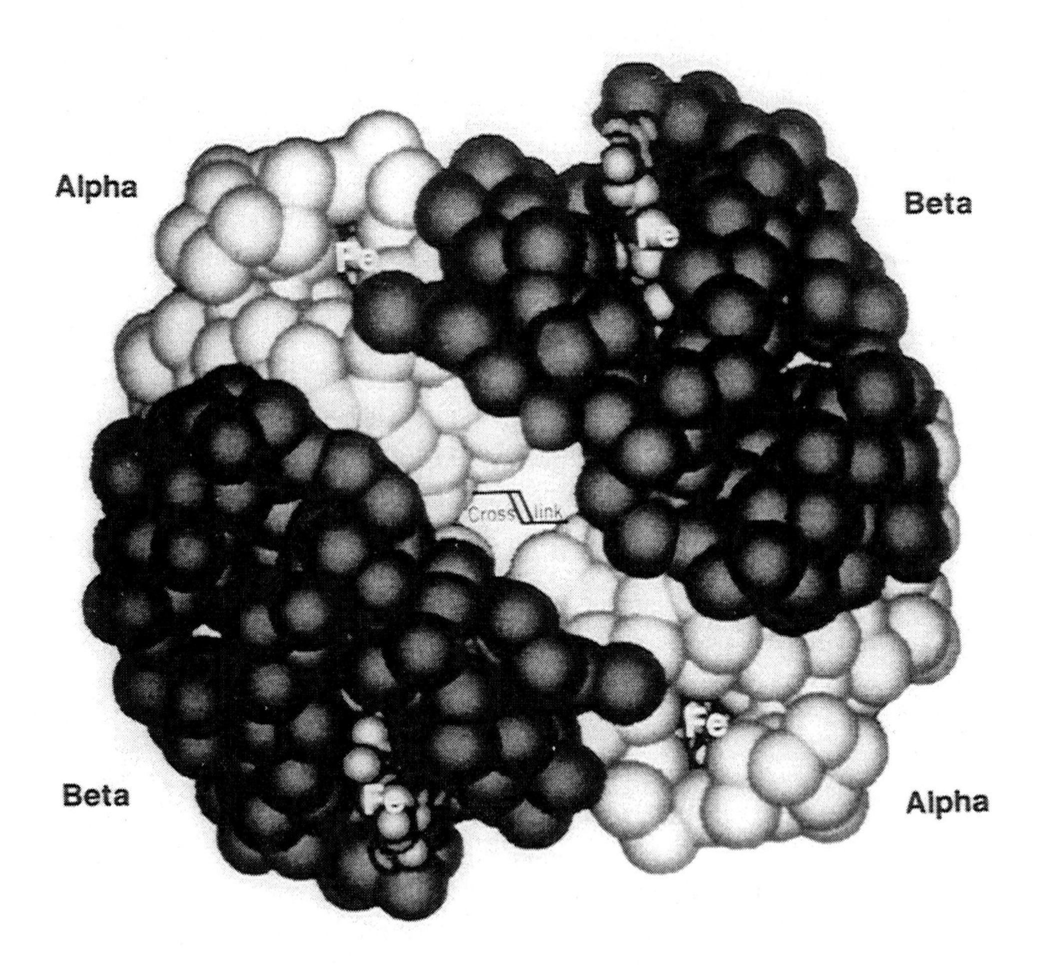

Figure 11-6 A Model of the DCLHb Molecule
Courtesy of Baxter Healthcare Corporation

(center) stabilizes the molecule, enabling it to deliver oxygen to tissues in a manner similar to whole blood.

Figure 11-7 is a photograph of technicians engaged in process engineering research for commercial scale manufacture of DCLHb.

LASERS IN SURGERY

In the later half of the 20th century the first successful use of lasers in surgery was in eye treatment, one of which was diabetic retinopathy. This condition is common in people who suffer from Juvenile Diabetes. In this form of diabetes networks of abnormal blood vessels spread across the surface of the retina in many victims. These blood

Figure 11-7 Engineering Research on DCLHb
Courtesy of Baxter Healthcare Corporation

vessels are very fragile, and they can leak blood into the normally clear liquid of the eye thus gradually dimming vision. Eye surgeons have found that illuminating the diseased retina with a continuous-wave visible light from an argon, krypton or dye laser can slow the spread of the abnormal blood vessels. This procedure has been used to forestall blindness in millions of diabetics.

After cataract surgery in about a third of all cases, the back membrane of the natural lens which remains becomes cloudy. This can again limit vision. An eye surgeon can send a series of short, intense pulses from a neodymium laser through the lens implant and break up the cloudy membrane. The laser pulse is focused with a lens having a very short focal length. This makes the laser light come to a very tight focus exactly at the membrane surface. Since the laser beam is focused very tightly, the power density at the implanted lens is too low to damage it. Behind the membrane, the laser light spreads out over a large area so it cannot damage the retina. The operation has been very successful.

The laser is also used to treat a condition called a detached retina. The detachment can be stopped by focusing a laser pulse onto the retina, so it causes a burn which forms scar tissue which "welds" the retina down to the back of the eyeball, so that it cannot break free.

Carbon-dioxide lasers operating at 10-micron wavelengths are used to remove tissues from the human body. The surgeon scans the beam from a carbon-dioxide laser across the tissue to be removed. Water in the cells absorbs the 10-micron beam strongly, vaporizing the top layers of cells, but doing little damage to the underlying tissue. This cauterizes small blood vessels which effectively stops bleeding.

Carbon-dioxide lasers became routine for some types of surgery. In gynecological surgery, it is very useful because it can remove abnormal cells from the surface of blood-tissues. Carbon-dioxide lasers can remove potentially precancerous cells on the surfaces of the vaginas of women. These lasers are also used to seal off blood vessels in endometris, a condition of abnormal bleeding which affects millions of women, causing infertility. Another common use of carbon-dioxide lasers is to perform delicate microsurgery on the larynx without having to open up the throat. The carbon-dioxide laser is passed down the throat in a small tube allowing the surgeon to remove small cancers from the vocal cords without damaging them.

A new laser prostate reduction operation was clinically tested in the United States and Europe in 1994 and 1995. Prostate enlargement is a common condition in old men resulting in difficulty in urinating. By age 75 nearly three out of four men require treatment, usually surgery. It is the commonest kind of in-patient surgery in the United States. The traditional prostate reduction surgery reaches the target through a cystoscope inserted through the penis. An electrocautery wire loop is used to carve away a portion of the prostate. The procedure requires a few days in the hospital and weeks for recovery. Patients risk sexual dysfunction and incontinence.

A small gallium aluminum arsenide laser diode has been used in the laser prostate reduction surgery. An optical fiber is inserted through a small puncture directly to the prostate. The laser delivers low-power energy through the optical fiber to the prostate for a few minutes. This thermally destroys a controlled volume of tissue reducing the size of the prostate. The system is designed to be used in clinics on outpatients under local anesthesia. It is a form of laparoscopic surgery.

LAPAROSCOPIC SURGERY

As early as 1910, long before video cameras were invented, doctors were performing minor surgery through a tube inserted in the abdomen. However, the method lacked the magnification provided by TV and it was not widely adopted.

Later in the 1960's gynecologists used a tube called a laparoscope to examine the ovaries. They inserted the tube through the vagina. The name laparoscope comes from the Greek word lapara which refers to the soft part of the body between the ribs and the hip.

In 1989 Dr. Phillipe Mouret, a French surgeon, removed a gallbladder using a video camera mounted on a laparoscope. Within four years 600,000 gallbladders have been removed in the United States alone by laparoscopy using a video camera. In a very short time this procedure for gallbladder removal became standardized and by the end of 1994 over 25,000 surgeons were performing this procedure.

The gallbladder laparoscopic surgery is performed under anesthetic. Four small punctures are made in the patient's abdomen. Each incision is no larger than a dime. The surgeon uses these holes to insert into the abdominal cavity tiny specialized instruments that are used to remove the gallbladder through the navel. A plastic tube called a trocar is first inserted through the puncture nearest the navel. A pipe carrying carbon dioxide is slipped through the trocar. The carbon dioxide causes the skin to rise making room for the other equipment. A very small video camera is inserted into one of the incisions. The camera is mounted at the end of a bundle of glass fibers that conduct a high intensity light to the area of the operation. The images from the camera travel up the fibers to a TV screen where the surgeon can observe the gallbladder and surrounding organs. The TV system enlarges the operating scene up to 18 times. This gives the surgeon a detailed view of the operating field.

The surgeon then inserts the tools for cutting, clamping, and stapling through the other small openings he has made in the patient's abdomen. The surgeon can use a tiny laser or an electrocautery to detach the gallbladder from the liver bed. Either instrument allows the surgeon to perform precise cutting without harming the other abdominal organs. The surgeon then removes the gallbladder through the navel.

The laparoscopic video gallbladder operation eliminates the need for major surgery that would require a large incision resulting in a visible scar up to eight inches long. Laparoscopy shortens the hospital stay from at least a week to perhaps one day. The patient can return to normal activities within a week or less compared to months for conventional gallbladder surgery.

Laparoscopic video surgery has spread to many other types of surgery. These include a complete hysterectomy through the vagina, removing sections of lung through the small space between the ribs, a procedure called arthroscopy for repairing a damaged knee, gastrointestinal, kidney, and other surgery. Towards the end of the 20th century laparoscopic video surgery began to dominate many surgical fields.

One of the problems with this type of surgery is the two dimensional picture presented on a TV screen. A three dimensional video picture would provide depth perception allowing surgeons to locate organs and do other tasks more easily and quickly than by watching a 2-D video monitor.

A three dimensional view of a flat video screen can be obtained by using the phenomenon of polarized light. If light is transmitted through certain crystals only the vibrations in one direction can pass through. The direction of polarization can be selected by the proper orientation of the polarizing crystal.

In three-dimensional laparoscopic surgery, two tiny tilted cameras are mounted in the laparoscope. One gives a left-sided view while the other gives a right-sided view. An electronic switch operating at 120 times a second switches back and forth between the two views.

A modulator causes the light from the liquid crystal display to change polarization 120 times a second in synchronization with the camera switching. The light from the left camera is polarized in one direction and the light from the right camera is polarized in another direction.

The surgeon wears special polarized glasses. The right eyepiece passes only the view from the right camera. Similarly the left eyepiece allows only the light from the left camera to pass. The brain integrates the two rapidly switched images into one three-dimensional picture.

One of the problems of laparoscopic surgery is the training of surgeons in the new technique. Simulators can solve this problem just as pilots are trained for new planes on a flight simulator. In the 1990's a number of companies in the United States designed simulators for different types of laparoscopic surgery for the purpose of training surgeons.

11.4 New Parts for Old in the Human Body

The 20th century saw the development of artificial parts that could replace or augment parts of the human body that became defective. This included hearts, heart valves, heart pacers, hips, knees, plastic lens in cataracts, cochlear implants, liver, teeth, and kidney transplants.

Heart Transplants

Surgeons in the early 1900's found that while they had the techniques to operate on the heart, there was one severe problem that prevented successful major heart surgery. They knew how to start and restart the heart but they had less than three minutes to avoid irreparable brain damage to the patient. Dr. John Gibbon of Philadelphia developed a machine that took over blood circulation. His machine was tested on animals in 1931. It was not until 1953 that Dr. Gibbon performed a successful operation on a human patient using total cardiopulmonary bypass. It was not until the mid-1970's that such machines were available. These bypass (heart/lung) machines can maintain a patient's complex circulatory system during surgery for hours without serious side effects. Another innovation allowing surgery for extended periods was the introduction of extreme cold to preserve a heart that has been stopped. These technologies made heart transplants possible.

On December 3, 1967, Dr. Christian Barnard performed the first human heart transplant in Capetown, South Africa on Louis Washkansky. This operation was followed three days later by a heart transplant in the United States. Dr. Adrian Kantrowitz, Professor of Surgery at Wayne State University College of Medicine in Detroit, performed the heart transplant at Maimonides Hospital in Brooklyn, New York. A large number of heart transplants soon followed. In 1994 the five-year survival rate was estimated at 50 percent. However, the life style quality of some patients improved remarkably.

New mechanical hearts were developed toward the end of the 20th century. Initially they worked best as temporary devices to sustain a patient waiting for a donor heart. In the future after much development they may possibly be used to replace the human heart. This would solve the problem of the shortage of human hearts for replacement. Animal hearts have been tried a number of times but the recipient rejected the organs and died.

HEART VALVE REPLACEMENT

The circulatory system is a continuous loop. The lower chambers of the heart, known as ventricles, pump blood into the arteries. It travels through the body and returns through veins, to the upper chambers, called the atria. The inlet valves, also known as mitral and tricuspid valves, lead from the atria to the ventricles. The outlet valves, also known as pulmonary and aortic valves, are gatekeepers between the heart and the arteries that carry blood away from the heart.

The valves consist of flaps of tissue, called cusps or leaflets, that open and close. The tricuspid valve, between the right atrium and right ventricle, has three cusps; the mitral valve on the left side of the heart has two cusps. At the entries to the aorta and the pulmonary artery the flaps are shaped like half-moons. These are sometimes called semilunar valves.

There are two general disorders of the valves which can lead to heart failure and death. One is stenosis or tightness where the valve does not open completely and blood backs up. The other disorder is called incompetence in which there is inadequate closing. The blood can move backward in the wrong direction. The valve becomes vulnerable to infection, leading to further damage.

Damaged heart valves are too narrow or fail to shut completely. Replacing them can save lives and relieve severe symptoms such as constant fatigue. The first operation to replace defective heart valves was performed in 1960. By 1994 some 50,000 operations were performed annually. Mechanical valves were made from metal and plastic while recycled valves came from pigs or corpses. Mechanical valves increase risks of clotting. Blood-thinning drugs are necessary with mechanical valves. The failure rate was estimated in 1994 to be two percent.

The Implantable Heart Pacemaker

The rate and rhythm of the heart are normally controlled by specialized cells within the heart in the sinus node of the right atrium. The sinus node generates an electrical impulse which results in the contraction of the atrium. The pulse is also carried by nerves down to the left and right ventricles causing them to contract. It does this via specialized conducting fibers.

The rate of discharge of the pacemaker cells is generally from 60 to 100 times per minute. The heart rate changes to match the situation. Sometimes the heart rhythm is not controlled properly, resulting in illness and even death. Physicians and engineers have long thought about using electrical pulses to take the place of defective electrical pulses generated by the heart.

The history of the idea of using electricity to stimulate the heart goes back to G. Aldini in 1819. He attempted to restart the heart of decapitated criminals by using the electricity from a voltaic pile. In 1932 Dr. A. S. Hyman invented an external cardiac pacemaker. He demonstrated the use of an electrode through the external jugular vein. Dr. Hyman was awarded a patent for his invention. Although portable his machine was quite large. Dr. Paul M. Zoll in 1952 developed an external pacemaker with two disk electrodes that were placed directly upon the skin. Rate and amplitude were variable and the operation of the unit was completely manual.

Wilson Greatbatch, a biomedical engineer, worked in the late 1950's on an implantable cardiac pacemaker in animals in a barn behind his house. On April 7, 1958, Greatbatch together with two medical doctors implanted the first successful self-powered implantable cardiac pacemaker in an experimental animal, a dog. Thus began a new era in controlling heart rate.

The first successful self-powered unit was installed in 1960. Figure 11-8 shows a consultation at the bedside of the first patient. From left to right are William C. Chardack, MD, Andrew Gage, MD, and Wilson Greatbatch, P.E. Figure 11-9 shows two encapsulated models, 1965.

Some problems limited the longevity of the pacemaker to two years, as late as 1970. The electronic components were not good enough for the "zero defect" requirements of pacemakers. Gradually by improved testing of all components this problem was resolved. However, the limiting factor became the original mercury-zinc battery. In 1970-72 Greatbatch after much research settled on a lithium battery. The pacemaker became a reliable long lasting device.

The original pacemaker was on all the time. In 1964 Barough Berkovitz published a series of papers describing a concept in which the pacemaker "listened" to the heart and worked only when the heart didn't. The pacemaker worked only "on demand." This came to be known as a demand pacemaker. Most modern pacemakers are of this type. While the original pacemaker had as few as two transistors, by 1983 there were as many as a hundred thousand transistors in one pacemaker. By 1983, 300,000 pacemakers were implanted every year.

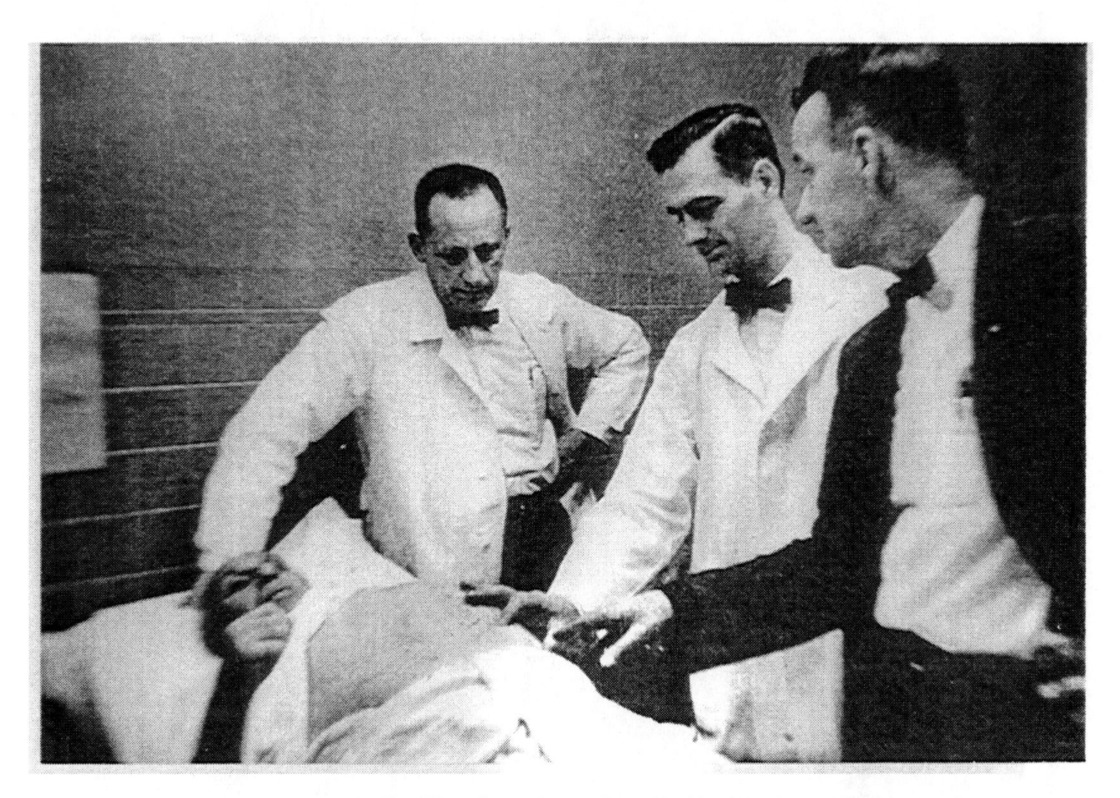

Figure 11-8 The First Heart Pacer Installed in a Patient, 1960
Courtesy of Wilson Greatbatch

REPLACEMENT OF HIPS

Surgeons replace damaged or arthritic hip joints with synthetic ones. The operation dramatically relieves pain. New ball and sockets are made from polyethylene, titanium, ceramics, chrome and cobalt. Cement is usually used to anchor the new joint in place. Patients walk a few days after the surgery. Figure 11-10 shows an artificial hip.

About 1 to 2 percent fail annually. Loosening is the biggest problem. The polyethylene socket lining may wear out and need replacement. There is a risk of dislocation possible during initial healing. Breakdown of cement and polyethylene can cause bone-damaging inflammation.

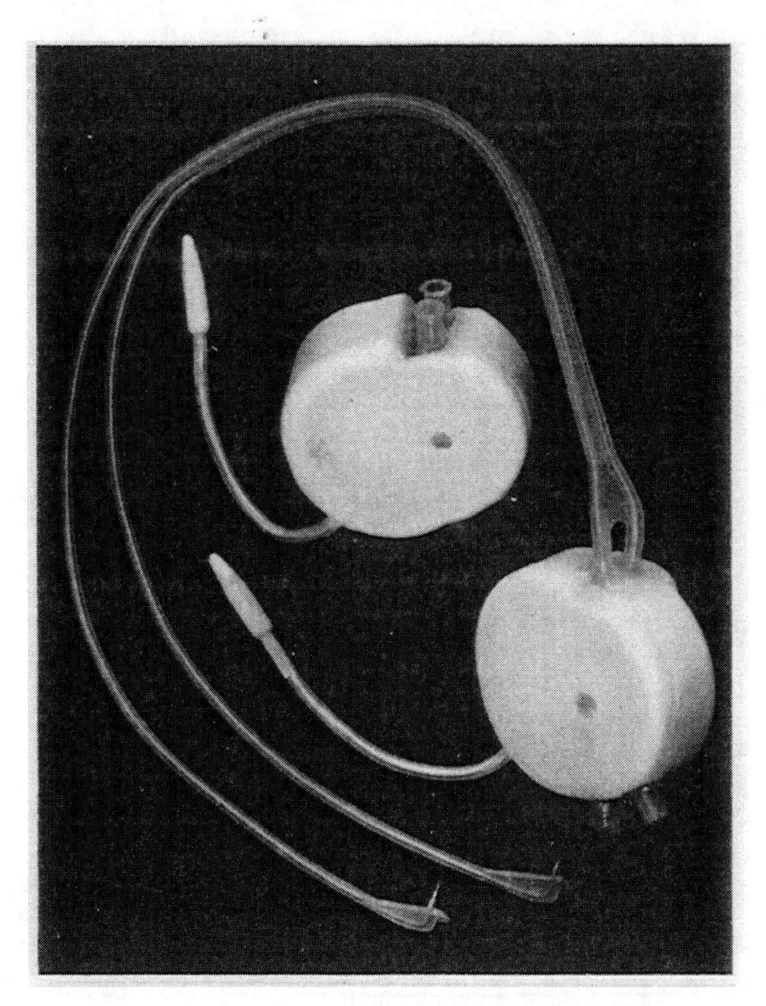

Figure 11-9 Early Models of Pacemaker, 1965
Courtesy of Wilson Greatbatch

A National Institute of Health panel in 1994 evaluated hip replacement. Since the early 1970's surgeons performed 800,000 hip replacements, 120,000 in 1993 alone. The panel was enthusiastic overall about the success of the procedures but pointed out a problem. This was the loss of the bone at the site. Physicians once attributed bone loss to cement used to bind the prosthesis to bones. However, the panel said 30 percent to 40

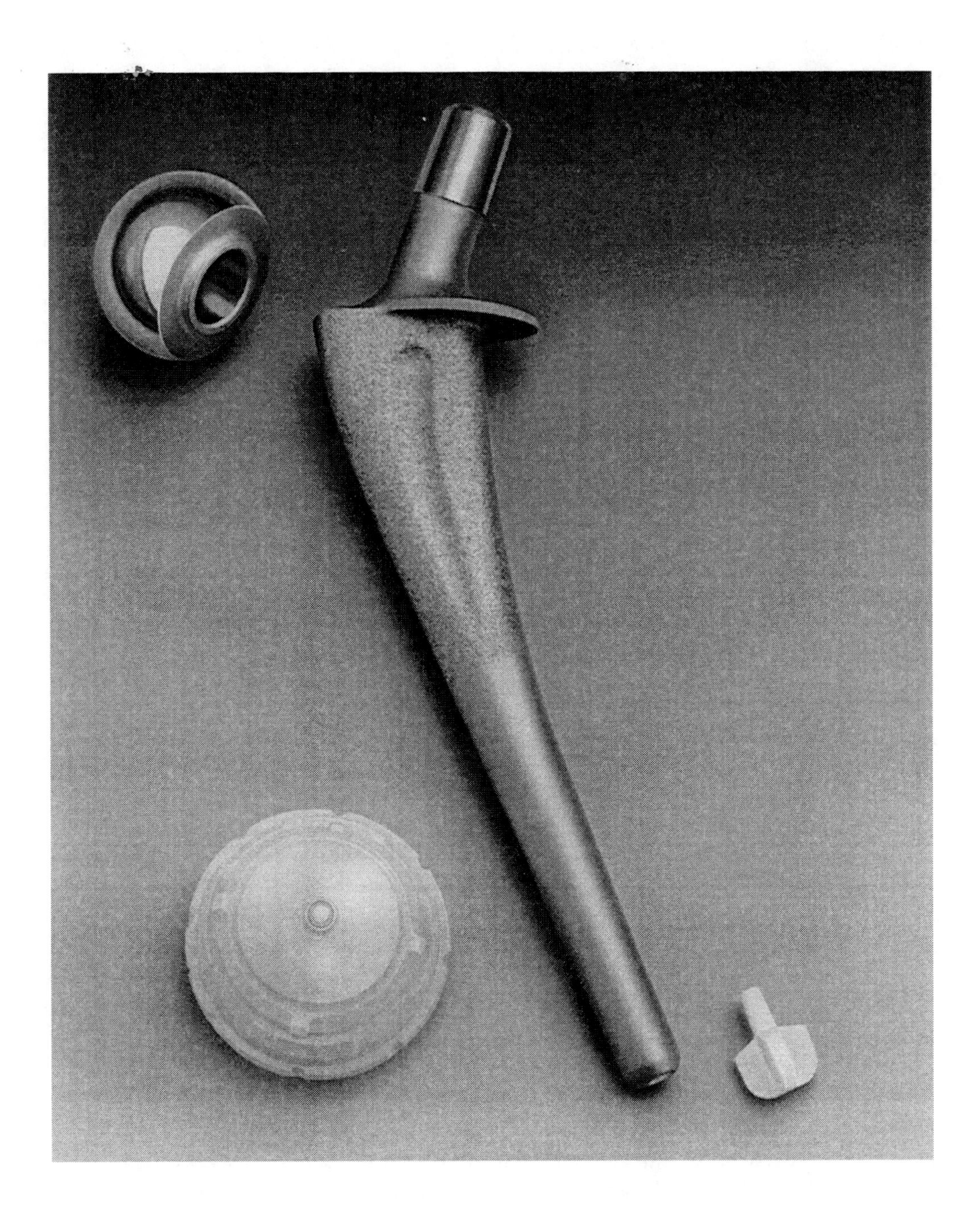

Figure 11-10 Artificial Hip Replacement
Courtesy of Howmedica Inc.

percent of cases where no cement was used in the implants also showed some degree of bone loss within 10 years of surgery.

Orthopedic surgeons have experimented both with attaching prosthesis devices without cement, relying on bone growth into porous surfaces for binding and with hybrids where cement is used on one side and not on the other. The panel concluded that the best procedure for first time hip replacement should use a cemented thighbone component and non-cemented pelvis attachment.

The panel found that hip replacements for men are generally performed when they are between 65 and 74. Replacements for women are most common from 75 to 84 years of age, when the disease would be more advanced.

Knees

The replacement of knee components is the most common joint replacement operation. It relieves pain usually resulting from arthritis. Surgeons replace the damaged ends of the bones that meet at the knee, as well as the underside of the kneecap. About 160,000 are performed annually. The most common problem is loosening. There is a slight risk of infection or blood clots. Disintegration of cement and polyethylene can cause inflammation, eroding bone. One percent or less of the operations failing was the estimation in 1992.

Replacing the Lens of the Eye

A cataract is the clouding of the normally clear lens of the eye. It is due mostly to aging although there are other causes like diabetes, steroids, or injury. As people are living longer at the end of the 20th century cataract surgery reached over 1.5 million people in the United States.

Up to 1965 cataract surgery was highly invasive and recovery meant lying in a darkened room for up to a week with heavy sandbags immobilizing the head. The patient's eye lens was replaced with cumbersome coke-bottle eyeglasses. The result was distorted vision so that the cure was almost as bad as the cause.

In the late 1960's Dr. Charles Kellman of Manhattan searched for a better, less traumatic way to excise a cataract. He tried a number of different devices without success. Then he had his teeth cleaned with an ultrasonic device. The story is that he asked to examine the instrument. Dr. Kellman leapt out of the chair--bib and all--and ran past the patients in the waiting room, shouting, "This is it!"

Dr. Kellman helped to develop a probe which breaks up the cataract with 10,000 ultrasound pulses per second, sucks out the debris and at the same time fills the eye with fresh saline-fluid to replace the matter being extracted. The incision to expose the lens may be as small as 3 millimeters.

Once the defective lens is removed, doctors can implant a new Lucite introcular lens whose shape is determined by a computer. The actual surgery takes anywhere from four to thirty minutes in a hospital's ambulatory surgery center. The patient arrives in the morning and goes home in the afternoon.

THE COCHLEAR IMPLANT

The cochlea is a small bony chamber in the inner ear. Normally, the fine hairs in the cochlea transform sound into electrical signals. The auditory nerve picks up those signals and conducts them into the hearing center of the brain. Profound deafness occurs when the cochlea no longer does its job.

A cochlear implant is a small device developed in the 1970's. It is inserted directly under the skin behind the ear and into the cochlea. The implant gets signals from a microphone behind the ear. A pocket-sized computer called a "speech processor" transforms those sounds into electrical pulses. These activate the implant to create electrical signals that travel down fine wires to the auditory nerve and then into the hearing center of the brain.

The device cannot restore normal hearing but allows users to understand some words without lip-reading and makes lip-reading much easier. Users hear sounds that are different from normal speech, but with training can interpret them properly. By 1994 about 10,000 formerly deaf children and adults could hear. The system is continuously being improved with more experience in use.

KIDNEY TRANSPLANTS

Kidney failure can be treated by dialysis or transplants between certain compatible individuals including close relatives. A dialysis machine can replace the action of nonfunctioning kidneys but the process is physically hard on the body as well as psychologically hard on the patient. It entails a five-hour treatment two to three times a week.

Kidney transplants must fight the rejection phenomenon. The human body attempts to destroy any foreign substance that tries to get into it. Special cells of the blood will attack any foreign body containing cells. In kidney transplants the recipient and the donor must have closely related tissues. This is true in transplants other than kidneys. Special drugs have been developed to suppress the rejection process.

Liver Transplants

Liver transplants got a great deal of publicity when on June 8, 1995 a liver from an unidentified donor was transplanted in Mickey Mantle, a 63 year old baseball Hall of Famer. His own liver had been ruined by cancer, hepatitis, and 40 years of heavy drinking. Unfortunately Mr. Mantle died two months later of cancer. In 1995, 65 percent of liver transplant patients encountered some form of foreign tissue rejection. While this problem exists in most transplants, it is particularly serious in liver transplants. Steroids are used to counteract the rejection in liver transplants.

Bone Marrow Transplants

This is used in certain cancers when the normal bone marrow is not functioning properly in making blood cells. However, the donor and recipient must be matched together. Searches often have to be made to find compatible donors in a short time. When a suitable donor is not found in time, the disease is often fatal. Bone marrow transplants are sometimes used after accidental exposure to ionizing radiation.

The Transplant Olympics

Athletic competitions among transplant recipients have been held since 1978. They are an Olympics involving people who have had transplants of hearts, kidneys, etc. The competition includes track and swimming. In 1995 the event took place in Manchester, England.

11.5 Battling Bacteria

In 1928 a British bacteriologist, Alexander Fleming (1881-1955), noticed that a culture plate of a penicillium mold was surrounded by dead staphylococci bacteria. Fleming found that the mold, Penicillium notatum, destroyed the bacteria and it was nontoxic to laboratory animals. In 1940 at Oxford, Ernst Chain, Howard Florey, and their co-workers isolated penicillin and showed that it was highly effective against many serious bacterial infections.

During World War II penicillin was very useful in treating wounds and infections. It was also effective in treating diseases caused by such bacteria as staphylococci, streptococci, the pneumococci of pneumonia, the gonococci of gonorrhea, and the spirochetes of syphilis. Thus began the era of antibiotics.

Antibiotics are chemical substances, originally produced by a living organism that is detrimental to other microorganisms. After the success of penicillin, many other antibiotics were discovered. Penicillin was not effective against tubercle bacillus. In 1943, streptomycin, isolated from *Streptomyces griseus* was found to be effective in treating tuberculosis. Later it was found that streptomycin also was very useful against typhoid fever and many other diseases. A large number of antibiotics were discovered over the years and became a very important part of medicine. However, soon it was learned that bacteria developed mutants that were highly resistant to antibiotics.

RESISTANT BACTERIA

Antibiotics attack germs by destroying the cell wall, breaking down their DNA and blocking instructions for protein synthesis. The antibiotics thus wipe out large numbers of bacteria. However, there are a few bacteria containing a gene for resistance to a specific antibiotic. These resistant bacteria have an opportunity to multiply rapidly in the absence of the bacteria that have been destroyed by the antibiotic. The resistant bacteria can quickly become dominant since one bacterium can produce almost 17 million offspring within 24 hours of dividing.

In addition, bacteria can transfer the resistant gene to other bacteria in a form of sexual reproduction. The resistant bacterium sends out a filament to a non-resistant bacterium pulling it in. The resistant individual makes a copy of a loop of DNA, called a plasmid, that contains the gene for resistance. When the two bacteria touch, the resistant one transfers the plasmid into the other bacterium. The second microbe is now resistant and can transfer the resistance to its own progeny and to other bacteria. One species of bacteria can even transfer immunity to another species in this manner.

Resistant bacteria have been documented since 1960. A resistant bacteria can secrete an enzyme that dismembers a number of antibiotics. This is the case in staphylococcus aureus, the germ that causes surgical infections.

Over prescription by doctors of antibiotics has contributed to the problem. Some doctors in the past have given antibiotics to patients who had viral infections which are not susceptible to those drugs. Physicians have even prescribed penicillin for pregnant women who did not have any disease. This unnecessarily increased the number of resistant bacteria in both the mother and child making it more difficult later if they have a bacterial infection.

A very large number of antibiotics became ineffective. A partial list includes those that originally were effective against blood poisoning and surgical infection, meningitis and ear infections and pneumonia, tuberculosis, gonorrhea, malaria, and diarrhea.

An additional major problem which produces resistant bacteria is the widespread use of antibiotics in livestock including cattle and poultry. Many bacteria immune to antibiotics came from patient's consuming hamburgers, milk, eggs, or chicken. This combined with the over prescribing of antibiotics have produced a serious problem. New antibiotics have become much more expensive as the cost of research required has increased.

There is another problem with the use of antibiotics which is also common to a large variety of pills. This is the question of side effects in certain people. For example, penicillin produces a very bad reaction in some patients that can even result in death. The risks versus the benefits have to be weighed carefully as in much of 20th century medicine. The wild stallion must be managed very carefully.

11.6 Vaccinating the World

The first vaccine was introduced in 1798 by the British physician Edward Jenner (1749-1823). Jenner knew that cows had a disease called cowpox. This produced a characteristic eruption on the teats of milk cows and was frequently transferred to people who milked cows. Milkmen and milkmaids believed that contraction of cowpox prevented subsequent susceptibility to small pox but this was not always the case.

Jenner's investigation of cowpox showed that the eruptions on the teats of infected cows differed. He found only one kind created a resistance to smallpox. He called this type "true cowpox." Jenner subsequently learned that true cowpox conferred immunity only when matter was taken from the cowpox pustules.

On May 14, 1796 Jenner inoculated an eight-year-old boy, James Phipps, with matter taken from a pustule on the arm of Sarah Nelmes, a milkmaid who had cowpox. The boy contracted cowpox and recovered within a few days. On July 1, 1796 Jenner inoculated him with smallpox which produced no effect. Jenner published the results at his own expense in 1798. He described the matter producing cowpox by the term "virus."

Smallpox vaccination spread slowly in the 19th century. Vaccination in the 20th century resulted in the eradication of the disease by the last decade of that century. In the 20th century vaccination was used to conquer many diseases mostly viral and some bacterial.

The virus is a small amount of genetic material with some proteins. A virus needs to infect a cell of a host in order to reproduce. In the 20th century a number of diseases were caused by one or more viruses. Pneumonia is caused by two viruses, AIDS - 2 viruses, dengue fever - 4 viruses, hepatitis - 1 virus, measles - 1 virus, meningitis and encephalitis - 6 viruses, plus other organisms, and polio - 3 viruses.

The Polio Vaccine

Infantile paralysis or polio was the scourge of the first half of the 20th century. Hundreds of thousands of young people were paralyzed and had to temporarily live in machines called iron lungs. Polio also affected adults including Franklin D. Roosevelt.

The iron lung was an example of technology that did not cure but allowed people to live in a crippled state.

In 1935 killed and attenuated viruses were tested on over 10,000 children. Not only were these vaccines ineffective but they were probably responsible for some cases of paralysis and deaths. In 1949, John Enders, a Harvard virologist, and his colleagues had shown how to culture the polio virus in test tubes. Enders won the Nobel Prize for this work. Another essential step was the discovery that there were three types of polio virus. Any vaccine had to work against all three.

Jonas Edward Salk, a Jewish American doctor and microbiologist, developed a successful polio vaccine in 1954. He used virus killed by formaldehyde. He first injected children who already had polio and tested their antibody levels after the injection. He found that high antibody levels were produced by the vaccine. Salk in 1954 administered either a placebo or killed vaccine to 1,829,916 children. An evaluation of the trial in 1955 reported that the vaccination was 80-90% effective. By the end of 1955, 7 million doses had been administered. In the period 1956-58, 200 million injections were administered. Some scientists like John Enders and Albert Sabin, a University of Cincinnati researcher, believed that a weakened virus would work better.

Albert Bruce Sabin, a Jewish American doctor and microbiologist, was born on August 26, 1906 in Bialystock, now in Poland. Sabin decided to use a live polio virus attenuated in monkey tissue. This vaccine unlike the Salk vaccine could be taken orally.

Sabin got the Russians to test the vaccine for him. In 1959 he was able to produce the results of 4.5 million vaccinations. The vaccine possessed a number of advantages over that of Salk. It gave a stronger, longer lasting immunity thus making it unnecessary to give more than a single injection. Great Britain changed over to Sabin's vaccine in 1962 with most other countries following soon after.

Access of Vaccines Internationally

In the developed world vaccines against polio and other viral diseases were quickly and successfully adopted. In the third world countries the new technology did not spread rapidly due to political and economic reasons. Millions of children died because they had no access to a proven technology. The World Health Organization in 1970's started an "Expanded Programme of Immunization." By 1990 immunization rates in the third world for major childhood diseases had grown from less than 5 percent to those under the age of l to at least 80 percent. Officials estimate that 3 million children's lives are saved each year.

A new vaccine technology is now being developed in a global effort, called the Children's Vaccine Initiative which will probably come to fruition early in the 21st century. It is an oral "supervaccine" that would provide lifetime immunity in a single dose. An oral liquid will contain microscopic pellets of vaccine-like time release capsules. Some dissolve right away, releasing the first dose of vaccine. Some of the other pellets take one to two months releasing another dose. Still other pellets take nine to twelve months to dissolve, releasing the third dose. The pellets are taken up by cells that live in the lining of the gut and are held there until they dissolve. This technology would provide lifetime immunity against all major infectious diseases in one dose without any boosters.

The goal is to develop a supervaccine that would keep its potency without refrigeration. This would eliminate a problem in poor rural tropics where refrigeration is expensive. Another goal is that the "supervaccine" would cost pennies per dose so that it could be used in poor sections of the world. Also since it would be given only once there would be no complicated schedules to remember.

11.7 Striking a Delicate Balance with Cancer Chemotherapy

Cancer chemotherapy became increasingly important since the 1970's. Chemicals that work by impairing cell division (alkylating agents) and agents that interfere with enzymes and thus block vital cell processes (antimetabolites) are used to attack rapidly multiplying cancer cells.

These chemicals have serious drawbacks that require very careful balancing of the good and bad effects. As they cannot distinguish between healthy and malignant cells, these drugs also attack rapidly multiplying noncancerous cells. They reduce the body's resistance to infection. Chemotherapy works on rapidly multiplying cells. Only about 5% of the cells in a tumor are multiplying at any one time. On the other hand, rapid normal multiplying cells that make hair and cells in the bone marrow are also attacked.

In spite of these problems chemotherapy has been spectacularly successful in one type of cancer, childhood leukemia. In 1965, 1% of children with leukemia survived. By 1995 over 85% of children with leukemia were cured of their illness.

The treatment consists of a combination of drugs made from a garden plant called periwinkle, and antibiotics and hormones. The periwinkle is an evergreen shrub that grows throughout the world. One extract of periwinkle is the drug vincristine which was found to be successful to some degree in treating childhood leukemia. The best antibiotic for the purpose was from bacteria found in sewage from the Island of Sardinia. Scientists also found that some hormone-like drugs were good at stopping cancer cells dividing. However, none of these substances cured cancer by themselves. It was learned that a combination of the three drugs was very effective.

The side effects are quite dramatic. The hair cells are destroyed and the patient loses his or her hair. However, the hair grows back after the treatment stops. The drugs are also toxic to the cells lining the intestine. There are other drugs that alleviate this problem. The idea is to balance the bad side effects with the curing effect.

11.8 CONTROLLING FERTILITY

The multiple biological functions of the female sex hormone progesterone became well known in the early part of the 20th century. These are maintenance of the proper uterine environment and inhibition of further ovulation during pregnancy. The latter function is in a way nature's contraceptive. The Austrian endocrinologist Ludwig Haberlandt was interested in the contraceptive potential of progesterone secreted by a woman's corpus luteum. It was found that the natural hormone displays only weak activity when given by mouth. It would work only with daily injections. Some chemists raised the possibility that artificial progesterone could work orally.

Carl Djerassi led a small chemical team in Mexico City which accomplished the first synthesis of a steroid oral contraceptive, Norethindrone, on October 15, 1951. This was accomplished starting with small inedible yams. Djerassi, together with co-workers, filed a patent application for Norethindrone on November 22, 1951. At that time Djerassi worked for a small company, Syntex, in Mexico and one of its competitors in the synthesis of oral contraceptives was Searle.

On August 31, 1953 Frank D. Colton of Searle filed a patent for a closely related oral steroid known as Norethynodrel. This was followed by a number of other oral steroids manufactured by different chemical companies. During 1953-54 these contraceptives were tested on animals by George Pincus of the Worcester Foundation for Experimental Biology in Shrewsbury, Massachusetts. The Harvard gynecologist Dr. John Rock and his colleagues performed the clinical studies to demonstrate contraceptive efficacy in humans. The tests showed that both the Syntex and Searle compounds were the best. Dr. Djerassi was recognized as the inventor of the first effective synthesized oral contraceptive. Later his patent was the first patent for a drug to be listed in the National Inventor's Hall of Fame in Akron, Ohio. Dr. Djerassi himself is very interesting both as a human being and a scientist.

Carl Djerassi was born in Vienna, Austria. His father was a Bulgarian Sephardic physician who was a descendant of Jews who had been exiled from Spain during the Inquisition. His mother was born of Jewish parents in Vienna. Djerassi's parents met in a medical school in Vienna where both were students.

When Carl Djerassi was 15 years old in December 1939, he and his mother fled Europe for America. He obtained a Ph.D. in chemistry from the University of Wisconsin in 1945. He went to Syntex (synt for synthesis and ex for Mexico) in Mexico City to do research on the synthesis of cortisone and then the synthesis of oral contraceptives, both starting with the wild Mexican yam. Djerassi went on to become a Professor of Chemistry at Stanford University. Later he founded one of the first environmentally aware pesticide companies. Then he became a successful writer of novels, poetry, and non-fiction.

The pill was approved by the Food and Drug Administration in 1960. By 1970 the pill was the most popular method of birth control with ten million American women users and many millions abroad. The 1960's were known as the decade of the pill. In

1990 it was estimated that over 100 million women were using the oral contraceptives or had used them in the past. The pill underwent changes over the years combining smaller doses of progestational and estrogenic components.

The pills had both side effects and benefits. Research has shown that older users of the pill are more likely than non-users to suffer from heart attacks, strokes, and blood clots in the veins. These effects are extremely rare in younger women but the occurrence is multiplied several times in all age groups among users who smoke. Benefits of the pill have been found to be reduction of pelvic inflammatory disease, ectopic pregnancy, benign breast disease, ovarian cyst, endometrial and ovarian carcinoma and some others. The Food and Drug Administration has added the positive effects to the list of side effects.

RU-486

Another chemical for birth control, Mifepristone, also known as RU-486, was developed by a French group of investigators in the early 1980's. Mifepristone counteracts progesterone, causing the uterus to let go of the fetal tissue. A single ingestion of RU-486 after confirmed pregnancy but not later than seven weeks after the last menstruation is used. A few days later, the woman returns to the clinic and is given Misoprostol, which causes uterine contractions. Within four hours, which must be spent at the clinic, the fetal tissue will pass in all but four percent of cases, in which surgery completes the abortion.

Norplant

The Food and Drug Administration around 1990 approved Norplant, an implant as a delivery system for steroid contraceptives over a five-year period. This eliminates the need to use a pill daily and makes it much easier to use.

Surgical Intervention

By 1992 over 30 percent of American women chose the surgical tying of tubes as a form of birth control. Men have a surgical method also, called vasectomy. It is interesting that the father of the pill had a vasectomy himself.

Assisted Reproductive Technology

The other side of controlling fertility is assisted reproductive technology. One method is in vitro fertilization. Eggs are harvested from a woman and fertilized in a laboratory glass dish with a man's sperm. After 48 hours, the embryo is implanted in the woman's uterus. This was first successfully used in 1978. After in vitro fertilization, a child was born, Baby Louise.

In order for a vitro fertilization to work, a drug must be used to increase the production of egg follicles. One such drug is Pergonal used by 30 to 50 percent of infertile couples undergoing treatment in 1994. The injectable drug is made from the urine of postmenopausal women. Another drug, Metrodin, does the same job but it is less costly. These developments have contributed to the marriage of technology to medicine together with all the other developments discussed in this chapter.

BIBLIOGRAPHY

Babayan, K. — *Retroperital Laparoscopic Nephrectomy Utilizing Three-Dimensional Camera Case Report,* Published in Journal of Endourology, Vol. 8, No. 2, 1994

Balkwill, Fran and Rolph, Mic — "Microbes, Bugs and Wonder Drugs," Portland Press Ltd., 59 Portland Place, London W1N3AJ, 1995

Djerassi, Carl — "The Pill, Pigmy Chimps and Degas' Horse," Basic Books, New York, 1992

Fischer, Jeffrey — "The Plague Makers," Simon & Schuster, 1994

Furman, Seymour and Escher, Doris — "Principles and Techniques of Cardiac Pacing," Harper & Row, 1970

Kleinfield, Sonny — "A Machine Called Indomitable," Random House, Inc., New York, 1985

Kremkau, Frederick W. — "Diagnostic Ultrasound, Principles and Instruments," 4th Edition, 1993

Melek, Jaques — "Cancer-Birth Control Pills, Cause & Effect Relationship," Sunbright Books, 1984

Smolan, Rick and Moffit, Phillip — "One Hundred Years of Healing, A Photographic Essay", Little, Brown and Company, 1992

Warren, Kenneth S. "Doing More Good Than Harm, The Evaluation of Health Care Interventions," Annals of the New York Academy of Sciences, Volume 703, 1993

Books without Listed Authors

"Ionizing Radiation," New York State Department of Health, Bureau of Environmental Radiation Protection, 2 University Place, Room 375, Albany, NY 12203-3399

"Oral Contraception in Perspective: 30 Years of Clinical Experience with the Pill," (History of Medicine Series), Parthenon Publications, 1987

CHAPTER 12

PLAYING GOD WITH GENETIC ENGINEERING

12.1 THE ROOTS OF THE GENETIC REVOLUTION

In 1865 an Austrian Monk Gregor Mendel published a paper, "Experiments in Plant Hybridization" which provided a basis for the mathematical analysis of inheritance. From the results of controlled crosses of garden peas, he showed that traits are inherited in a predictable manner as discrete bits of information or factors.

Mendel's work remained forgotten until the beginning of the 20th century. In 1900 the Dutch scientist Hugo de Vries and the German scientist Carl Correns rediscovered Mendel's paper and published research confirming his earlier work. Walter Sutton, a student at Columbia University, in 1902 substituted the word "gene" for Mendel's "factor." Sutton showed that genes are physically located on chromosomes. He studied chromosome behavior in grasshoppers and discovered that the grasshopper genetic material consists of 11 pairs of chromosomes and that Mendel's factors or genes resided in the chromosomes.

In 1908 a British physician, Archibald Garrod, proposed a theory that some human diseases result from the lack of a specific enzyme to perform a biochemical reaction. He put forth the then novel idea that lack of enzyme function resulted from a defective gene inherited at birth. It would take 30 years before his theory was tested and proved.

George Wells Beadle and Edward Tatum at Stanford University mutated or changed a gene of the red bread mold by X-ray radiation. After many trials they succeeded in mutating a gene that normally produced an enzyme necessary for the synthesis of vitamin B6. The mold with the mutated gene no longer produced B6 which was essential for the metabolism of the mold. Numerous metabolic mutant strains then were isolated that failed to produce the amino acid arginine. Beadle and Tatum found that each mutant strain lacked a different enzyme needed at different points along the arginine synthesis pathway. Beadle and Tatum published their work in 1941 confirming Garrod's hypothesis, now stated as "one gene/one enzyme." This was later broadened to "one gene/one protein."

An important investigation that led to the discovery of the chemistry of the gene itself was the work of the English microbiologist Fred Griffith who experimented with diplococcus *pneumoniae*, the ball-shaped bacterium that causes pneumonia. He discovered that there were two naturally occurring strains of this bacterium. The virulent smooth (S) strain possesses a smooth polysaccharide capsule that is essential for infection. The nonvirulent rough (R) strain lacks this outer capsule and has a rough surface appearance.

Griffith found that mice injected with the S strain succumbed to pneumonia. Neither the living R strain nor heat-killed S strain caused illness when injected alone. However, he found that when the R strain and the killed S strain were injected together, an infection resulted. In addition, he was able to retrieve the S strain bacteria from the infected mice. Griffith published his hypotheses in 1928 stating that some "principle" transferred from the killed S strain converted the R strain to virulence by enabling it to synthesize a new polysaccharide coat.

For the next 15 years a search for this "transforming principle" was undertaken by Oswald T. Avery and his team at Rockefeller Institute. Avery, Colin Macleod, and Maclyn McCarty reported in 1944 that they had isolated and purified the transformation principle. They found that it was deoxyribonucleic acid or DNA. Avery then concluded that a gene was made up of DNA which in this case made a gene product, the capsular antigen. Later it was found that several genes were required to produce the capsular antigen.

Avery's work on DNA published on February 1, 1944 did not get immediate acceptance because there were only four different nucleotides in its structure. It was not apparent how combinations of four components could result in an incredible diversity of large and complex protein molecules.

In 1952, Alfred Hershey and his assistant Martha Chase performed experiments at Cold Spring Harbor with viruses called bacteriophages which infect bacteria. The bacteriophage has an inner core of DNA surrounded by an outer capsule of protein. They more clearly made the connection between DNA and heredity. Using radioactive tracers and a blender they performed a series of experiments that convinced most biologists that DNA was the stuff of heredity.

There began an effort to connect the chemistry of DNA with the mechanism of heredity. The chemical base of DNA had actively been obtained many years before. In 1869 a German doctor Friedrich Miescher isolated a substance he called "nuclein" from the large nuclei of white blood cells. Chemical analysis of nuclein in 1900 showed that it was a long molecule composed of three distinct chemical subunits. These were a sugar, acidic phosphate, and five types of bases (Adenine, Thymine, Guanine, Cytosine and Uracil).

In the 1920s it was discovered that there were two forms of nucleic acid, ribonucleic acid (RNA) and deoxyribonucleic acid (DNA). It was found that DNA contained Adenine, Cytosine, Guanine and Thymine and that RNA contained Adenine, Cytosine, Guanine and Uracil. RNA and DNA also differed in their sugar composition. Avery and others discovered that DNA was the basis of heredity and investigations were begun to connect the chemistry and structure of DNA with the mechanism of heredity.

The study of the physical arrangement of atoms within the molecule was made possible by two developments that happened before World War II. The first was the law of chemical bonding that governs the arrangement of atoms within molecules. These laws had been described by Linus Pauling in the "Nature of the Chemical Bond," a series of monographs he wrote between 1928 and 1935. By 1950 the molecular structure of the individual subunits of DNA were known. These subunits were deoxyribose sugar, phosphoric acid and each of the four nucleotides, Adenine, Cytosine, Guanine, and Thymine. However, Pauling's laws could not determine the three-dimensional arrangement of the subunits of the very large DNA macromolecule. This was left to the second great prior development, the use of X-ray crystallography.

In 1912, the German physicist Max von Laue made the important discovery that X-rays are diffracted by the regularly arranged atoms of a crystal. It was also in that same year that the Australian physicist, William Henry Bragg, and his son, William Lawrence, worked out the mathematical equations to interpret diffraction patterns. They used the equations to determine the molecular structure of table salt. In 1934 the first X-ray photograph of an important biological molecule, the protein pepsin, was obtained by John Desmond Bernal. Linus Pauling and R.B. Corey in 1951 obtained precise atomic measurements of a helical *polypeptide* structure, the alpha helix.

12.2 THE RACE TO FIND THE STRUCTURE OF DNA

It was realized by 1950 that the structure of DNA could explain the replication of identical cells and the chemical basis of heredity. It was also understood that the first one to reach that goal would probably win the Nobel Prize. A number of people were involved: Rosalind Franklin, Figure 12-1, working in King's College in London, Linus Pauling of the California Institute of Technology, and a team of James Watson and Francis Crick, both working at Cambridge University. In order to have some understanding of what happened, it is necessary to look into the background and personality of the scientists.

Rosalind Franklin was born on July 25, 1920 to a prominent Jewish British family who traced their roots in England to 1763. Her uncle, Hugh was such an advocate of women's suffrage that he attacked Winston Churchill, a prominent antisuffragist, with a dog whip. Rosalind was born to a family that treated girls much more equal to boys than was the custom at the time. Her father brought a carpenter's workshop home in order that his children learn some useful skills. He included Rosalind who learned mechanical skills which later were useful in machine shop work in scientific laboratories.

By the time Rosalind Franklin was fifteen, she had decided to be a full-time professional scientist. At that time, around 1935, there were very few women scientists working in England. Her father thought she should go into social work but Rosalind was

Figure 12-1 Rosalind Franklin
Courtesy Cold Spring Harbor Laboratory Archives

determined and she entered Cambridge in 1938 to study physical chemistry. This university at that time allowed women only "titular" degrees but withheld ones that were identical to men's degrees. This was not changed until 1947.

Rosalind Franklin graduated and in 1942 during World War II she was appointed

Assistant Research Officer of the British Coal Utilization Association. Coal fuel was a war-related subject of importance. Between 1942 and 1946 Rosalind Franklin was the sole author of three research papers and a co-author of two. She also wrote her doctorate thesis which she submitted to Cambridge. She was awarded this degree in 1945.

In February 1947, Rosalind Franklin received an appointment to the Laboratoire Central des Services Chimiques de L'Etat in Paris. Women scientists in France had long been recognized as prominent contributors in the intellectual fields unlike England. Here she learned X-ray diffraction. This was the beginning of her work in crystallography which was the study of the structure of crystals by X-ray diffraction.

Rosalind Franklin learned the techniques of applying crystallography to non-crystalline biological substances which could be crystallized. She stayed in Paris until the end of 1950. She left Paris and went to King's college in London to study the structure of biological substances by X-ray diffraction. Later, people referred to her as being hired as the Assistant to Maurice Wilkins the Director of the laboratory at King's College. Wilkins himself denied this but it has been repeated as the truth by certain scientists. She was offered a fellowship on the understanding that she would be put in charge of building up an X-ray diffraction unit within the laboratory, which at that time lacked one. She set about doing just that.

Although she was not hired specifically to do research on DNA that was the most interesting project underway in the Biophysics Department of King's College in 1951. Rosalind Franklin was fascinated by DNA because it was the stuff of heredity. Previous attempts to delineate the structure of the DNA molecule by X-ray diffraction failed. Franklin was determined to find the answer through her background in that area.

King's College in 1950 did not welcome women, according to Ann Sayre in her book, *Rosalind Franklin and DNA*. It has been said as an excuse that King's College as a Church of England foundation cultivated male traditions which did not change appreciably when the college ceased to be theological or began to admit women. Rosalind Franklin did not feel welcome at the College. The male staff lunched in a comfortable large dining room while the female staff all lunched in either the students' hall or off the premises. There were seven women scientists on the laboratory staff other than Rosalind Franklin according to Horace F. Judson in his book, *The Eighth Day of Creation.*

In 1951 Rosalind Franklin set up a laboratory for the X-ray diffraction of DNA. She applied an improved method of hydration to DNA fibers and used new appropriate X-ray apparatus to obtain diffraction photographs of what was called the B-form of DNA, Figure 12-2. She gave a symposium on her work in November 1951 and discussed these photographs. According to Anne Sayre* Franklin's notes for the lecture described a big helix in several chains. Somehow Wilkins came away with the impression that she was definitely "anti-helical." Another participant in the symposium, James Watson, did not take notes.

---

*See Bibliography

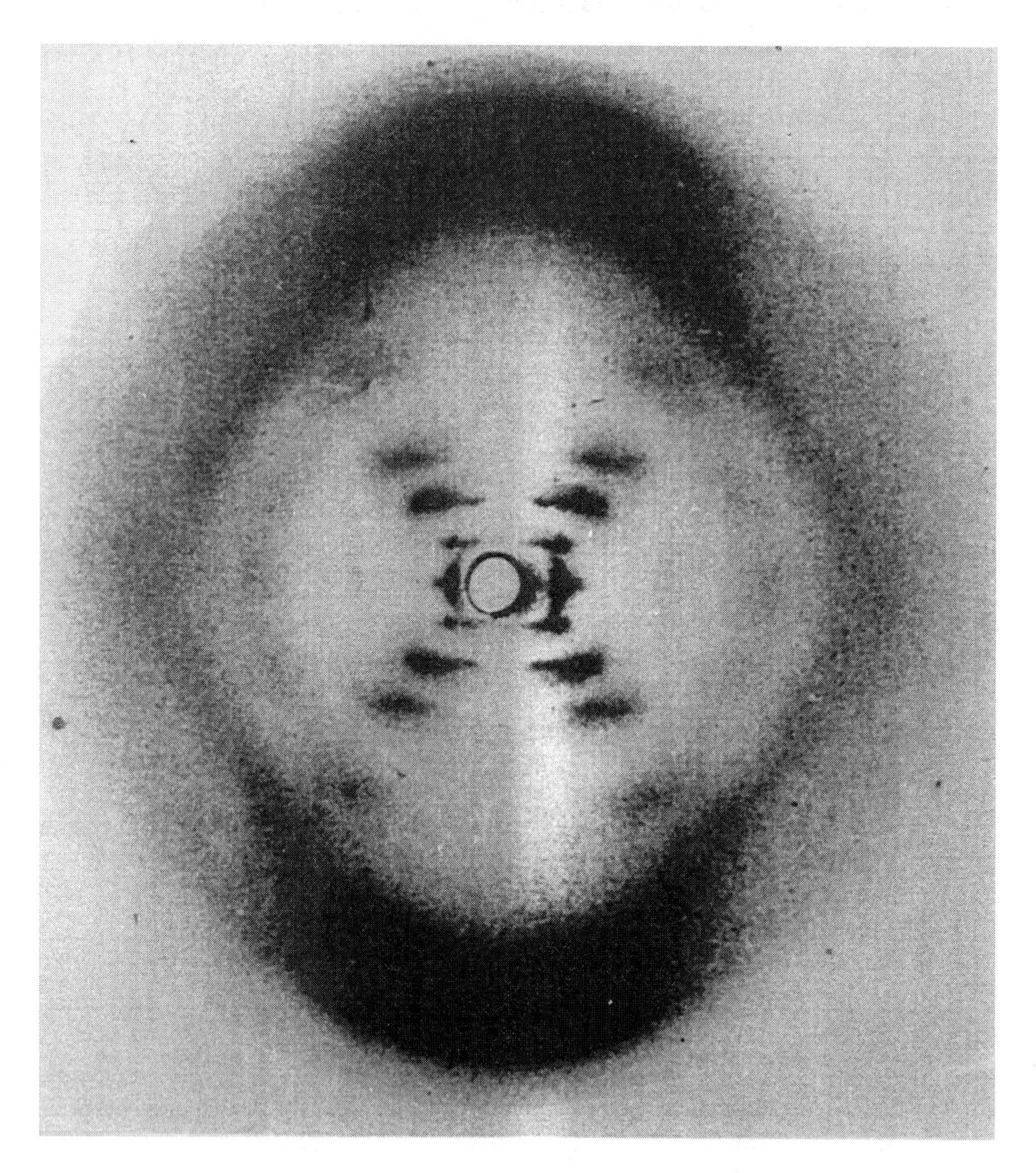

Figure 12-2 Diffraction X-Ray of B-Form of DNA, 1951
Courtesy of Cold Spring Harbor Laboratory Archives

This leads to the second team working on the structure of DNA, James Watson, an American geneticist, and Francis Crick, a British physicist, both at the Cavendish Laboratory in Cambridge. Watson and Crick worked with model building, using first paper models and then metal ones.

At about this time in 1951, Linus Pauling obtained precise atomic measurements of a helical polypeptide structure, the alpha helix. Watson, according to his book, *The Double Helix,* thought that Pauling was some six weeks away from discovering the structure of DNA. When Pauling's son visited Watson and Crick at Cavendish he received information designed, according to Watson, to put his father off the track.

In the summer of 1952 Wilkins showed Franklin's X-ray diffraction photograph of the B-form of DNA to Watson without her permission. He commented that the picture showed the helix but, "that damned woman just won't see it."* Actually Rosalind Franklin, as her notes showed, had already back in 1951 stated that DNA had a helical structure. There was a lack of communication and a clash of personality between Franklin and Wilkins. Watson and Franklin also had a decided dislike of each other. Part of this may be because she was a woman working in what male scientists considered to be a man's world.

Watson and Crick used information from Franklin's B-form X-ray photographs and her statement that the sugar-phosphate chains of DNA must be on the outside of the helical structure. There was another vital piece of information that enabled Watson and Crick to arrive at the correct structure of DNA. This was the work of Erwin Chargaff in 1949. He found that in DNA the number of molecules of adenine equals the number of molecules of thymine while the number of guanine molecules was equal to the number of cytosine molecules. Crick realized that a complementary pairing of adenine and thymine and a complementary pairing of guanine and cytosine would explain Chargaff's data. This implied a chemical bonding of complementary pairs.

The laws of chemical bonding that govern the arrangement of atoms were laid out by Linus Pauling of the California Institute of Technology in *The Nature of the Chemical Bond,* a series of monographs he wrote between 1928 and 1935. An expert in hydrogen bonds, Jerry Donohue, visited Watson and Crick at the Cavendish Laboratory. Donohue was considered to be second only to Pauling in his knowledge of hydrogen bonds. He provided some useful hydrogen bonding advice that Watson and Crick used in putting together their model.

Watson and Crick brilliantly synthesized all of the above into a correct structure of DNA. They built and manipulated a series of wire and metal models. On February 28, 1953, Watson and Crick arrived at a structure (Figure 12-3) which explained replication and heredity in a straightforward and convincing way. They published their results in the journal Nature on April 1953. This issue included among others an article by Rosalind Franklin and her student, Raymond Gosling, which offered experimental confirming evidence.

The DNA molecule that Watson and Crick proposed is composed of two alpha helices resembling a gently twisted ladder, Figure 12-4. The rails of the ladder, run in opposite directions containing alternating units of deoxyribose sugar and phosphate. The

* From page 151 of *Rosalind and DNA*

Figure 12-3 Watson (On left) and Crick with DNA Model
Photographer, Science Source/Photo Researchers

planar nucleotides stack tightly on top of one another, forming the rungs of the helical ladder. Each rung is composed of a pair of nucleotides held together by relatively weak hydrogen bonds. Adenine always pairs with thymine and cytosine always pairs with guanine. In that way the nucleotide alphabet on one half of the DNA double helix determines the alphabet of the other half. The planar nucleotides stack tightly on top of one another forming the rungs of the helical ladder. There are 10 bases per turn of helix.

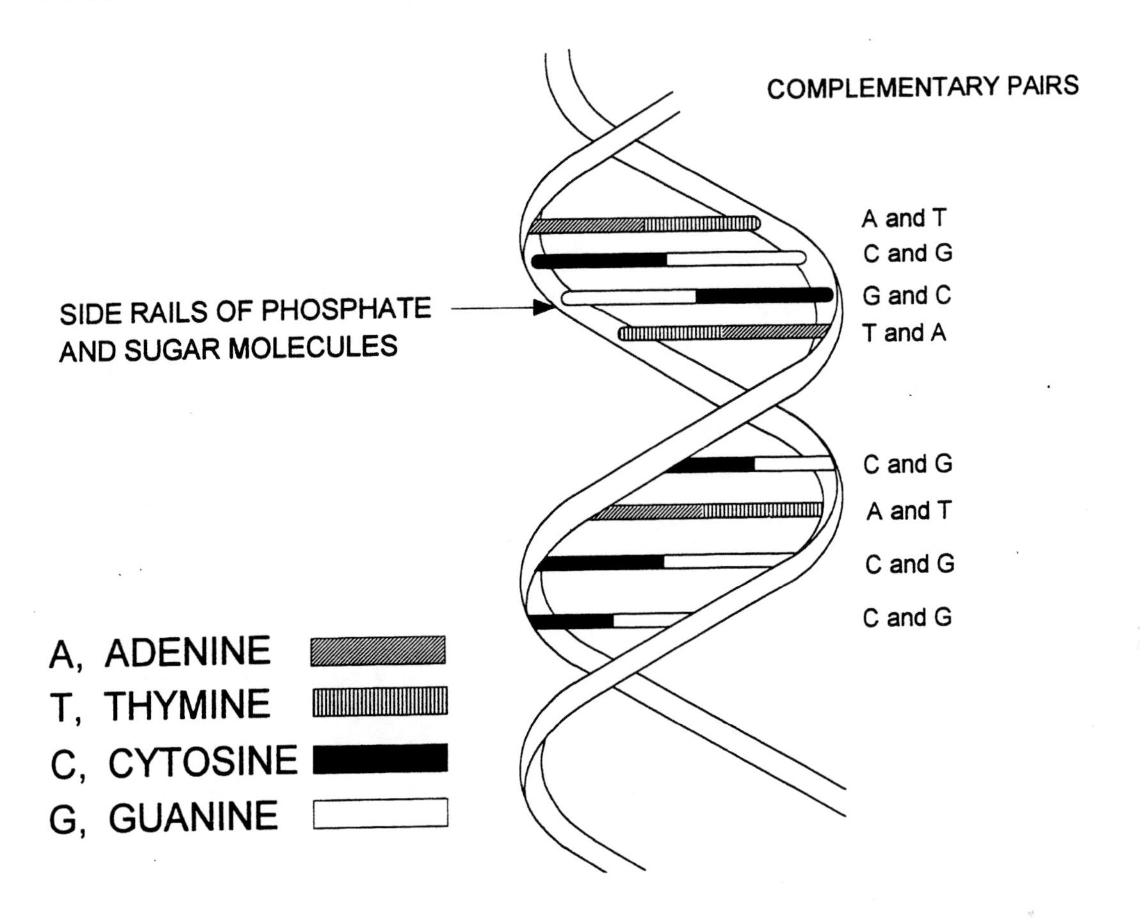

Figure 12-4 DNA, The Chemical Ladder of Life

The DNA structure described above shows how the molecule precisely replicates during cell division, so that each daughter cell receives an identical hereditary instruction. The hydrogen bonds between nucleotides break up which allows the DNA ladder to separate into two equal helices. Each half serves as a template for the reconstruction of the other half. The result would be two identical DNA molecules. This explanation of replication was confirmed experimentally by Matthew Meselson and Franklin Stahl at the California Institute of Technology in 1958. Arthur Kornberg at Washington University showed in the same year that a particular enzyme was synthesized only in the presence of a DNA template.

Rosalind Franklin died of cancer on April 16, 1958. In 1962 the Nobel Prize for Medicine and Physiology was shared among Francis Crick, James Watson, and Maurice Wilkins. The Nobel Prize is not awarded posthumously so Rosalind Franklin was not included. Many scientists believe that she would have been if she had not died.

12.3 Breaking the Genetic Code

The next great question was how the genetic language translated into the synthesis of specific proteins. In 1957 Francis Crick laid down two basic principles that led to the cracking of the genetic code:

1. Genetic information is arrayed in a strictly linear fashion along the length of the DNA molecule.
2. The "central dogma" is that genetic information stored in DNA flows through RNA which is the intermediate translator of the genetic code.

In 1961 Crick and Sydney Brenner performed a series of experiments that suggested that the genetic code consists of three-letter sequences of nucleotides called codons in RNA. The different codons code for one of the 20 amino acids. Amino acids in turn are the building blocks of protein.

The genetic code was cracked in 1966 by teams led by Marshall Nirenberg at the National Institutes of Health and H. Gobind Khorana at the University of Wisconsin. They confirmed that triplet RNA codons specify each of the 20 amino acids. Since all proteins begin with the amino acid methionine, its codon AUG represents the start for protein synthesis. Three codons UAA, UAG and UGA are stop signals that terminate translation.

Only methionine and tryptophan are specified by a single codon. All other amino acids are specified by two or more different codons. Because of this redundancy, single-base changes in RNA often do not change the amino acid coded. An example of this is that any codon beginning with GG specifies the amino acid glycine regardless of the nucleotide in the third position (GGU, GGC, GGA, or GGG).

All possible RNA combinations were tried, which resulted in a complete dictionary for the translation of RNA into one of the 20 amino acids.

As time went on there were a number of changes made to Crick's central dogma. The flow of genetic information is not one-way exclusively. Elements exert feedback regulation. A single gene may through rearrangement and biochemical editing be responsible for the production of several different proteins.

One major change is that some organisms do not use DNA for storing genetic codes. RNA viruses, or retroviruses, store genetic information as RNA. Their genetic information is converted to DNA by the enzyme reverse transcriptase. This is a backflow of genetic information. Another major change is that genes are not immutably fixed on the chromosomes. Physical rearrangement of genes permits a large array of proteins from a relatively limited amount of DNA code. These are just some of the changes that have been made in Crick's central dogma.

12.4 THE DEVELOPMENT OF GENETIC MANIPULATION

After the elucidation of the DNA molecular structure a series of discoveries were made. These included the restriction enzymes which act as biological scissors operating on DNA, and recombinant DNA which can combine genes of different species.

RESTRICTION ENZYMES

Around 1953 Salvador Luria, Giuseppe Bertani of the University of Illinois and Jean Weigle discovered a primitive immune system in bacteria. Bacteria are normally attacked by viruses called bacteriophages. This immune system seemed to be a property of the bacterial cell which is able to restrict the reproduction of bacteriophages. In 1962 Werner Arber of the University of Geneva showed that the resistant bacteria possess an enzyme system that selectively recognizes and destroys bacteriophage DNA within the bacterial membrane. In addition, the enzyme modifies the DNA of the bacterium to prevent self-destruction.

However, these enzymes in themselves were of no practical value as tools for manipulating DNA. They cut the DNA molecule randomly at different positions. In addition, the restrictions or cutting function was combined with a modification function. In order for this technique to be useful for manipulating DNA the cutting must be precisely at one point on the DNA chain. Also the restriction function must be separate from the modification.

It was not until 1970 that the technology of restriction enzymes became practical. In that year Hamilton Smith and his student Kent Wilcox at Johns Hopkins University isolated a new restriction endonuclease from Haemophilus influenzae called Hind II. This cleaves DNA predictably and the restriction activity is separate from the modification activity.

A colleague of Smith's at Johns Hopkins, Daniel Nathans, then showed that restriction endonucleases had very broad applications. Nathans used the purified DNA of Simian virus 40, a small virus that infects monkeys. He separated the resulting fragments by size in an electrical field. He deduced the order of the fragments and the corresponding restriction sites in the 5000 nucleotide circular chromosome. This created a restriction map that was then related to the existing map of SV-40.

12.5 THE DEVELOPMENT OF RECOMBINANT DNA

In 1972 Paul Berg used a restriction enzyme, EcoRI to cut DNA in a predictable manner and showed that DNA fragments from different organisms could be merged together. He joined the DNA from two organisms into a plasmid, a small circular chromosome. He planned to introduce the recombinant plasmid as a vector into animal cells. Concern in both the scientific and general communities halted research on gene transfer into mammalian cells for a number of years.

A vector is an autonomously replicating DNA molecule into which a foreign DNA fragment is inserted and then propagated in a host cell. Plasmids are the simplest bacterial vector. They are circular DNA molecules that exist separate from the main bacterial chromosome. One type of plasmid, called the relaxed plasmid, replicates autonomously of the main chromosome. A restriction enzyme cuts the plasmid in a specific location. A gene is inserted into the plasmid vector which is then introduced into an appropriate host cell. Subsequent mitosis of the host cell creates a population of clones, each containing the gene of interest.

In 1973 Stanley Cohen and Annie Chang of Stanford University showed that a recombinant DNA molecule can be maintained and replicated with E. coli, a common bacterium found in the human intestine. Cohen and Chang purified from E. coli a plasmid that incorporated a number of important features. The first was a unique restriction site, to allow the molecule to be cut at a single location by the restriction enzyme *Eco*RI. The second feature of this plasmid is that it contains a nucleotide sequence, the origin of replication, to allow the plasmid to be replicated within a host bacterial cell. The third feature was a gene which codes for resistance to the antibiotic tetracycline. The new plasmid from E. coli was named pSC101 after Stanley Cohen.

Cohen's and Chang's first step was to devise a method to introduce this plasmid into E. coli efficiently. They succeeded in this and the transformed cells were then spread on nutrient agar plates containing the antibiotic tetracycline. The bacteria grew showing that the plasmid had been taken up and expressed.

The next step was to construct a recombinant plasmid. A second plasmid, pSC102, was isolated that contained a gene coded for resistance to another antibiotic, kanamycin. Cohen and Chang cut pSC101 and pSC102 with the restriction enzyme, mixed together and rejoined with DNA ligase. The resultant recombinant plasmid was transformed into E. coli cells, which were plated on media containing both tetracycline and kanamycin. The appearance of colonies with resistance to antibiotics, tetracycline and kanamycin, showed that a recombinant DNA molecule had been introduced successfully into living bacterial cells.

Around 1974 Stanley Cohen joined with Herb Boyer to produce a recombinant molecule containing DNA from two different species. These accomplishments and others enabled scientists to enter the era of genetic engineering. The recombinant DNA techniques enabled human beings for the first time to change the genetic constitution of a living thing in a controlled manner and to transcend established species barriers. Each novel DNA combination creates a new biological entity with altered genetic and biochemical characteristics. Recombinant DNA depends heavily on restriction enzymes.

Types of Restriction Enzymes

Restriction enzymes are used as molecular scissors to cut DNA in a precise and predictable way. They are members of the class of nucleases, enzymes which break the phosphodiester bonds that link adjacent nucleotides in DNA and RNA molecules. A phosphodiester bond is one in which a phosphate joins adjacent carbons. There are two basic types of restriction enzymes, endonucleases and exonucleases. Endonucleases cleave nucleic acids at internal positions. Exonucleases progressively digest from the ends of nucleic acid molecules.

There are three major classes of restriction endonucleases, type I, type II, and type III. The restriction enzyme used in DNA work are type II. These have only restriction activity while modification is caused by a separate enzyme. Also each cuts in a predictable and consistent manner, at a site within or adjacent to the recognition sequence. Also, type II restriction enzymes unlike the other types requires only Mg^{++} as a cofactor.

More than 1200 type II enzymes have been isolated. More than 70 types are commercially available. Restriction endonucleases are named according to a specific nomenclature: The first letter is the initial letter of the genus name of the organism from which the enzyme is isolated. The second and third letters are usually the initial letters of the organism's species name. The first three letters of the endonuclease name are italicized. A fourth letter indicates a strain of an organism. A Roman numeral usually indicates the order of discovery. An example is *Eco*RI where E is the genus Escherichia, co is the species coli, R represents the strain RY and I the first one found.

Various enzymes leave cohesive or sticky ends which are extremely important in making recombinant DNA molecules. These exposed nucleotides serve as a template for realignment. This allows the complementary nucleotides, adenosine-thymine and cytosine-guanine, or two like restriction fragments to hydrogen bond to one another. A given restriction enzyme cuts all DNA in exactly the same way, whether the source is a bacterium, a plant, or a human being. Thus any sticky-ended fragment is generated by the same restriction enzyme.

One of the tools of DNA recombination is the plasmid in a bacteria called E. coli. E. coli is one of the normal bacteria that inhabits the human colon. Its primary genetic complement is contained on a single chromosome of approximately 5 million base pairs. A single E. coli replicates by binary fission to create a clone of identical daughter cells. One cell of E. coli produces more than 1 billion cells in 11 hours.

Because the genetic code is nearly universal, a plasmid of E. coli can accept foreign DNA derived from any organism. The DNA of bacterium, a plant, a fruit fly, an animal, or a human is constructed of the same four nucleotides, adenine, cytosine, guanine, and thymine assembled in the same structure. DNA is replicated by the same basic mechanism for all organisms. Each organism transposes DNA into messenger RNA, which in turn is translated into proteins according to the genetic code. A foreign gene inside the plasmid of E. coli is replicated in exactly the same manner as the native bacterial DNA.

12.6 Changing Plants by Genetic Engineering

In 1994 the U.S. Agriculture Department estimated that there were almost 300 projects underway to develop genetically engineered plants through recombinant DNA. One example is the genetically engineered flavr savr tomato. The idea is to turn off the genes that control softening so that these tomatoes can be allowed to ripen on the vine but still not turn mushy when shipped. With ordinary tomatoes they are picked green and turned red with ethylene gas. Another genetic change is to insert a gene that would increase their solid content. Since tomatoes are mostly skin and water, increasing the amount of solid could have a big economic advantage for processors. Also a gene modeled after a flounder gene has been added to tomatoes to produce a cold resistant variety.

Similarly, potatoes are being genetically changed to produce higher solid potatoes. A potato with more starch and less water would hold up better during processing. Since potatoes are almost exclusively mechanically harvested and many are bruised as they bump into conveyor belts, the smaller fluid content of the genetically altered potato would have a distinct advantage. Secondly, a potato with less water would absorb less oil when it is deep fried so French fries potentially could be less fatty. Also, pest-resistant potatoes have been developed by genetic manipulation.

Corn is being modified by implanting a gene into sweet corn that would prevent the sugar from being converted to starch, thus keeping the vegetable sweet long after it is picked.

Cooking oils that contain saturated fat pose health hazards. Genetic changes are being made to lower the amount of saturated fat in soybeans and corn oils.

Genetic changes are being made to turn off the gene in coffee plants that cause caffeine to form. This would decrease the processing cost of decaffeination and could also result in better tasting decaffeinated coffee.

A soybean has been genetically altered to be more nutritious. Ordinary unmodified soybeans have six of the eight amino acids that humans require from food. The genetically altered soybean has all eight amino acids. Also soybeans naturally are moderately high in saturated fats, which are linked to heart disease. The modified soybean has much less saturated fat and is therefore healthier.

One of the problems with ordinary soybeans is that they have compounds that produce flatulence. Traditional breeding techniques produced a small decrease in this unwanted compound. Genetic engineering decreased it much further and even improved the taste. All of this made a much superior soybean.

12.7 Altering Animal Genes

Animals have also been altered by adding the genes of humans and other animals. These changed animals are called transgenic. One of the uses of this technology is to produce valuable human proteins in their milk. This turns sheep, goats, and cows and other animals into living pharmaceutical factories which produce large copies of rare human natural substances. One of the animals, a sheep, has produced in milk about 2-1/2 ounces of a protein, human alpha-1-antitrypsin, used to treat a form of emphysema.

Another example is a transgenic goat that produces tissue plasminogen activator, or TPA, a protein widely used to treat heart attacks. This animal factory produces 3 grams of TPA per liter. Dairy transgenic cows have also been made to produce milk with human lactoferin, an iron binding protein.

As part of this research, Dr. Ian Wilmut of Scotland, cloned a sheep in 1996 from the DNA of a cell in the udder of a six-year-old ewe. The next step is to create clones of transgenic sheep which produce human proteins.

The cloning generated a storm of controversy because of the possibility that the same technology could be used to clone human beings.

One genetically engineered product that has aroused controversy is the bovine somatotropin (BST) that can be injected into cows to make them produce more milk. All cows produce BST naturally. Cows that produce more BST yield more milk. With synthetic BST, farmers can turn an ordinary cow into a high-yielding one. The most productive cows get the most mastitis (an udder disease commonly treated with antibiotics). BST-injected cows may get even more mastitis, and other diseases. This will require more antibiotics which raises the possibility that milk drinkers have drug-resistant bacteria. Milk produced by genetically engineered BST is undetectable from milk produced by natural BST.

12.8 Laying the Foundation of Gene Therapy

In 1982 the Banbury Conference Center of Cold Spring Harbor Laboratory held a meeting on human genetic diseases. Up to that time gene therapy, the curing of diseases by changing genes, did not have the tools to perform experimental testing because of the inefficient gene transfer methods. Before that time gene transfer only transferred one foreign gene into one cell out of many hundreds of thousands exposed cells.

At about the time of the meeting Robert Weinberg, Howard Temin, Edward Scolnick, and their colleagues made a great advance by using retroviral vectors to produce cDNA (DNA from RNA). This made it possible to transfer foreign genes into virtually all exposed cells. This led to experiments to test the feasibility of efficient and large-scale transfer of wild-type genes into defective cells and to determine whether such a transgene could complement a genetic defect and correct a disease phenotype.

The first successful experiment in vitro was in cells deficient in the purine biosynthetic enzyme hypoxanthine guanine phosphoribosyltransferase (HPRT).

Defects in HPRT underlie Lesch-Nyhan disease which is one of over 3,000 that are due to genetic defects. Another later in vitro experiment was in cells defective for adenosine deaminase, another purine biosynthetic enzyme whose absence is responsible for severe combined immunodeficiency disease (SCID). Also it was demonstrated in vitro that many types of cells such as bone marrow progenitor and stem cells, cultured hepatocytes, epithelial cells, fibroblasts, keratinocytes, lymphocytes, myoblasts, smooth muscle cells, and others were susceptible to efficient gene transfer with retroviral vectors. This implied that disorders of these cell types might be suitable targets for gene therapy.

The above experiments were of the type performed in test tubes in the laboratory. These experiments led to a study involving a retrovirally transferred foreign gene in humans. One of the basic problems was to meet the requirements of the National Institute of Health and the FDA to regulate the application of molecular genetics to human experimentation. In 1989 W. French Anderson and Steve Rosenberg and their colleagues undertook the first study involving a retrovirally transferred foreign gene in humans. The study involved the introduction of the neomycin resistant gene into tumor infiltrating lymphocytes (TIL) as a way of marking the cells and determining their fate and stability after return to tumor-bearing patients. This experiment was performed basically to satisfy a set of criteria for human gene transfer experiments, called "points of criteria." The study also showed that genetically modified TIL cells could make their way back to an existing tumor in human beings and therefore could potentially deliver an antitumor gene to a tumor.

In 1990 Michael Blaese, W. French Anderson, Ken Culver and their NIH colleagues performed the first therapeutic genetic study. This experiment involved two little Ohio girls with a genetic disease called ADA deficiency. The girls were born without functioning immune systems. This meant facing life in isolation and early death because of extreme vulnerability to infection. ADA deficiency means the unending fear of contacts with disease. The experiment required taking samples of the girls' white blood cells, infecting the cells with a specially tailored retrovirus carrying a healthy copy of the necessary gene, and then putting the blood cells back into the girls. One of the girls was four years old and the other was nine years old.

In 1992, W. French Anderson and Kenneth Culver working at the National Heart, Lung, and Blood Institute in Bethesda, Maryland, announced that the first authorized gene-therapy experiment was successful. A year after the experiment the two girls were able to go to school. Both of them have functioning immune systems and their doctors said that they are leading near-normal lives. This has encouraged experimenters to expand gene therapy to more widespread diseases such as malignant melanoma, a deadly skin cancer. Preliminary tests for safety of genetically altered white cells showed no adverse effects.

Another gene therapy advance is in cystic fibrosis. This disease is one of the most prevalent hereditary illnesses among whites. One in 20 Americans is a silent carrier of the defect, and if two carriers have children, they have a one-in-four chance of bearing a child with the disease.

The disease results when a child lacks a working version of the cystic fibrosis protein which controls the flow of chloride molecules across the cell membrane. Because of the lack of the cystic fibrosis protein the passage of chloride and sodium molecules through the body is impeded. The water balance is disrupted and mucus builds up in the lungs. Infectious bacteria thrive in the mucus and destroy the fragile tissue there, eventually leading to pulmonary failure and death.

In 1989 Dr. Lap-Chee Tsui of the Hospital for Sick Children in Toronto and others discovered the cystic fibrosis gene. Then scientists performed an experiment in a test tube, adding the gene to diseased lung cells of cystic fibrosis patients. The cellular defect was corrected and the test tube experiment was followed by animal testing in 1992.

Dr. Ronald G. Crystal of the National Heart, Lung and Blood Institute and his colleagues infected laboratory rats with an adenovirus vector. This virus causes colds, bronchitis and other respiratory diseases. Dr. Crystal and colleagues removed the gene of the adenovirus that allows it to multiply and kill body cells. Then they used recombinant DNA to insert a copy of the human cystic fibrosis gene into the adenovirus. The altered adenovirus vector was then placed in a liquid which was dripped down a brachial tube of a rat.

The viruses flourished for six weeks, causing the rats' lung cells to produce human cystic fibrosis proteins. The rats themselves do not have cystic fibrosis but the experiment showed that the human gene addition in the adenovirus vector could cause lung cells to produce the missing human cystic fibrosis protein. However, there remained the possibility that the crippled adenovirus could regain its ability to multiply and then begin dividing wildly and generating uncontrollable amounts of cystic protein. Soon experiments showed that excess cystic fibrosis protein was not harmful to the body. Another question is that in human patients the immune system must over time rally a defense against the altered adenoviruses, killing the vector and rendering the method useless. To answer these and other questions, the next step is to use the method in monkeys and then slowly test it in humans first for safety. However, the foundations for an eventual gene therapy for cystic fibrosis have been laid down.

In order to use gene therapy successfully it is necessary to know the sequence of the four nucleotides, adenine, cytosine, guanine, and thymine. They are referred to as A, C,G, and T. Their order in a long sequence determines the gene and its function in the body. Automated methods of determining the DNA sequence has become very important.

12.9 AUTOMATED DNA SEQUENCING

An early DNA sequencing technique developed by Applied Biosystems of California in the early 1980's used fluorescent dyes to determine the order of the A^s, C^s, G^s, and T^s in a cloned sample of DNA. The DNA is first heated to separate the strands. The single strands are then cut at various points, and the resulting fragments are added to a mixture of fluorescent dyes in which strands that end in a specific letter are all tagged with the same fluorescent dye. The mixture is next run through an agarose gel which in turn is passed through a laser beam. The shorter fragments which travel farther in the gel are exposed to the beam first, followed by successively longer fragments. Each base (A, C, G, or T) emits a different fluorescent color corresponding to its position in the DNA sequence. The DNA sequence is obtained from the sequence of fluorescing colors emitted as the fragments in the gel go past the laser beam. This can be used to determine the structure of a gene.

Every gene begins with a sequence of three nucleotides known as an initiating codon and ends with a sequence of three nucleotides known as a stop codon. With automated DNA sequencing and a computer a scientist can scan a long DNA sequence for initiating and stop codons, marking off the genes between them. The complete determination of all of the genes of a species is known as the genome of that species.

Other countries also initiated an automated DNA sequencing system. In Japan the Science and Technology Agency began in 1981 to support a project to automate DNA sequencing. In 1989 a Japanese DNA sequencer was produced by Hitachi and marketed only in Japan. In Europe an automation project was begun in the early 1980's at the European Molecular Biology Laboratory, EMBL, in Heidelberg, Germany. They used a fluorescent-dye detection system that was somewhat different from American designs. The EMBL design served as the basis for the ALF DNA sequencing system marketed by LKB-Pharmacia in Sweden beginning in 1989.

A new technique to produce enormous amounts of short stretches of DNA, called the polymerase chain reaction (PCR), was developed by Karl Mullis and published in late 1987. Copying DNA had become possible with the advent of cloning in the 1970's. However, cloning had limitations since it required bacteria, yeast, or other organisms and also long sequences of DNA. PCR made it possible to repeat DNA fragments from much smaller samples of DNA even down to the theoretical limit, a single molecule. The PCR reaction built on what was known about how certain enzymes known as DNA polymerase enzymes synthesized new DNA strands from existing ones. The PCR reaction was carried out in a test tube. The ingredients included the DNA to be analyzed, the nucleotide precursors to make new DNA and a DNA polymerase enzyme.

Starting with a very small amount of DNA a fragment could be copied hundreds of billions of times. This was accomplished without having to clone it in bacteria, yeast or other organisms. All that was necessary was a means to heat and cool the reaction mix and the use of an enzyme and chemical reagents. PCR was used in the detection of the

AIDS infection and also genetic diseases. PCR enabled many laboratories to make billions of copies of small DNA samples cheaply and efficiently. PCR made genome projects practical by generating billions of copies of short DNA sequences.

12.10 FINDING THE GENOME OF MICROSCOPIC ORGANISMS

BACTERIAL GENOMES

Bacteria cells do not have a nucleus but they do have genes. Two bacterial genomes, *Haemophilus influenzae* and *Mycoplasma genitalium* were mapped in the early 1990s. This was followed by Esherichia coli, a very prevalent bacteria in the human body. All of these are relatively simple compared to the eukaryotes or cells with a nucleus. Here there are a number of chromosomes, each with thousands of genes, and millions of nucleotide pairs.

THE FIRST EUKARYOTE GENOME

The first eukaryote genome to be mapped was *Sacchromyces cerevisiae* or yeast, in 1996. Yeast has sixteen chromosomes with a total of 6,000 genes and 12.5 million nucleotide pairs. The work done on the yeast genome will help to map the human genome. Yeast has a genome three times the size of the genome of the bacterium, *Esherichia coli.* However, a few yeast genes are similar to human genes.

The yeast genome project was initiated around 1983 by Stephen G. Oliver, a geneticist at the University of Manchester Institute for Science and Technology in England. The work spread to nearly 1,000 laboratories around the world. Pieces of chromosomes were parceled out to different laboratories for analysis.

The work on yeast genome helped in the realization of the human genome program. The scale of the human genome with its 70,000 genes is, of course, much greater than the completed yeast genome.

12.11 THE HUMAN GENOME PROJECT

Charles DeLisi, a mathematical biologist, was the originator of the Department of Energy Human Genome Project. The human genome project is the recording of the linear sequence in the order of the nucleotides, adenine, cytosine, guanine, and thymine. In addition, it includes the linear sequence of information in the order of amino acids in human beings.

A meeting held at Rockefeller University in March 1979 brought out the need for computers and data banks to determine and store sequence data. It was pointed out that transmitting a long, seemingly random sequence of four letters, A, C, G, T from one person to another without errors is hardly possible except by putting the information on a computer-readable medium. There were people at the meeting who were enthusiastic about the idea of a database and DNA sequence center at the Los Alamos DOE Laboratory. There were also scientists from Cambridge, England, and a group from the European Molecular Biology Laboratory (EMBL).

A workshop was held at the National Institute of Health (NIH) on July 14, 1979. A plan was presented a little later by NIH for a human genome project. Phase I was the establishment of a public national database for DNA sequence information. Phase II was to develop mathematical and computational methods to analyze the information. Late in 1981 Phase II was dropped because of cost, politics, and lack of clarity about what a DNA analysis should be. A contract was awarded by NIH on June 30, 1982 to a private firm, Bolt, Beranek & Newman (BBN) together with a group at Los Alamos and the database began to operate in October.

It was soon found that automatic sequencing instruments were necessary because of the magnitude of the problem. J. Craig Venter and Leroy Hood were instrumental in devising methods to automate the sequencing of DNA. Venter's work was originally done at the National Institute of Health and later as the head of the Institute for Genomic Sciences, Inc. in 1992. The Institute used state of the art automation including 50 DNA sequencing machines, computer workstations, many small computers, and much automated equipment. This organization and the French Genethon became leaders in automating the human genome project.

In the meantime a struggle broke out between James Watson and Bernadine Healy. In October 1988 James Watson had become the head of the Human Genome Project at the National Institute of Health. In 1991 Bernadine Healy was appointed NIH Director. Watson was the most famous geneticist at this time and Healy was the most powerful biomedical research administrator. Watson in the 1950s had been involved in a controversy with a female scientist, Rosalind Franklin. Some years later he became immersed in another controversy with a female scientist. However, in forty years the situation had changed drastically. Dr. Bernadine Healy had experienced sexism in her class at Harvard Medical School and had also been a victim of sexual jokes later while working at Johns Hopkins.

Their battle began back in 1985 when Dr. Healy was Deputy Director for Biomedical Affairs in the Office of Science and Technology Policy. Watson complained about the Reagan White House administration of genetic technologies. He stated that "The person in charge of biology is either a woman or unimportant. They had to put a woman some place." Healy took offense at this although Watson claimed that he meant it as a slap only at the Reagan administration. The two struggled over a number of issues including the question of patenting genes and gene products. Healy was for the patents while Watson thought the patents would impede international research. On April 10, 1992, after Healy was appointed as Director of NIH in 1991, Watson resigned as head of the Human Genome Project.

The Human Genome Project has continued despite all of the above political wrangling. The technical problems have been alleviated by spreading out the work. It is scheduled to be completed by 2001.

12.12 DNA POINTS THE FINGER AT CRIME

Scattered through human DNA are regions in which very short adjacent nucleotide sequences repeat over and over up to thirty times. Such regions are called Variable Number of Tandem Repeats (VNTR). For a specific region of repeats, that is a particular VNTR, the number of repeats differs from person to person. This led to a system of identifying a criminal suspect.

DNA from semen and blood is collected from a crime scene and compared with the DNA from suspects. Technicians prepare the sample DNA and cut it with an enzyme, a restriction endonuclease, to produce fragments of discrete length. For this, an endonuclease is chosen that cuts outside the VNTR. Technicians then compare the lengths of the DNA fragments from different samples by gel electrophoresis. In this method a slab of gelatin material is prepared. The gel slab contains a row of small wells or holes near one edge, and each DNA sample, dissolved in water, is squirted into a different well. An electric current is then run the length of the gel to drive the DNA samples through the walls of the wells and into the gel. The current is turned off before the DNA runs out the other end of the slab, so the DNA fragments are left distributed throughout the gel. Since shorter fragments travel faster than longer ones, the position of each fragment reflects its length. Each sample loaded into a well has many DNA molecules in it so that there are many copies of each fragment size. For each fragment size the copies move together as a band that can be seen if the DNA is stained.

The next step is to place a sheet of nylon on the gel, and all the DNA bands are driven up and out of the gel onto the nylon. The bands stick tightly to the nylon and maintain the same relative positions as in the gel. Then the nylon sheet is treated with a radioactive DNA, a probe that binds only to the DNA in the bands containing the VNTR of interest. That makes the VNTR bands radioactive. The exact position of those radioactive bands on the nylon is then determined by placing the nylon against a piece of X-ray film. The radioactive emissions from the probe bound to the bands of interest expose the film. After development, thin dark bars, about 1/4-inch long and l/32 inch wide are seen on the film. These bars are called bands and are what jurors see.

THE STATISTICS OF DNA IDENTIFICATION

The numbers given here are only for the purpose of illustration. A particular VTNR that has nineteen repeats occurs in only one in 4,000 people. Then the chance that a random person in the population would also have nineteen repeats is 1 in 4,000. If a DNA sample from a crime scene and a DNA sample from a suspect both had VNTRs that repeated nineteen times then the chance that both VNTRs would match coincidentally is one in 4,000.

When two VNTRs are used a much higher probability can be obtained. If two DNA samples have fragments that match and if the odds for the first VNTR were 1 in 4,000 and the second 1 in 1,000 then the chance that both VNTRs would match coincidentally is 1 in 4,000 x 1000 or 1 in 4 million.

If three VNTRs are used, the chance that all VNTRs would match is extremely low making it highly probable that the suspect was at the scene of the crime. For example, suppose VNTR one occurs in 1 in 4,000, VNTR two occurs in 1 in 1,000, and VNTR three occurs in 1 in 2,000 then the chance that all three VNTRs from two DNA samples would match coincidentally is 1 in 4,000 x 1,000 x 2,000 or 1 in 80 million. In this case, a DNA sample from the crime scene and a DNA sample from a suspect would point to the latter with a very high probability.

Specialized DNA laboratories sprung up in the latter part of the 20th century and have been used to both exonerate and convict people accused of committing crimes. One famous case occurred in Leicester County, England. A fifteen year old girl was found raped and murdered in 1983 and another fifteen year old girl was found raped and murdered in the same locality in 1986. A worker at a local psychiatric facility was arrested. However, a DNA expert, Alec Jeffreys at Leicester University, took DNA material from vaginal swabs of each victim and compared them to the suspect's DNA. Jeffreys concluded that each girl was the victim of the same criminal. However, Jeffreys stated that the DNA tests showed that the suspect was innocent.

In January 1987 the police in Leicester County took DNA samples from all young males in the vicinity of the crime. More than 3,600 DNA typings were performed by May 1987 without a match with the samples from the victims. In August 1987 a man admitted substituting his blood samples for that of a coworker of his, Colin Pitchfork. Pitchfork was arrested and he confessed to both murders and he was convicted.

12.13 THE CLASH OF FEAR AND GENETIC TECHNOLOGY

Jeremy Rifkin in 1977 published a book with Ted Howard, *Who Should Play God,* attacking genetic research. Rifkin in 1982 joined a coalition of religious leaders against human genetic experimentation even to cure diseases. Rifkin followed this in 1983 with a book *Algeny,* against gene therapy. Rifkin has played on the fear of the public. He has even come out against the genetically altered tomatoes. Some of the fears are irrational but some have a basis in fact.

The Human Genome Project will probably lead to tests for most genetic diseases. The geneticist Paul Billings of the California Pacific Medical Center presented a study of genetic testing discrimination published in the March 1992 issue of the American Journal of Human Genetics. This was an initial pilot study of forty-one people who said they had experienced discrimination because of genetic test results. Thirty-two cases involved discrimination by insurance companies. Most of the other cases dealt with employers and adoption agencies. In this pilot study many of the people had no symptoms whatever but this had no influence on the decisions of the insurance companies. Even though a disease could be treated successfully the insurance companies still canceled the policy. This included cases where the syndrome did not require treatment but the policy was canceled.

When genetic tests become available to detect a person's predisposition to heart disease, cancer or other serious diseases, an insurance coverage problem may arise. In order to forestall this the state of Wisconsin passed a law in 1992 prohibiting insurance companies from using genetic test results. Insurance companies share information through the Medical Information Bureau in Westwood, Massachusetts. Another fear is that employers with access to insurance medical records may discriminate against employees with genetic conditions.

Among the most common inherited genetic health problems are sickle-cell anemia, Tay-Sachs disease, and cystic fibrosis. Sickle-cell anemia is a disease that severely impairs the blood's ability to carry oxygen. Blacks have a 1-in-10 chance of carrying the gene that causes sickle-cell anemia. Tay-Sachs is a brain disorder that usually results in death before a child reaches the age of 3. Ashkenazi Jews have a 1-in-25 chance of carrying the gene for Tay-Sachs disease. Cystic fibrosis is characterized by chronic lung infections and inability to absorb fats and other nutrients. One in 22 whites is a carrier for cystic fibrosis.

In 1993 there were three major gene discoveries in three serious diseases. These involved colon cancer, Lou Gehrig's disease and Huntington's disease. The colon cancer gene, called MSH2 may be responsible for as many as one in every six colon cancers. A blood test can tell whether an individual carries the gene. While this can result in early treatment it could also result in problems with insurance companies and possible employers.

Lou Gehrig's disease is amyotrophic lateral sclerosis (ALS). It causes progressive paralysis by destroying muscle-controlling nerves in the brain and spinal cord. It usually begins in middle adulthood, generally causing total paralysis and death within two to five years. The gene was a well-known one that oversees production of a substance called superoxide dismutase 1 (SOD1) which in turn helps the body destroy toxic substances called free radicals. If defects in the gene reduces the effectiveness of SOD1, at least some ALS might be caused by buildups of free radicals. This leads to the possibility that some antioxidant vitamins may be useful against ALS. On the other hand, the early detection of the defective gene can cause problems in insurance and employment.

The discovery of the Huntington disease gene and a test for it poses a different problem. There is no treatment for the terrible disease as yet. Huntington's Chorea attacks brain cells in later life, impairing movement and intellect. It runs in families and many individuals in fear of becoming a victim would rather not know in advance and hope that they have not inherited it.

Up to 1994 genetic manipulation effected changes only in a specific individual and could not be passed along to future generations. However, in November of 1994 researchers at the University of Pennsylvania announced that they succeeded in modifying genes in mice and then tracked them along to two generations of offspring. Dr. Ralph Brinster, whose work led to this announcement, was of the opinion that the genie is out of the bottle and there is no putting it back. The technology for permanently altering an animal's genes and passing the changes along to its offspring can have profound consequences for the future.

Gene manipulation of this kind might be able to ward off birth defects, some forms of cancer and heart disease, and many other presently incurable maladies and even some personality disorders. However, this technology can stray into the field of eugenics. Will it be acceptable to enhance intelligence or artistic talent, or to have everybody have blond hair? There is the fear that such a technology could fall into the hands of a future Hitler. Will humanity be able to so alter the gene pool in a way that brings disaster down upon humanity?

12.14 THE P53 GENE AND CANCER IN THE 21ST CENTURY

In 1979, Dr. Arnold Levene of Princeton and Dr. David Lane of the University of Dundee independently discovered a protein which they called p53. In 1982, scientists isolated the p53 gene which produced the p53 protein. In 1989, Dr. Bert Vogelstein found that 80 percent of colon cancers involved the p53 gene.

Further work in 1989 by Dr. Vogelstein and separately by Dr. Levene revealed that p.53 instead of causing cancer actually prevented tumors in 60% of different types of cancers.

It was discovered that the healthy p53 gene produces a mechanism that inhibits the abnormal growth of cells. Sometimes p53 causes the cancerous cells to commit suicide. However, mutations, either inherited or caused by carcinogens, can cripple the cancer fighting properties of the p53 gene. These mutations consist of changes in the order of the nucleotides, A, C, G, T.

An example of this action of a carcinogen is benzopyrene in cigarette smoke which changes G to T in p53. Other chemicals in cigarette smoke change C to A. These mutations can allow lung cancer to spread and eventually kill the smoker. Ultraviolet light in skin cancer can change CC in p53 to TT. All of these mutations will prevent p53 from crippling the growth of cancer cells. In the 1990's scientists experimented with the healthy p53 gene to treat cancer. In 1996, researchers placed healthy p53 genes in viruses which were injected into patients with lung cancer. In a small sample, in one third of the patients the tumors shrunk and in another one-third the tumors ceased growing. This is only one example of using the p53 genes to stop some cancers.

However, there is a great deal of work before this type of therapy can be used on a large scale. If this promising genetic therapy comes about, it will be in the 21st century after more research and extensive trials.

Bibliography

Ballantyne, Jack and Sensabaugh, George and Witkowski, Jan — *DNA Technology and Forensic Science.* Cold Spring Harbor Laboratory Press, 1989

Cook-Degan, Robert — *The Gene Wars: Science, Politics and the Human Genome.* W.W. Norton & Co., Inc.1994

Crick, Francis — *What Mad Pursuit.* Basic Books, Inc. 1988

Davis, Kay E. and Tilghman, Shirley M. — *Genetic and Physical Mapping, Vol. 1.* Cold Spring Harbor Laboratory Press, 1990

Dickerson, Richard E. — "The DNA Helix and How it is Read." Science American 249/6: 94-111, 1983

Drlica, Karl A. — *Double-Edged Sword: The Promises and Risks of the Genetic Revolution.* Addison Wesley, 1994

Friedmann, Theodore — *Gene Therapy, Fact and Fiction in Biology's New Approach to Disease.* Cold Spring Harbor Laboratory Press, 1994

Judson, Horace J. — *The Eighth Day of Creation.* Cold Spring Harbor Laboratory Press, 1996

Kevles, Daniel J. and Leroy Hood, Editors — *The Code of Codes: Scientific and Social Issues in the Human Genome Project.*

Harvard University Press, 1992

Liu, Margaret A., Maurice R. Hilleman and Reinhard Kurth, Editors — *DNA VACCINES, A New Era in Vaccinology,* Annals of the New York Academy of Sciences, Vol. 772, 1995

McCarty, Maclyn — *The Transforming Principles: Discovering that Genes are made of DNA.* W.W. Norton and Co., 1985

Micklos, David A. and Greg A. Freyer — *DNA Science, A First Course in Recombinant DNA Technology.* Cold Spring Harbor Press, 1990

Sayre, Anne — *Rosalind Franklin and DNA.* W.W. Norton and Co., 1975

Watson, James D. — *The Double Helix: A Personal Account of the Structure of DNA.* W.W. Norton and Co., 1980

CHAPTER 13

20TH CENTURY FOOD DEVELOPMENTS

13. 1 A REVOLUTION IN THE GROWING OF FOOD

The 20th century saw the mechanization of the farm rapidly in industrialized countries and then very slowly in the rest of the world. The substitution of internal combustion machines for animal power reduced the need for labor. In 1910 feeding the horses and mules on farms required more than l/4 of the output of the world's farms and nearly l/10th of the work on farms involved caring for draft animals. As late as 1922, 32 mule teams in the state of Washington, Figure 13-1, took a month to harvest 1,200 acres of wheat. Later a gas-powered combine, Figure 13-2, took l/3rd the time to harvest the same number of acres. By 1950 almost all farms in the developed world were using tractors, mechanical plows, harvesters, and trucks powered by internal combustion engines.

Another great change was the development of artificial insemination in farm animals. The practical application of artificial insemination in domesticated farm animals was developed in Russia during the first decades of the 20th century. Semen was collected and methods were developed for the dilution of semen and its preservation for several days outside the body. In the 1950's a great advance in this technology was made when it was demonstrated that bull semen could be frozen at -79° C, provided that glycerol was added before freezing. Later it was shown that liquid nitrogen made it possible to store semen at -196° C for an even longer time. Semen frozen for a number of years has been shown to be viable after being defrosted. Artificial insemination produced more than 10,000 calves from one bull. By 1995 most cattle in the developed world were produced by artificial insemination. In addition to cattle artificial insemination was introduced in the growing of sheep, pigs, and poultry.

Another development in the raising of farm animals was embryo transplantation. The production of many eggs or ova can be induced in one female by hormone treatment. After artificial insemination, the fertilized eggs or embryos can be collected from the uterus of the donor and transferred to the uteri of recipient females for development into normal fetuses. At the time of transfer, the recipient's sexual cycle must be synchronized with the donor's sexual cycle by means of the injection of hormones. It is possible to harvest a very large number of embryos from superior animals and implant them in many inferior females

Figure 13-1 32 Mule Team Harvester, 1922
Courtesy of National Geographic Society

that serve as factories to produce highly selected animals for the production of milk or meat.

One of the problems of artificial insemination in farm animals is inbreeding. Inbreeding increases the chances of unfavorable as well as favorable genes. As a result there is a segregation of various kinds of congenital defects and a general decline in fertility and viability of the inbred animals. In order to avoid this occurring in artificial insemination, farmers changed to the practice of using tested sires for only relatively short periods of time. Then they are replaced with younger sires to help avoid too much inbreeding. A rapid turnover serves also to reduce the length of the generation interval in breeding programs.

Many experiments have been carried on in the 20th century on the crossing of inbred lines of the same breed or of different breeds. It has been possible with this technique to produce commercial hybrid chicks with superior egg-laying performance. Because the individual inbred lines are poor producers, the hybrid chicks are usually developed by a four-way cross. This is carried out by mating the offspring from crossing of lines A and B with the offspring from crosses between lines C and D, producing in effect a double hybrid. In order to obtain the best possible result, a large number of lines are tested in

Figure 13-2 Combine Grain Harvester 1996
Courtesy Deere and Company, Moline, Illinois

various crosses. By the last decade of the 20th century most of the eggs marketed in the industrialized world were produced by hybrid chicks.

In the 1950s and '60s an agricultural scientist, Norman Borlaug, produced the "green revolution" of high-yield seed, fertilizers and irrigation that transformed Asian agriculture. In 1970 Borlaug received the Nobel Peace Prize for his work. The "green revolution" produced rapid results. For example, in Bangladesh, high yield varieties of rice were first introduced in the 1968-69 growing season. One year later a fourth of the total crop came from the new strains. Countries in Asia that had imported food became exporters after the transformation to the agricultural technology of the "green revolution."

In the 1990s agronomists working in five African nations started to apply the lessons of agricultural technology that had increased food output dramatically in Asia. They were employed by Sasakawa-Global 2000 known as SG-2000. This was a joint development project of Japanese philanthropist Ryoichi Sasakawa and the Carter Center, former U.S. President Carter's organization in Atlanta. Norman Borlaug who in the 1990s was a professor at Texas A & M University, was the chairperson of SG-2000. In 1990 agronomists working for SG-2000 targeted Tanzania, a fertile country in Africa, that had to import 300,000 tons of grain because of its primitive agricultural technology. The project sponsored 10,000 one acre management-training plots across Tanzania. Each plot was meant to be a model for the neighbors. SG-2000 gave interest loans so that the farmers on the demonstration plots could buy the fertilizers and high-yield seed. SG-2000 farm extension agents advised the farmers on planting techniques and weed and insect control. Corn yields on the demonstration acres tripled or quadrupled.

However, problems have arisen in using the technology over the whole country of Tanzania which has 12 million acres planted with grain compared to the 10,000 acres of demonstration acres sponsored by the $1 million SG-2000 project. The farmers and the government are too poor to afford the new technology. The World Bank, the multibillion-dollar Washington institution, that dominated Third World development has its own program in Tanzania and other African nations. It also sends agents to farms in Tanzania to instruct peasants and establish demonstration plots. However, it leaves out hybrid seeds and special fertilizers as too expensive for less developed African nations. Instead they advocate improving traditional agricultural methods.

In the United States in the last quarter of the 20th century there was an interest on the part of some people in organic farming* as a sort of protest against the "green revolution." Organic farming uses no chemical pesticides or fertilizers. Techniques include the use of cover crops which goes back to the turn of the century in the United States.

Part of the revolution in agricultural technology in the 20th century has been the widespread use of chemical pesticides. Alternatives have been and are being developed.

13.2 Chemical Pesticides and Alternatives

DDT or dichlorodiphenyltrichlorethane was synthesized in 1874 but its insecticidal properties were not discovered until 1939. Its usefulness was immediately apparent, with the results that it was soon being used on a large scale. Other insecticides of a similar

*See Wolkomir in Bibliography

organochlorine chemical structure soon followed. Lindane, Aldrin, Dieldrin, Endrin and Chlordane were all introduced in the 1940's.

During World War II, American troops and refugees alike were infested by body lice which carried typhus. The U.S. Army sent out a call to chemical companies for an answer. The Geigy Co. of Switzerland sent in a packet of DDT which was tested on 25 volunteers at an Agriculture Department facility in Orlando, Florida. It proved to be twice as effective as anything else. Millions of two-ounce cans were used on millions of men, women, and children.

DDT also at first seemed to be an ideal agricultural pesticide. It was toxic to insects at low doses and appeared to be relatively nontoxic to humans and animals. DDT was very stable and repeated applications were unnecessary making them cheap to use. It killed mosquitoes, boll weevils and other agricultural pests. DDT killed not only pests, but the natural enemies of pests. However, DDT and other organochlorine chemicals were long lasting in the soil where they lodged in earthworms. Robins and other birds consumed large amounts of worms and absorbed the chemicals. Eagles devoured smaller prey and were contaminated with the pesticide chemicals which weakened the eagles' eggs. The chemicals lodged in the fat tissue of immeasurable species because they are soluble in fat. They lodged in the fatty organs, like the liver, of human beings.

In the 1960's, a team of scientists, sponsored by the World Health Organization and the International Agency for Research on Cancer, conducted a worldwide survey to determine the extent to which organochlorine insecticides were getting into the chemicals in the body fat of people all around the world. They concluded that DDT had become a constituent of the human body.

Rachel Carson, Figure 13-3, a former staff biologist of the American Fish and Wild Life Bureau, had observed the effects of DDT and the other organochlorine chemicals on wild life. Fish in lakes and streams died suddenly after heavy rains caused runoff of pesticides from treated croplands. Birds were found dead after DDT was applied aerially to control forest insects. The shells of birds' eggs became so thin that nesting parents crushed the weakened shells under their weight and ospreys, eagles, robins and others were threatened with extinction.

Rachel Carson's book, *Silent Spring,* was published in 1962 and focused American attention on the use of chemical pesticides which could destroy the environment. Chemical companies threatened lawsuits to block publication. They failed and the book became an international best seller despite the hysteria against her criticisms. Carson stated that indiscriminate pesticides were killing off many bird species directly and triggering genetic resistance among insects that attack crops. She predicted that pesticides might destroy the earthworm which would break the food chain of many birds.

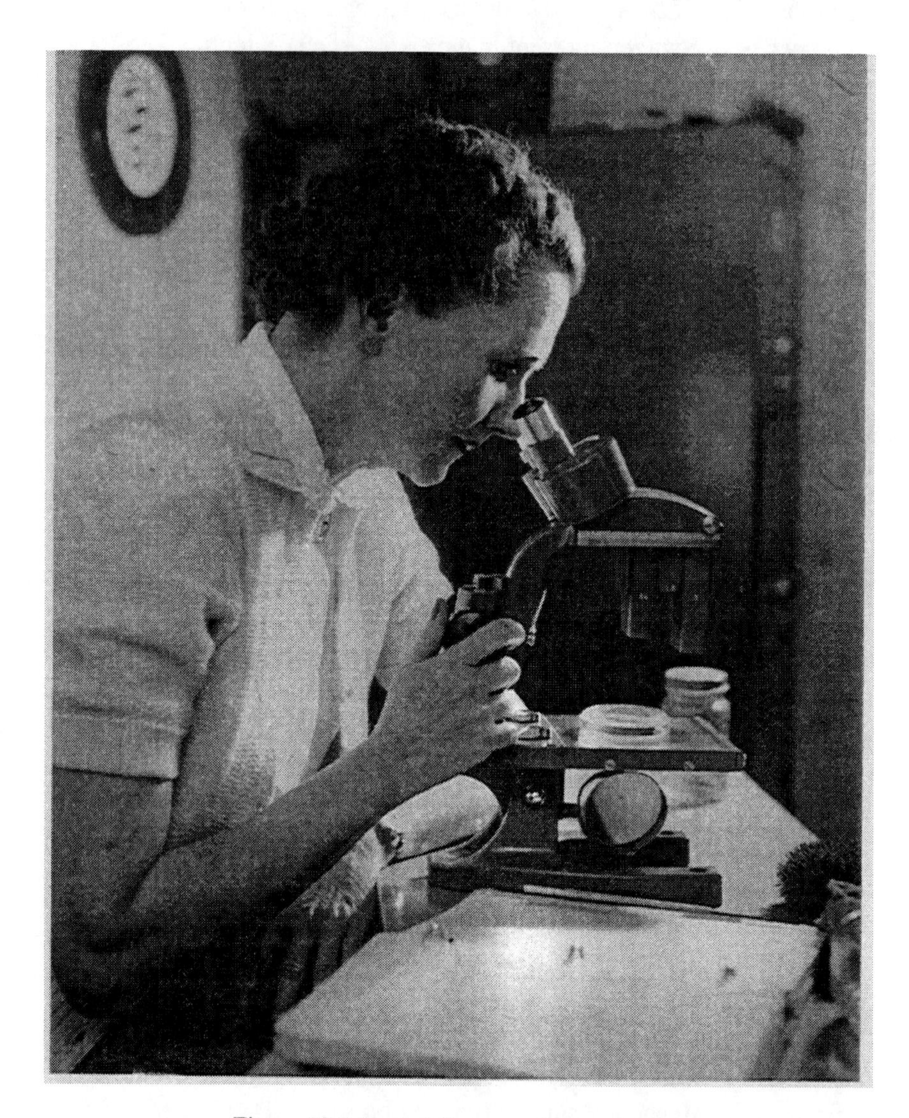

Figure 13-3 Rachel Carson at Microscope, 1951
By Permission of Rachel Carson History Project
Brooks Studio Photographer

Carson wrote that mutated insects might inherit the earth, made stronger by the very poisons chemists designed to destroy them.

Rachel Carson's book had a profound effect. In the 1970's the use of DDT and other organochlorine insecticides were banned in the United States. In 1970 the mean concentration of DDT in a human fat sample was 7.88 parts per million. By 1974 it had fallen to 4.99 parts per million. Bird populations including the Ospreys and Eagles slowly came back. In the United States organochlorine insecticides were replaced by organophosphorus insecticides, including parathion and malathion, and carbamates. Some of these are highly toxic to humans and require great care during application. They are much less stable, however, and degrade rapidly in the environment. Unfortunately, DDT and other organochlorine insecticides are still exported in large quantities. Their economic advantages weigh heavily where the demand for food is most urgent. Thus they continue to be a threat to the world environment.

BIOCONTROL

At the time when Rachel Carson wrote *Silent Spring* in 1962 most of the pest control research at the Agricultural Research Service (ARS) of the U.S. Department of Agriculture was devoted to improving chemical pesticides. By 1994 at least 80 percent of the ARS budget for pest control was for alternatives to chemical pesticides. Those include making plants more pest resistant, developing viruses and finding the natural diseases of insects. For a long time the Department of Agriculture has been importing parasites and predators of the elm leaf beetle, the alfalfa and clover leaf beetles and the brown tail and gypsy moths. The bacillus papillae, or milky spore disease which suppresses the beetle is an example. Another nonchemical agent is bacillus thuringiensis which produces a toxin effective against the sugar beet maggot and the Colorado potato beetle that attacks tomato plants and eggplants. There are also species-specific parasites for grape insects, for the oriental fruit moth in peaches, the tomato pinworm in Mexico, the pink bollworm in Arizona.

One technique of using species-specific parasites in gardens is by adding micro-organisms to compost produced from yard wastes. The microorganisms are specifically designed to attack the fungi and bacterial diseases that were previously fought with chemicals such as methyl bromide which will be banned by the end of the century. This customized compost is being made commercially and sold to nurseries and orchards.

A pioneer in the use of compost with special anti-pest microorganisms was Land Recovery, Inc. in Purdy, Washington. This company uses naturally occurring organisms to break wastes down into a dark organic material that can improve poor soils. However, the compost has special microorganisms added to act as bio pesticides. The ground-up waste is mixed, soaked with water, and stored in piles in rows about eight feet high. A gridwork of pipes in the pad blows air into the piles to keep temperatures down as the organisms multiply and give off heat. The piles are turned periodically to keep them loose and to expose new surfaces. After 45 days the material is allowed to dry for a month and then is screened to remove large twigs. Any needed microorganisms are added at this stage, by spraying the compost with a microbe-laden solution. The microorganisms, which are commercially available, are natural enemies of fungi or microbes that cause bacterial plant diseases. This material is then allowed to cure in small mounds for another month. The temperature is kept under 70 degrees to keep the beneficial organisms healthy. The big piles get too hot for the beneficial microbes.

An additional technique for biocontrol is male sterilization of insects by radiation. This was pioneered by Dr. Edward Knipling of the Agricultural Research Service. He and his team developed the sterile fly technology which eradicated the screwworm in the United States, Mexico, and parts of Central America.

Another technology is the use of large amounts of sex hormones. Around 1970 the coddling moth was one of the first insects for which a sex hormone was identified but in tiny quantities. Chemical synthesis produced very large amounts of the coddling moth sex hormone. This allowed for the sowing of the hormone in thousands of acres befuddling the sexual appetites of male moths and which greatly lessens procreation, a progressive experience generation after generation. In California the coddling moth developed resistance to chemical insecticides and the sex hormone technology became available as a substitute.

13.3 The Loss of Genetic Agricultural Diversity

In the 20th century farmers used technology for maximum efficiency and yields. This meant concentration on the best, most productive few animals and plants. This leaves more and more species of plants and animals to fall into disuse and obscurity. Since the beginning of the 20th century, 75 percent of the genetic diversity of the world's agricultural crops has been lost. In Europe about half of all breeds that existed in 1900 are extinct.

About 6,000 apple varieties that grew on American farms 100 years ago are gone. Globally, domesticated animals are disappearing at the rate of one breed a week. The United Nations' Food and Agriculture Organization (FAO) has been monitoring the loss of genetic agricultural diversity. A FAO World Watch List published in November 1993 showed that of about 4,000 known breeds of farm animals, 1,000 were threatened by extinction. The FAO listed a number of endangered breeds. They named, for example, California's San Clemente goat, Santa Cruz sheep, Florida cracker cattle, the Rocky Mountain horse, the Navajo-Churro sheep, Maine's Katahdin sheep, Iowa's Cream Draft horse, the Tennessee Fainting goat, the Red Minnesota No. 1 pig, and the Black Montana No. 1 pig.

A protective International Convention on Biological Diversity has produced an accord which establishes global priorities and policies for preserving biodiversity. Biodiversity has been defined by the National Academy of Sciences as "the variety and variability among living organisms, and the ecological complexes in which they occur." As of 1994, 34 nations ratified the accord with an additional 120 nations that signed it but did not ratify it.

Cary Fowler is the co-author of a book, *Shattering: Food, Politics and the Loss of Genetic Diversity.* He is stationed in Rome where he is helping to develop a first overview of the state of the world's genetic plant resources and a companion global action plan. The problem with the loss of genetic diversity in food plants is the susceptibility of one particular plant type to a new type of virus. This could wipe out a food source in the future.

An example of the vulnerability of modern food plants is the 1970 U.S. devastating loss of a great deal of the corn crop. Almost all corn plants had been uniformly bred into one vulnerable strain. The southern corn leaf blight attacked the corn and destroyed 700 million bushels.

In order to avoid such problems, seed banks of different varieties have been established in different parts of the world. However, some of these are susceptible to wars and revolutions.

13.4 PRESERVING FOOD

CANNING

The roots of the canning process go back to the Napoleonic wars. As Napoleon prepared for his invasion of Russia, he searched for a new and better means of preserving food for his troops. He offered a prize of 12,000 francs to anyone who could come up with a solution. Nicolas Appert, a Parisian candy maker, was awarded the prize in 1809. Appert had observed that storing wine in airtight glass bottles kept it from spoiling. He filled wide-mouth glass bottles with food, carefully corked them and then heated the bottles in boiling water.

This was followed by the tin can developed by an Englishman, Peter Durand. Soon the British Army used the tin cans on campaigns. The use of canning slowly spread among civilians in the last half of the 19th century. In 1865 the steamboat Bertrand was heavily laden with canned provisions when it set out on the Missouri River for the gold mining camps in Fort Benton, Montana. The boat sunk under the weight and went down to the bottom of the river. It was found a century later, under 30 feet of silt, a little north of Omaha, Nebraska.

Among the canned food items retrieved from the Bertrand in 1968 were brandied peaches, oysters, plum tomatoes, honey, and mixed vegetables. In 1974, chemists at the National Food Processors Association (NFPA) found that although the food had lost its fresh smell and appearance, there was no microbial growth. They determined that the foods were as safe to eat as they had been more than 100 years before. The cans were made of three pieces, a top, a bottom, and a body formed from a plate lead soldered into a cylinder. With the lead soldering there was a concern about lead leaching into metal canned foods.

In the 20th century, canning was greatly improved and the food was packaged in materials other than metal cans. Foods packaged in glass jars, paperboard cans, and plastics are considered canned by food processing specialists if the food undergoes the canning preservation process.

The canning process deals with food-spoiling bacteria, yeast, and molds that are naturally present in foods. These microorganisms need moisture, a low-acid environment (acid prevents bacterial growth), nutrients, and an appropriate (usually room) temperature. In canning food processing the foods are preserved from food spoilage by controlling one or more of the above factors. For example, frozen foods are stored at temperatures too low for organisms (bacteria, yeast, and molds) to grow. When foods are dried, sufficient moisture is not available to promote growth. It is the preservation process that distinguishes canned from other packaged foods. During canning, regardless of the package, the food is placed in an airtight, hermetically sealed, container and heated to destroy microorganisms. The hermetic seal is essential to ensure that microorganisms do not contaminate the product after it is sterilized through heating.

In the 20th century, in the United States, some canned foods come with a peel-off metal top and plastic lid ready for the microwave. Barriers (made of sophisticated synthetic materials) that provide an airtight seal are sandwiched in these plastic layered containers. They are used for applesauce, pudding, and other foods that can be stored on supermarket or home shelves for years. Containers were also developed for new transparent plastic materials like polyethylene terephthalate used for peanut butter and catsup. Packages made of paperboard layers have been designed in the shapes of boxes to contain such foods as fruit juices, tomato sauce, and milk.

In the 20th century the can itself has changed. In the traditional three-piece cans, a welded side seam has replaced the lead soldered side seam in most American cans. There were also two-piece cans which eliminate the side seam and one seamed end. These cans are made by feeding metal into a press that forms the can body and one end into a single piece. However, many imported cans in the last decade of the 20th century still bore lead-soldered seals. They could be detected by peeling back the label to expose the seam. The edges along the joint of a lead-soldered seam are folded over and silver-gray metal is smeared on the outside of the seam. A welded seam is flat with a thin, dark, sharply defined line along the joint. Lead soldered cans in small numbers are still used in the United States for dry foods such as coffee packaged in cans. Leaching here is not a concern in such dry material.

It was learned in the mid-20th century that the acid content of food has a great deal to do with the required temperature for canning. Foods with a naturally high acid content such as tomatoes, citrus juices, and fruits will not support the growth of food poisoning bacteria. Foods that have a high acid content, therefore, do not receive as extreme a heat treatment as low-acid foods. They are heated sufficiently to destroy bacteria, yeast and molds that could cause food to spoil.

Foods with low acid content, such as mushrooms and green beans, require a much higher temperature for canning. The deadly Clostridium botulism bacterium, which causes botulism poisoning, produces a toxin in these foods that is highly heat-resistant. In February 1990, some 22 students at Mississippi State University became seriously ill after eating omelets made with canned mushrooms imported from China. Similar outbreaks followed in New York and Pennsylvania, affecting more than 100 people. The Food and Drug Administration (FDA) of the United States identified the bacteria causing the problem as *Staphylococcus aureus* which produces a poison, staphylococcal enterotoxin. A FDA investigation suggested that the mushrooms were contaminated with staphylococcal enterotoxin even before they were canned. The canning process could not destroy the highly heat-resistant toxin. The U.S. banned mushrooms canned in China from entering the United States.

FROZEN FOODS

Ice and salt mixtures were used in the 19th century to preserve fish. However, slow freezing of food damages it. Freezing between 31° and 25 °F causes crystal formation. It was realized early in the 20th century that damage to frozen food could be minimized if a product was brought below 25° as quickly as possible. A number of innovators attempted to develop such a process. Two of the best systems were invented by Clarence Birdseye.

In 1912, Birdseye went to Labrador as a fur trader. While there he saw natives catching fish in fifty below zero weather. As soon as the fish were taken out of the water, they quickly froze solid. Months later some of the fish were still alive after they were thawed out. Birdseye stayed in Labrador until 1917 but he never forgot what he saw in the fisheries there.

In 1920 he became an Assistant to the President of the U.S. Fisheries Association. This organization was interested in quick freezing processes which would better preserve their highly perishable products and keep prices steady during fluctuations in the catch.

In 1923 Birdseye began working full time on a rapid frozen-food process. The next year he moved to Gloucester, Massachusetts where he founded a small firm, General Seafoods Corporation, which later became the General Foods Company. Birdseye used two processes for quick freezing.

In his first method he packed the food in a rectangular package. This package was held between two belts that were chilled by a very cold calcium chloride. Birdseye's second method was simpler and soon took over. Packaged food was held under pressure between two hollow metal plates chilled by the vaporization of ammonia.

While Birdseye's ideas seemed simple, they were covered by 168 patents to make the whole process practical. One of the problems with quick-freezing vegetables was called autolysis, a destructive enzymatic action which results in unpleasant flavors. This was solved in 1930 by H. C. Diehl and C. A. Magoon who found that if vegetables were briefly scalded before freezing then autolysis would not take place.

Another problem was that private homes had to have freezers to keep the food frozen. In 1939 refrigerators came out with a separate freezer compartment with its own thermostat. The new refrigerators had two separate thermostats. One controlled the temperature of the cold compartment for ordinary foods. The other thermostat kept frozen foods in that state.

Another problem was the long distance transport of frozen food. In the late 1930's refrigerated trucks were produced. In 1949 refrigerator train cars were developed but by the last decade of the 20th century refrigerated trucks carried most of the frozen food in the United States.

High Pressure Food Processing

In 1899 Bert Hite of the University of West Virginia Agricultural Center showed that certain fruits could be preserved by subjecting them to high pressure. The pressure kills the microorganisms in a fruit while preserving nearly all its innate flavor, color, and nutritional value. This is not the same as home pressure-cooking. Hite's idea was not used on any scale until the late 1980's when the Japanese began to apply the technology to jam, yogurt, and fruit sauce. In 1990 the Meidi-Ya Food Corp. in Osaka began selling premium jams, jellies, fruit sauces and yogurts manufactured under the process.

In conventional manufacturing of packaged liquid products, food makers typically boil the food in batches to kill bacteria. However, heat destroys a substantial amount of vitamins and mineral and slightly alters the taste. In high-pressure food processing, water pressure is used instead of heat. The pressure, which far exceeds that found in the deepest parts of the ocean, kills the bacteria but leave the food's vitamins and natural flavor practically unscathed. The pressure also bursts the cell membranes in the fruit, releasing its flavor without pulverizing the chunks.

In Japan, although the price of food sterilized through pressure is three times more expensive than conventional heat, sterilized products sales have been very brisk because they taste better. In the United States, American food companies have avoided the technology because of their doubts about the market and the cost of building special pressurizers.

Preserving Milk

In the last part of the 20th century, a new method of preserving milk was developed by killing the bacteria in it much more effectively.

Parmalat milk goes through an ultra high temperature which exposes it to a much more complete pasteurization for a few seconds. The high temperature destroys bacteria normally present in milk. It is in an aseptic package protected from air, light, and bacteria.

Parmalat milk does not require refrigeration before opening the package and will last a considerable time. After opening it must be refrigerated and has to be used within 10 days.

Radiation

Gamma rays, emitted by radioactive cobalt-60, have been used mostly on an experimental basis to kill bugs and fungus on fruits and vegetables, and to extend shelf life. Irradiation kills salmonella which taints half of the chickens sold in the United States. The amount of radiation necessary for this process is fewer than 100,000 rads which is equivalent of about 10 million chest X-rays.

Opponents of irradiated food state that when the gamma rays zap food they destroy some nutrients including vitamins. Molecules found in zapped food include carcinogenic formaldehyde in irradiated starch, benzene from starch, mutagenic peroxides from plant tissue, and formic acid from sucrose. These are referred to as "radiolytic products." Proponents of irradiated foods state all such products total no more than 30 parts per million in irradiated food. Individual compounds are present at levels of parts per billion. There is more benzene in some nonradiated dairy products and produce than in zapped meat. The FDA studied the pros and cons of the food irradiation process and decided that it was worth doing.

The FDA in 1963 approved wheat and wheat flour irradiation; in 1964 white potatoes, pork in 1985, and fruit and vegetables in 1986. However, no large amounts of irradiated food was sold until 1992 when Vindicator of Florida, Inc. shipped 1,000 pints of strawberries to north Miami Beach groceries. Citrus and other fruits followed. The Florida Department of Citrus endorsed irradiation as a safe alternative to the banned ethylene dibromide as a way to kill the Caribbean fruit fly. However, Maine, New York, and New Jersey prohibit radiated food. In addition, H.J. Heinz, Quaker Oats, Ralston Purina, McDonald's, Campbell's and many poultry producers have all stated they will not sell irradiated food or use irradiated ingredients. They feared the wrath of consumers who did not want the process to be used.

Chemical Preservative Additives

In the 20th century various chemical additives were added to food to preserve it. Some examples were formaldehyde, monochloroacetic acids, borate, nitrite, and sulfite. Other examples were the organic acids such as sorbic, acetic, benzoic, lactic, and propionic acid. In addition, gases were used such as carbon monoxide, carbon dioxide, ethylene oxide, propylene oxide, sulphur dioxide, and ozone.

There was a reaction against addition of chemical preservatives to food among many consumers. There is much interest in developing preservation techniques which use natural processes. These include enzyme/protein. This avoids some of the problems that arise with the use of chemicals. Enzyme/proteins can function as anti-microbials in several ways.

1. Depriving spoilage organisms of an essential nutrient.

2. Generating substances toxic to spoilage organisms.
3. Attacking a cell/wall membrane component, thereby physically disrupting the cell or changing the permeability of the cell wall membrane. This is referred to as microbicidial substances.

Depriving spoilage organisms of an essential nutrient is illustrated by adding an enzyme such as glucose oxidase-catalase which in the presence of glucose depletes oxygen when food is stored in a closed container. This prevents the growth of bacteria that require oxygen. Besides the prevention of microbial growth by the removal of oxygen, the glucose oxidase catalase system prevents oxidation reactions which lead to browning and off-flavors. The enzymes are applied to fresh orange juice and mayonnaise dressings.

An example of generating substances toxic to spoilage organisms is the coupling of lactoperoxidase with a very small amount of peroxide. This converts SCN^- to $SCNO^-$ which are very lethal ions for microorganisms. SCN^- as well as lactoperoxidase are indigenous to milk. Low levels of hydrogen peroxide added to milk is an effective preservative. Preservation on soft ice cream mix and pastry cream has been demonstrated using this method.

Examples of microbicidal substances are lysozymes. Lysozyme is a natural occurring enzyme which destroys bacteria cell walls. Special lysozymes from different sources attack different bacteria. Lysozyme derived from hen egg white has been added to preserve certain hard cheese. This lysozyme prevents gas production by a special bacteria which is a contaminant in cows milk. The same lysozyme from hen egg white also has been reported to be effective against the bacteria that causes botulism. There is more work that has to be done to enable this lysozyme to have widespread application. Another class of microbicidal substances is chitinase which act on enzymes as a component of anti-fungal systems in food. The enzymes can be made from a special bacteria. Another example is bacteriocins which are proteins produced by bacteria to kill other bacteria. Nisin is a bacteriocin used as a food preservation agent in cheese, tomato juice, cream-style corn, chow mein, meat slurries and beer.

13.5 Adding Flavors

Historically many food flavors have been generated as a side effect of enhancing the stability of the food. This has been the case for vinegar, cheese, yogurt, beer and wine in which the preservative effects of complex molecules generated by bacteria were accompanied by desirable flavors. The presence of complex mixtures of acids, alcohol, esters and other chemicals gives the individual character and identity to the food or beverage. In addition, many flavors and fragrance raw materials have for thousands of years been derived from plants growing in areas such as Indonesia, the Spice Islands, India, and Africa.

In the 20th century with the advent of sophisticated chemistry, complex flavor mixtures have been analyzed and broken down into various chemical compounds which can be synthesized in the laboratory. Added chemicals for flavoring food raised some objections from consumer groups. In the last half of the 20th century there was a growing interest in using biotechnological processes to create flavors for food. These use either cells or enzymes and can be divided up into those that produce either flavoring complexes or single flavor compounds.

Flavoring complexes are multi-component flavor systems. An example is the use of specific strains of lactic acid bacteria and fungi to arrive at certain cheese flavors. Another example is enzyme modified cheese short-chain fatty acid-specific lipases and protease to obtain multi-component flavors. A blue cheese type flavor is generated by a group of fungi, particularly Penicillium roqueforti. The free fatty acids that are generated by lipolysis are relatively toxic to P. roquerforti which as a protective measure converts the fatty acids into products called methylketones which are generally considered the key flavor components of blue cheese.

Single flavor compounds are produced by the biosynthesis, isolation, and purification of individual flavor compounds or synthesis by either microbial fermentation or by using specific enzyme systems. Examples are esters, pyrazines, and pungent tastes. Esters are responsible for fruity, flavored aromas. They can be produced by different species of bacteria or yeast. Pyrazines are nitrogen-containing compounds with roasted nutty flavor. They are produced by specific species of bacteria. Pungent tastes like those of mustard and horseradish can be made by producing isothiocyanates from odorless precursors known as glucosinolates by the action of the enzyme myrosinase.

FLAVOR ENHANCERS

These materials are added to certain foods in order to enhance sensory responses. They can be produced by microbial fermentation or by enzyme action, in combination with chemical processes. Examples are monophosphate, MSG, yeast extracts, amino acids, and citric acids.

There are two monophosphate flavor enhancers, inosine monophosphate (IMP) and guanosine monophosphate (GMP). They can be produced by the action of an enzyme phosphodiesterase on ribonucleic acid (RNA). The RNA used is from microorganisms such as the yeast Candida utilis.

MSG, another flavor enhancer, is the monosodium salt of L-Glutamic acid. Hundreds of thousands of tons of MSG are used annually around the world. It is produced by fermentation of sugars with a specific bacteria, Corynebacterium glutamicum. Strain selection, classical genetics and optimization of fermentation have resulted in improved production. Yeast extracts are used as savory flavor enhancers and are widely produced for this purpose. Amino acids have many uses one of which is as a flavor enhancer.

Citric acid is also used as a flavor enhancer as well as for many other purposes. It is produced in large quantities by fermentation of molasses using the fungus Aspergillus niger or the yeast Candida lipolytica.

SWEETENERS

The bulk of the sweeteners consumed in the 1960's was in the form of cane and beet sugar sucrose. In the early 1970's enzyme technology opened up the way to a new class of sweeteners based on starch instead of sucrose. An example is high fructose corn syrup (HFCS) where enzyme technology developed this sweetener from raw materials. HFCS by 1985 had decreased sucrose consumption appreciably.

From 1965 on a large number of low-calorie sweeteners such as cyclamate, aspartame, saccharin, and thaumatin were developed for use particularly in the soft drinks industry. One of the most popular of the low calorie sweeteners is aspartame (trade name Nutra-sweet) which was approved for use in soft drinks by the FDA in the United States in 1983. It has gained a rapidly increasing market share. Aspartame was produced originally by a chemical process but later an enzymatic synthesis was introduced. Aspartame rapidly replaced saccharin after the latter was found to produce cancer in animals. Aspartame has a relative sweetness of 150 compared to sucrose, a natural sugar.

On a weight basis the sweetest product is thaumatin, a protein derived from the berries of the plant Thaumatococcus danielli. It is 3,000 times sweeter than sucrose. Thaumatin is used to a considerable extent in Japan and also in the United Kingdom. Several laboratories have undertaken to develop a process for production of thaumatin by genetically engineered microorganisms.

13.6 USING AND MANUFACTURING VITAMINS

Vitamins are an essential part of the food that humans need. They are required in minute quantities for the maintenance of normal metabolic functions. Unlike proteins, carbohydrates and fats, they do not provide energy or serve as building units. The value of certain foods in maintaining health was recognized centuries before the first vitamins were identified. Long ship voyages resulted in scurvy in the crews until in the 18th century the British found that citrus fruits prevented the disease. It was discovered in the 19th century that substituting unpolished rice for polished rice prevented the disease beriberi.

In 1906 the British biochemist Sir Frederick Hopkins showed that foods contained what he called "accessory factors" in addition to proteins, carbohydrates, fats, minerals, and water. In 1912 a chemist, Casimir Funk, demonstrated that the anti-beriberi substance in unpolished rice was an amine, a type of nitrogen containing component.

Funk named it a vitamine which stood for vital amine. This term soon came to be applied to the "accessory factors" in general, all of which were originally thought to be closely related. Later it was discovered that different vitamins have different chemical properties and different functions and that many of them do not contain amines at all. However, because of its wide use, the term continued to be used. Later, the final e was dropped resulting in the "vitamin."

In 1912 Hopkins and Funk postulated the theory that the absence of sufficient amounts of a particular vitamin in a system may lead to certain diseases, such as scurvy or beriberi. This started a wide search for vitamins that were necessary for the human body. In the early days of research vitamins were assigned letters in the order of their discovery. Later they were assigned specific names after their chemical structure was revealed by analysis. Some of the many vitamins will be described.

The existence of vitamin A was first clearly recognized in 1913. The chemical structure was discovered in 1933 and it was first synthesized in 1947. Vitamin A combines with proteins in the retina of the eye to aid in night vision. Vitamin A is a component of a pigment, called visual present in the retina of the eye. A deficiency results in night blindness, the ability to see in the dark. Vitamin A deficiency may also result in defective bone and teeth formation and in poor general growth. Vitamin A is found in all animal livers, in milk fat and fish oils. In addition, many yellow and green leafy vegetables contain carotenes that are converted in the body to vitamin A.

Vitamin A is a fat-soluble alcohol that is not excreted in the urine. Only very small amounts are required by the body. The recommended dietary allowance for adult humans is 1.0 milligrams. This can be provided for by 6 milligrams of carotene in yellow and green vegetables. A high intake of vitamin A in the order of 150 milligrams daily over a period of months may result in a toxic condition. The excess will not be eliminated in the urine but will be stored in the liver. This can cause nausea, blurred vision, bone pain, and headache. In infants it can cause growth failure, enlargement of the liver and nervous irritability. Vitamin A concentrates are sold and it is routinely added to milk. Care must be taken to avoid excesses of this vitamin. Carotenes, the yellowish pigment found in carrots, yellow vegetables, spinach and broccoli that produces vitamin A is not toxic. However, a high concentration may impart a yellow condition of the skin.

The vitamin B complex are several vitamins that have been grouped together because of loose similarities in their properties, their distribution in natural sources, and their physiological functions. All of the B vitamins are soluble in water which means excesses can be excreted in the urine. All of the vitamins in the B complex are essential in facilitating the metabolic processes. B_1(thiamin) is found in pork, liver and whole grains such as unpolished rice. The body needs 1.5 mg daily and a deficiency causes beriberi. B_2 (riboflavin) is found in milk, egg white, liver and leafy vegetables. The body needs 1.7 milligrams daily and a deficiency causes conditions known as cheilosis and glossitis. There are a number of others in the B complex such as B_6 (pyridoxine) and B_{12} (cobalamin). B_6 is found in whole grains, yeast, egg yolk, and liver. The body requires 2 milligrams daily and a deficiency causes skin disorders and convulsion in infants. B_{12} is found in liver and meats. The body requires 2 milligrams daily and a deficiency results in pernicious anemia.

Vitamin C (ascorbic acid) is water-soluble and an excess can be excreted in the urine. It is found naturally in citrus fruits, fresh vegetables and potatoes. A deficiency results in scurvy. Starting in the 1940s, vitamin C was produced synthetically by the seven step Reichstein-Grussner process. In this method glucose, a natural sugar, was reduced to sorbitor which was converted to L-sorbose by the microorganism, *Acetobacter suboxydans.* Then a number of chemical steps convert L-sorbose to L-ascorbic acid. By the late 1980's over 40,000 tons of synthesized vitamin C were manufactured. Claims were made for Vitamin C as a method of preventing colds but this has not been generally accepted in the medical establishment.

Vitamin D (calciferol) is essential for the proper formation of the skeleton. It is fat-soluble and a surplus is not excreted. It is found in fish oils, and fortified dairy products. The body requires only 5 micrograms daily. Exposure of the skin to ultraviolet light catalyzes the synthesis of vitamin D_3 (cholecalciferol). A deficiency of the vitamin causes rickets. In the United States the major source of dietary vitamin D is fortified foods such as milk. The vitamin is stable in foods and storage and cooking does not appear to affect its activity. Even moderate amounts of exposure of the skin to sunlight can provide adequate synthesis of the vitamin, although dark-skinned persons and the elderly have more limited capacity to synthesize the vitamin. Vitamin D is reasonably well tolerated in adults but may be potentially toxic in young children.

Vitamin E is found in at least eight naturally occurring forms. Alpha-Tocopherol is the most common form and also has the highest biological activity. The richest sources of vitamin E are seed oils and margarine and shortenings made from these oils. Nuts, whole grains, leafy green vegetables, eggs, and milk are also good sources.

Vitamin E is fat-soluble and an excess is not excreted in the urine. Vitamin E functions primarily as an antioxidant protecting the body tissues from damaging reactions that arise from many normal metabolic processes and toxic agents. It protects biological membranes such as those found in the nerves, muscle and cardiovascular system. It helps to prolong the life of red blood cells and helps the body to use vitamin A optimally.

The requirement for vitamin E is related to the amount of polyunsaturated fat consumed in the diet. The higher the amount of this type of fat, the more vitamin E is required. However, sources of polyunsaturated fat are usually good sources of vitamin E. Examples are seed and vegetable oil. Dietary deficiency of vitamin E is rare in human beings. However, a progressive neuromuscular disease has been observed in children and adults who have a malabsorption of fat resulting in a severe vitamin E deficiency.

Vitamin K is another fat-soluble vitamin and is also unstable to light. There are two naturally occurring forms, K_1 which is synthesized by bacteria and K_2 which is made by plants. A third form K_3 (menadione) is a manufactured version. Vitamin K is essential for a number of blood clotting proteins. Defective coagulation and the resultant bleeding is the main sign of vitamin K deficiency. In some newborn infants vitamin K deficiency has occurred. This is because the intestinal bacteria do not begin to synthesize adequate amounts of the vitamin until about one week after birth. Because of this vitamin K is given to infants right after birth to prevent bleeding. All infant formulas are fortified with vitamin K. In adults prolonged treatment with antibiotics can inhibit the growth of intestinal bacteria resulting in a deficiency of vitamin K.

Niacin is a white water soluble powder which is stable to heat, acid and alkali. Humans are capable of synthesizing niacin from the amino acid tryptophan. Niacin is widely distributed in yeast, wheat germs, and organ meats. Milk is a poor source of niacin but contains large amounts of tryptophan from which the human body can synthesize niacin. A deficiency of niacin results in a disease known as pellagra. It is particularly prevalent among people whose diet is largely untreated corn. Very high levels of niacin are used to lower blood cholesterol. However, this type of treatment must be monitored for side effects such as liver damage.

Vitamin Supplements

These manufactured vitamins have been sold by the millions in the last quarter of the century. Many doctors believe that diverse foods are the best natural sources of vitamins but that in some cases vitamin supplements are useful.

13.7 Changes in Cooking Technology

Two changes in cooking technology occurred in the 20th century. The major one is the microwave oven and another change is self-cooking food packages.

The Microwave Oven

The magnetron, the heart of the microwave oven, was invented in England during the early years of World War II. It generated microwaves for a radar with a high resolution to detect enemy aircraft. After the war it was found that microwave energy is absorbed in most food and could be a source of quick, uniform heating or cooking. After considerable research, especially in the safety aspects, the microwave oven was adopted as a cooking technology in fast food restaurants and in private homes.

The microwave power, supplied by a magnetron, is at a frequency of 2.45 GHz (2.45 billion Hertz) with rated power from 350 to 750 watts. The heating level is adjusted by a circuit that turns power off and on over a selected time base, usually 20 seconds. Since the oven size is generally much larger than the 12 centimeter wavelength, standing waves could form, resulting in heat nonuniformities. This problem is avoided by using either a rotating mechanism or a moving platform to shift the food through the standing-wave pattern.

Safety is a primary concern in the design of a microwave oven. Exposure to microwave power of 100 milliwatts per square centimeter for several minutes can lead to patho-physiological effects in laboratory animals. The microwaves penetrate beneath the skin and heat the tissues, and tissue destruction can result if the temperature rise is faster than the control mechanisms of the body can handle. It was assumed that human tissue reacted to microwave exposure in the same manner and a safety factor of 10 was proposed. In the United States this resulted in the first maximum recommended exposure of 10 milliwatts per square centimeter for durations greater than 6 minutes. Other countries used much stricter limits. The limit in the United States is still being studied. In microwave ovens interlocks are provided so that the power is off when the door is open, and federal regulations require a backup system should the primary interlock system fail. Leakage levels of microwave energy from the units must also meet strict requirements. All safety features must be operable for at least 10,000 times.

The microwave oven became very popular in the last two decades of the 20th century. By 1993 more than 70 percent of U.S. households had a microwave oven. Many Americans considered the microwave their favorite household product. Computers were incorporated into the ovens to control the cooking time. Many foods are now packaged specifically for microwave cooking. Some of these packages are made of vinyl chloride. This is fine for microwave cooking but some of these packages have instructions for alternative cooking in boiling water. This may cause problems as the boiling water may cause some of the packaging material to be absorbed into the food in minute amounts.

SELF COOKING FOOD CONTAINERS

A Super Soup has been introduced in Japan. A compartment in the bottom of the cans contain special chemicals. A chemical heat reaction is started by turning a key. This boils the soup in the top compartment. The soup can then be poured directly into a plate, eliminating in the process a cooking pot.

In 1986 the U.S. military used an iron-magnesium tablet and water to supply heat for cooking. In 1995 this was adapted for commercial use, known as HeaterMeals. A Heater Meals box contains food inside a sealed container sitting on a Styrofoam tray. At the base of the tray, there is a tablet of iron and magnesium. To cook the meal the food is lifted up and some water is poured onto the tablet. The entire contents is then slid back into the HeaterMeals box where it cooks for fourteen minutes using the heat of the iron-magnesium-water process.

BIBLIOGRAPHY

Altieri, Miguel A. — *Agroecology: The Science of Sustainable Agriculture,* 2nd Edition, Westview, 1995

Burger, Anna — *The Agriculture of the World,* Avebury Publications (UK), 1994

Carson, Rachel — *Silent Spring,* Houghton Mifflin Co., 1962

Currel, B.R. et al., Editors — *Biotechnical Innovations in Food Processing.* Butterworth-Heineman, Ltd., 1991

Fowler, Cary and Pat Mooney — *Shattering: Food, Politics and the Loss of Genetic Diversity,* University of Arizona Press, 1990

Fowler, Cary — *Unnatural Selection: Technology, Politics and Plant Evolution,* Gordon & Breach, 1994

Gore, Al — *Ecology and the Human Spirit,* Houghton Mifflin, 1992

Mather, Robin — *A Garden of Unearthly Delights, Bioengineering and the Future of Food, Dutton,1993*

Wolford, J. Editor — *Developments in Food Colour.* Elsevier Applied Science, 1984

Wolkomir, Richard — "Bringing Ancient Ways to Our Farmers' Fields," Smithsonian, November 1955

CHAPTER 14

USING ARTIFICIAL MATERIALS

The 20th century saw the development of a large number of artificial materials which changed civilization in both good and bad ways. Included are plastics, non-natural textile fibers, advanced composite materials and powder technology.

14.1 POLYMERIZATION

Many of these technologies depend on polymerization. It starts with monomers which are relatively small molecules. Thousands of monomers are combined chemically to form a very large molecule called a polymer. Polymerization is characterized by the formation of stable covalent chemical bonds between the monomers.

There are two general types of polymers classified according to the way the monomers are linked together. Linear polymers are connected in a long chain. These soften or melt when heated. Cross-linked polymers are connected in a network. These form when heated but once formed they do not melt or soften with heat. They are known as thermosetting resins. Cross-linked polymers, unlike linear polymers, do not dissolve in solvents. Most synthetic polymers are made of several different monomers. The monomers are combined to make a product that has certain unique physical properties such as elasticity, high tensile strength, and the ability to form fibers.

14.2 PLASTICS

A plastic is a large polymer developed from a combination of synthetic chemicals extracted from raw materials such as oil, wood, natural gas and coal. The first completely synthetic plastic was Bakelite produced from phenol and formaldehyde by Leo Baekeland in 1909.

There are two general types of plastics, thermosetting and thermoplastic. Thermosetting plastics are plastic resins that cannot be remelted once formed. Thermoplastics on the other hand can be recycled after melting. Examples of thermosetting plastics are Polyvinyl Chloride (PVC) and Acrylonitrile-Butadiene-Styrene (ABS), both used in residential plumbing. Examples of thermoplastics are polyethylene and acrylics.

The chief source for plastics is petroleum. Secondary sources are coal and cellulose. Plastics are shaped by different means, including molding, and casting. Foam plastics are produced by forming gas bubbles in the molten material. Plastic products are further shaped and finished by methods ranging from mechanical through laser machining and ultrasonic welding.

Plastic Processing

As plastics took over a vast array of products in the last half of the 20th century, plastic processing became very sophisticated. Advanced methods and techniques were used to convert plastic materials in the form of sheets, granules, powders, fluids or pellets into parts or formed shapes. The plastics material may contain a variety of additives which influence the method of manufacture as well as the properties of the plastics. After forming, the part may be subjected to a variety of operations such as machining, adhesive bonding, welding, painting, and metallizing.

One of the more popular forms of plastic processing is injection molding. This process consists of heating and homogenizing plastics pellets in a cylinder until they are sufficiently fluid to allow for pressure injection into a relatively cold mold where they solidify and take the shape of the mold cavity. For thermoplastics, no chemical changes occur within the plastic and therefore the process is repeatable.

Injection molding of thermosetting resins has the cylinder heated to homogenize and preheat the reactive materials, and the mold is heated to complete the chemical cross-linking reaction to form an intractable solid.

The development of Reaction Injection Molding (RIM) allowed the rapid molding of liquid materials. Two highly reactive resin systems are first injected into a mixing head and from there into a heated mold where polymerization and cross-linking occur. This process has proven particularly effective for high-speed injection molding of such materials as polyurethane, epoxies, polyesters, and nylon.

Another type of plasticizing process is extrusion. In this process plastic pellets are fluidized, homogenized, and continuously formed. Extrusion products include pipe sheet, wire coating, and tubing. This process is used to form very long shapes which can be cut into smaller shapes. The extruder provides a continuous length resembling a toothpaste tube. A long barrel containing a screw mechanism feeds the pellets from a hopper through several heated compartments to a cylindrical die. The die is used to establish the particular design form. The extrusion process produces pipe and tubing by forcing the melt through a specific cylindrical die.

A third form of plastics processing is the forming of plastic sheets into parts through the application of heat and pressure. The pressure can be obtained through the use of pneumatics or compression or vacuum. Sometimes radio frequencies are used to apply heat. Tooling for this general process is the most inexpensive plastic process. It can accommodate very large as well as very small parts.

A fourth form of plastic processing is rotational molding. In this process, finely ground powders are heated in a rotating mold until melting occurs. The process can be performed by relatively unskilled labor. However, the finely ground plastics powder are more expensive than pellets or sheets. In addition, the process is not suited for large production runs of small parts.

A fifth type of plastic processing is the making of foamed plastic materials. The basic processes used include an agent that generates gas in the polymer liquid or melt. Structural foam plastics differ from other foams in that the part is produced with a hard integral skin on the outside and a cellular core in the interior.

14.3 SYNTHETIC FIBERS

Synthetic textile fibers are made from very long chain-like molecules called linear polymers which consist of small monomer units. Fiber forming polymers must possess at least 200 monomer units joined in a chain. In addition, there must be a high degree of intramolecular and intermolecular attraction and the polymer must have the ability to be oriented along the axis of the fiber.

The synthetic fibers may be formed either by addition polymerization or by condensation polymerization. In the addition polymerization process, small monomers containing unsaturated carbon-carbon bonds are built into polymers by using catalysts. These open the double bonds joining the carbon atoms of each molecule. The reactive molecule subsequently attacks a second molecule, joins with it, and propagates the reactive entity. This results in a number of monomer units joined together in a single chain by the addition of links of monomers.

In condensation polymerization, molecules containing reactive groups at each end are combined with the elimination of water, forming larger molecules, which still contain reactive groups at each end. The reactive groups can continue to combine so as to build, step by step, a large polymer.

Fibers are produced from polymers according to a number of procedures. First the fiber-forming material must be made fluid. Then the fluid is forced under pressure through tiny holes into a medium which causes it to solidify. Finally the fibers are further processed to obtain their optimum properties.

The materials are spun in one of three methods to form fibers. In wet spinning the polymer is dissolved in an applicable reagent to form the fluid. The fluid is then pumped through metal plates (spinnerets) containing many small holes into a liquid bath. A chemical reaction between the spinning material and the bath causes the fiber to solidify. An example is the production of rayon by the viscose process.

In dry spinning the polymer is dissolved in a solvent and extruded through a spinneret like the wet spinning process. However, the liquid bath is replaced by a stream of warm air which evaporates the solvent causing the polymer to solidify as a filament. An example is the manufacture of cellulose diacetate.

In melt spinning the polymer is melted into a hot molten fluid which is extruded into a stream of cold air which solidifies it into a filament. Filaments may be produced in various thickness and different cross-sectional shapes, such as round, squares or lobed. Examples of melt spinning are nylon, polyester, and other thermoplastic fibers.

There are two main sources of synthetic textile fibers. One is plant cellulose and the other is from organic intermediates, derived from petroleum, coal, and natural gas. There are a number of ways of classifying synthetic fibers. Some are known by the manufacturer's name or trademark or by terms referring to characteristic properties. In the United States the Textile Fiber Products Identification Act (TFPIA) standardized the nomenclature.

RAYON

One of the first artificial fibers was rayon. Rayon may be defined as any man-made textile fiber produced from the plant substance cellulose. Viscose rayon, the most widely used type was developed in 1892 in an attempt to produce silk chemically. Originally rayon was known as artificial silk and wood silk but in 1924 it was given the name rayon.

To make rayon, cellulose is obtained from the short fibers adhering to cottonseeds. The cellulose is chemically treated to form a solution that is forced through tiny holes in a spinneret. Rayon emerges in the form of filament, a fiber of great length, and is hardened by chemical means. The filament is cut into short pieces of uniform length and twisted together to form yarn.

ACETATE AND TRIACETATE

Acetate became an important artificial textile fiber in the 1920's. Its original name was acetate rayon and it is also called cellulose acetate. The official name by the Federal Trade Commission is acetate. It is a textile fiber produced from the plant substance cellulose, which is obtained from soft woods or the short fibers adhering to cottonseeds. The material is treated with acetic acid and acetic anhydride and then partially hydrolyzed so that it is soluble in acetone to form cellulose acetate. The cellulose acetate is extruded through a spinneret forming a fiber. During the manufacturing process the fiber is slightly stretched, increasing its strength.

Acetate is produced in the form of a long filament but is frequently cut into short lengths to be spun into yarn. Acetate fiber is much weaker than nylon. It does not wrinkle easily when worn, and, because of its low moisture absorption, does not easily retain types of stains. Acetate garments launder well, retaining their original size and shape and drying in a short time but have a tendency to retain creases imparted when wet. The fiber is used, alone or in blends, in such apparel as dresses, sportswear, underwear, shirts and ties.

Cellulose triacetate, known in the United States by the trade name Arnel, is chemically different to but related to cellulose acetate. It has greater heat resistance than normal acetate, is faster drying, and can be permanently pleated by heat setting. It is used alone or in blends, for knitted and woven fabrics especially for women's clothes.

NYLON

Nylons were developed in the 1930s by Wallace H. Carothers, a U.S. chemist, and his research team, working for E.I. du Pont de Nemours & Company. Nylons are defined as any synthetic plastic material composed of polyamides of high molecular weight and are usually, but not always, manufactured as a fiber. Polyamide resins are whitish translucent, high-melting polymers. They may be made from various chemicals such as dicarboxylic acid and amino acids. In the most common nylon adipic acid and hexamethylenediamine are used.

Nylon can be drawn, or extruded through spinnerets from a melt or solution to form fibers, or sheets to be manufactured into yarn, fabric and cordage. It has high resistance to wear, heat, and chemicals.

POLYESTER

Polyester was introduced in 1951. It is a class of organic substances composed of polymers formed from monomers by establishment of ester linkages between the monomers. An ester is a class of compounds formed by the reaction between the acid and an alcohol usually with the elimination of water. Polyesters are most commonly prepared from equivalent amounts of glycol (an alcohol containing two hydroxyl groups) and dibasic acids. The long-chain polyester made from ethylene glycol and terephthalic acid is the basis of the fiber called Dacron and the film Mylar.

Polyester fibers are generally similar in performance and properties. They recover quickly after extension. They melt at about 260° Celsius. Prolonged exposure to light reduces their strength but does not affect their color. Polyesters have good resistance to chemicals. They can be washed or dry-cleaned with most common cleaning solvents. Their low moisture content makes them likely to accumulate static charges unless treated with antistatic agents.

Polyester soon got a bad name because it was stiff, uncomfortable and made people sweat. By 1970 polyester was considered not usable. However, by 1990 engineers produced very fine holes to make very thin microfibers. A microfiber is less than one denier per filament. A denier is a unit of fineness equal to the fineness of a yarn weighing one gram for each 9,000 meters. The less the number of deniers the finer the yarn. Microfiber polyester is very soft, comfortable and luxurious like silk. The new polyester meets the objections of the old polyester. Microfibers are also used in nylon, acrylic, rayon and other fibers.

ACRYLIC

Acrylic is a manufactured fiber composed of at least 85% by weight of acrylonitrile (C_3H_3N) units. Its successful use in fiber form depended upon the discovery of appropriate solvents for dry or wet spinning, and satisfactory methods to dye the fiber. The first commercial acrylic fiber was Orlon which has attractive properties and has a great deal of versatility.

Polyacrylonitrile (PAN) was developed in 1954 in Japan from the addition polymerization of acrylonitrile. However, the pure polymer is a brittle, glassy substance at room temperature and is not well suited for use as a textile fiber. It was found that by forming copolymers of acrylonitrile and other vinyl monomers the mechanical properties of the fiber would be greatly improved. The fiber is produced from either the wet-spun or dry-spun process from solution in appropriate solvents. After drawing and texturing, the product is a soft yarn with characteristics similar to wool. It takes dyes well, is extremely resistant to sunlight, is durable, and is relatively easy to maintain. It is widely used by itself and in blends with wool or nylon in sweaters, socks, and robes. Because it does not absorb moisture, acrylic has an annoying static cling and shock especially in dry weather.

MODACRYLIC

This is a manufactured fiber composed of at least 35% by weight of acrylonitrile but less than 85%. The modacrylics are modified acrylic fibers. These fibers are produced by partial dissolution of the copolymer in appropriate solutions followed by either wet or dry spinning. The most important use of modacrylics has been in imitation furs and in imitation hair for wigs.

SPANDEX

Spandex is a manufactured fiber used as a replacement for rubber, both natural and synthetic. It is very useful in intimate apparel and sportswear, where its high elasticity give both figure control and a close fit without restricting the movements of the wearer.

Spandex is composed of at least 85% of segmented polyurethane. In Spandex, a long-chain polyester is combined with a short-chain diisocyanate to produce a polymer containing long lengths of a relatively soft material joined by short lengths of a relatively hard material. This is the segmented polyurethane. By appropriate combinations of hard and soft segments, these fibers can be engineered to provide any required degree of stretch and strength.

Kevlar

Kevlar is a polymer fiber introduced by du Pont in 1961 for use in body armor for policemen and others who needed protection from bullets. A wet solution is fed through spinnerets to form a fiber which is wound on drums and then woven to make a layer of very strong material. Many layers are used to obtain a bulletproof garment.

Spectra

Spectra is a polyethylene fiber manufactured by Allied Signal for body armor. It is not woven like Kevlar but a plastic resin holds fibers together. Both Spectra and Kevlar are combined to form a very strong body armor. On May 8, 1991 a New York City policeman was hit 13 times by a 9-millimeter handgun. He was wearing body armor which saved his life.

Hydrophilic Fibers

Hydrophilic fibers are used to absorb moisture both in the form of sweat and water. A fiber called hydrophil is used in Alaska dog races to keep the musher dry even though he or she is soaking wet from falling in the water. The fiber absorbs moisture of all kinds and keeps the wearer dry.

14.4 Composites

A composite material results when two or more distinctive materials are combined to show definite new improved characteristics. The use of composites goes back thousands of years to the Bronze Age when a small amount of tin was added to copper to obtain a much stronger material that found its first use in weapons.

In the 1960's a composite was made of fiberglass reinforced plastics or FRP. Later polymer-matrix composites using carbon fibers were developed as well as others like ceramic-matrix and metal-matrix composites. These are known as advanced composites to distinguish them from FRP.

In advanced composite systems, fibers of a particular material are imbedded in a matrix of a different material. The fibers are bonded and held in position by the matrix. The fibers improve the qualities of the matrix material. Continuous fibers from 0.1 to 0.2 millimeters are used as reinforcing constituents. These are carbon or organic fibers, inorganic fibers, ceramic fibers, and metal wires. Inorganic materials are also used in the form of discontinuous fibers and whiskers. Whiskers are single crystals that exhibit fibrous characteristics. Whiskers are extremely strong and stiff compared to continuous fibers. Silicon carbide and silicon nitrides are examples of whiskers.

Different Types of Fibers

Carbon fibers were invented for composite materials in 1954 in both the United States and Japan. In the United States rayon-based carbon fiber was developed while the Japanese used polyacrylonitrile-based fibers (PAN). By the 1970s PAN-based carbon fibers dominated organic fibers for composites. After 1985 pitch-based carbon fibers produced in Japan were found to be stronger and cheaper than PAN-based carbon fibers.

Inorganic Fibers

The most important inorganic continuous fibers for advanced composites are boron and silicon carbide. Both exhibit high stiffness, high strength, and low density. Stiff strong discontinuous fibers that predominate as reinforcements for metal matrix composites are silicon carbide, alumina, and aluminosilicates.

Ceramic Fibers

An example of a commercial continuous ceramic fiber is polycrystalline aluminum oxide. This fiber exhibits high stiffness, high strength, a high melting point, and good resistance to corrosive environments. The fibers are produced by the dry-spinning technique from various solutions. This is followed by heat treatment of the fibers.

Strong, stiff continuous metal fibers are prepared from tungsten and tungsten alloy. Their primary use is in the reinforcement of matrices of nickel and iron-base alloys for high-temperature applications.

Matrices

Matrices can be made of organic material, ceramics, or metal. The organic matrix is used extensively in advanced structural composites. The continuous reinforcing fibers for organic matrices are available in several forms. These are monofilaments, multifilament fiber bundles, and unidirectional ribbons.

A number of processes are used for fabrication of an organic matrix composite. The first step involves fiber placement to orient the unidirectional layers at discrete angles to one another in order to achieve the desired load distribution. Hand or machine lay-up techniques and filament winding are used frequently. In machine lay-up the continuous reinforcing fibers and matrix resins are combined into a nonfinal form called a prepreg. The prepreg is cut and laid up, layer by layer, to produce a laminate of the desired thickness, number of plies, and ply orientations. In filament winding, a fiber bundle or ribbon is impregnated with resin and wound on a mandrel to produce various shapes. Consolidation into the final composite structure is achieved by means of heat and pressure.

The ceramic matrix is usually a glass or a ceramic produced by fine-scale crystallization from glass. Ceramic matrix composites have high-temperature strength and wear resistance. Different kinds of fibers are used including continuous fibers, whiskers and particles. Carbon fibers, silicon carbide fibers, alumina (form of aluminum oxide) fibers and silicon carbide whiskers are employed in ceramics matrices.

An example of a metal matrix composite is a continuous graphite fiber reinforced metal matrix made of aluminum or magnesium. One method of manufacturing uses precursors. First a composite precursor is made by infiltrating liquid aluminum or magnesium into a multifilament graphite untwisted strand or tow. Precursor layers are then bonded together under high pressure and temperature. A different method of manufacture without using precursors is to form the fibers in a mold and then molten metal is poured in to form a shaped composite.

Another example of a metal matrix composite is the use of continuous boron fibers in aluminum foil. Also Silicon Carbide continuous fibers are used in an aluminum matrix. They may be made by low-pressure hot molding.

An additional example is the use of continuous tungsten fibers incorporated in a nickel or iron base alloy matrix. This can be manufactured by one of a number of techniques such as liquid metal infiltration, powder metallurgy, and casting. Discontinuous fibers or whiskers are also used with metal matrices.

AUTOMOTIVE APPLICATIONS

Advanced composites of different types are used in auto components to reduce weight, and make more fuel efficient engines. Ceramic matrix and metal matrix advanced composites are both used in automobiles. Ceramic matrix composites are used in turbocharged diesel engines in valve seats and inserts, piston rings, liners, and exhaust manifolds. Silicon carbide fibers are used with a matrix of aluminum silicate glass.

Metal matrix advanced composites are used in diesel engine pistons to improve thermal fatigue and wear. The composite consists of short alumina fibers in an aluminum alloy matrix. Connecting rods using metal matrix composite connecting rods have been developed using aluminum reinforced with stainless steel fibers and aluminum reinforced with polycrystalline alumina fibers.

Aircraft Composites

Advanced composite materials are used in both commercial production aircraft and military planes to obtain increased strength with lighter weight. Metals and alloys have been replaced with different types of composite materials. Advanced composite materials in aircraft exhibit wear resistance or strength retention at high temperatures.

Commercial aircraft such as the Boeing 777, the Concorde and the Airbus A310 all have used organic matrix fiber reinforced composites. Most of these composite components are of a honeycomb sandwich construction. The Airbus and the Concorde aircraft used composite materials embedded in a carbonaceous matrix for brake materials. This type of composite has low weight, excellent friction behavior, and wear resistance.

Military aircraft make a great deal of use of carbon-reinforced composites. About a quarter of the weight of the U.S. Navy's AV-8B is carbon fiber-epoxy. An epoxy is a compound containing a 3-member ring containing one oxygen and two carbon atoms. On the F-18 aircraft, carbon fiber-reinforced composites make up about 10% of the structural weight and more than 50% of the plane's surface area. Uses include wing skins, horizontal and vertical tail surfaces, and doors.

The Voyager, an aircraft that flew non-stop around the world, was constructed of carbon fiber and epoxy resin bonded together to form a composite. The B-2 stealth bomber also used composites bonded to epoxy resin.

Sports Product Applications

Advanced composites in the 20th century have found widespread use in sports equipment. Boron fibers or carbon fiber-reinforced epoxies are used in the frames of rackets for tennis, squash, racquetball, and badminton. Other products utilizing composites are fishing rods, skis, and golf club shafts. Golf club heads of hot forged aluminum silicon carbide are also used. Bicycle frames have been fabricated from boron-aluminum composites, and wheel rims for bicycles have been made from an aluminum-aluminum oxide composite which is stiffer and lighter than a rim fashioned from an aluminum alloy. High performance sailing craft made from honeycomb sandwich panels of carbon fiber-reinforced plastic have strong rigid, lightweight hulls.

Electronic Applications

Integrated electronic circuits generate heat which must be removed to avoid damage to components. Metal heat sinks to remove the heat from integrated circuits have a coefficient of thermal expansion different from that of the circuit leading to the failure of the component. Heat sinks fabricated from metal matrix composites with exactly matched thermal properties have solved this problem. Precision radar systems require antennas with low inertial and structural rigidity. Metal matrix composites with high specific stiffness and strength and a low coefficient of thermal expansion can meet these requirements. Graphite-reinforced copper is replacing alloys of iron, nickel and cobalt in some microwave circuits.

MACHINING TOOL APPLICATIONS

A 20th century cutting tool material for rough-machining superalloys is a composite of a matrix of aluminum reinforced with silicon carbide whiskers. The whisker reinforcement enhances toughness and it makes it possible to operate at higher speeds with improved reliability.

BONE REPLACEMENT

One biocompatible composite that can replace bones and joints is a three-dimensional structure of graphite fibers imbedded in a hydrophilic matrix material. Carbon-carbon composites consisting of carbon fibers imbedded in a carbonaceous matrix is biocompatible with living tissue and have physical characteristics close to bone.

14.5 EXTREMELY FINE POWDER MATERIALS

Over the centuries inventors have come up with techniques for grinding all types of materials to produce small particles for foodstuffs, cosmetics, cement, medicines, washing powders and many other products.

In the 20th century it became possible to make dry materials with particle diameters of 100 microns (a micron is a millionth of a meter) down to 10 microns. By the early 1980s more than a billion tons of solid material was reduced in size, crushed or ground annually in the United States. An example of 20th century production of fine powders is the use of a toner powder in copying machines. In order to make perfect copies every time, the powder particles must be of small uniform size. Another example is the use of dry coatings in which powdered ingredients with specific and uniform particle sizes are necessary to achieve the desired appearance and long-term resistance to wear of modern products.

New powder technologies are used in many areas in 20th century civilization. Foods such as spices, freeze-dried coffee, and confectionery are produced by using special powder ingredients. Powder technology is used in certain chemicals, pharmaceuticals, industrial minerals, and in aircraft engines. Aircraft metal engine parts are subject to heat. Bonding a heat resistant ceramic coating to the metal would solve the heat problem except that the different responses of the ceramic and the metal to heat would weaken the bond. Using metal powder and ceramic powder to form a multiple layer coating for the metal part solves the problem.

Powder Production

There are two basic methods of making powders, mechanical and non-mechanical. In the mechanical process grinding, cutting, friction, and compression are used to produce very small particles. One common mechanical method uses a grinding process called ball milling. Basically the ball mill is a hollow cylinder filled with balls. The material to be reduced is fed in pieces to the ball mill. The cylinder is rotated so that the balls and fed material tumble together. The sizes of chunks of material are reduced by pounding, squeezing and rubbing.

Another mechanical method is the use of the pin-disk mill. Spinning pin-studded disks are used in this process. Particles fed into the center of the mill are forced against the whirling pins. The impact forces reduce the size of the particles.

A third method of a mechanical process is the use of a high compression roller mill. One of the rollers is fixed. The material is fed between two rollers. The second roller pushes towards the first applying considerable force against the material.

Non-mechanical methods of producing powders include processes to remove solid particle dissolved or suspended in liquids. These processes are precipitation, filtration and sedimentation. Powders may also be produced from gases by chemical vapor deposition. Another method is the jet mill which can get smaller particle size than mechanical methods. Jet mills use rapid streams of gas to force particles to collide with each other. However, the bulk of most powder production is still done by mechanical methods.

14.6 Artificial Material Poisons

PCBs

In June 1968 a group of people were brought to Kyushu University Hospital in Japan with a peculiar set of symptoms. These symptoms included skin rash, headaches, and discharge from the eyes, and a dark pigmentation of the nails and skin. Soon the disease spread to 1200 people in the western part of Japan.

A search started for the source of the problem. It was found that all the sick families had used the same rice oil from a particular factory for cooking. An analysis of the oil showed that it had been contaminated with heat-exchange fluid during processing at the factory. This latter fluid contained PCBs.

Extensive medical examinations revealed problems with the liver, immune system and reproduction. The skin rashes tended to disappear but the other symptoms got worse. The Japanese called PCB poisoning Yusho or oil disease.

PCBs are any of a number of highly stable organic compounds that are prepared by the reaction of chlorine with biphenyl. A commercial mixture provides a colorless, viscous liquid that is relatively insoluble in water, does not degrade under high temperature and is a good insulator. They are particularly useful as lubricants and insulators in transformers and capacitors.

PCBs were discovered in the late 19th century and were introduced into U.S. industry in 1929 and have been in use from that time in most industrial nations. However, in the mid-1970s the production and application of these chemicals have been restricted because they have been found to be injurious to living organisms. Although PCBs were never intended to be released into the environment, they found their way into the air, water, and soil via industrial and municipal waste disposal and leaks from equipment. Although electrical equipment containing PCBs is no longer produced in the United States, still in 1981 40% of all electrical equipment in use in the U.S. contained PCBs.

The high resistance of PCBs to decomposition ensures that they remain in soils and bodies of water for many years, enabling them to accumulate and enter the food chain. PCBs are toxic to fishes and invertebrates. In humans PCBs cause liver dysfunction, dermatitis and dizziness and are suspected of causing cancer.

PCBs have been reported in fishes, eels and other organisms in the North Sea near the Netherlands. Fish products from the U.S. waters and the Baltic Sea have been found to contain several parts per million of PCB and have been declared unfit for human consumption. PCBs have been found in all organisms analyzed from the North and South Atlantic, even in animals living in the deep ocean, two miles down.

PCBs have been detected in human fatty tissues and in the milk of cows and humans. The estimated percentage of the U.S. population with detectable levels of PCBs was nearly 100% in 1981. Except for occupational contact, human exposure is mainly through food and is stored in fatty tissue. PCBs can cause skin discoloration, liver dysfunction, reproductive effects, and toxicity.

The incineration of solid and liquid PCB materials is a common and highly efficient method of destruction. PCB contaminated fluids have been decontaminated by using chemical reagents such as sodium to remove the chlorine atoms from the biphenyl molecule.

In 1973 researchers discovered that sunlight hitting a chemical catalyst destroys certain compounds. In 1992 a group of scientists at the State University of New York at Oswego cleaned up PCBs containing water and sludge by adding titanium dioxide and setting the mixture on the roof for a few hours. The sunlight hit the titanium dioxide which stimulated formation of active molecules of oxygen and hydrogen that destroy the PCB.

The photodecomposition method is promising for certain situations and is less costly than incineration. It is also much quicker than breaking down PCBs with bacteria. However, there are some limitations. It is less effective on multiple decontaminates and it is not useful if the contaminant is down at the bottom of 10 feet of cloudy water.

PBBs

In 1973 a mysterious illness struck whole herds of dairy cattle in the state of Michigan. The animals lost their appetite, grew thin and weak, became sterile or gave birth to dead calves, produced dwindling quantities of milk, and developed open sores that would not heal. An investigation showed that the toxic chemical had found its way into the cattle feed. The chemicals were being excreted in the milk and contaminated the food supply of Michigan.

The chemicals were polybrominated biphenyls or PBBs, ingredients of a fire retardant manufactured by Michigan Chemical Corporation. This company also manufactured a feed supplement for dairy cattle. A mix-up at the factory resulted in shipping the fire retardant to farms all around the state. The PBBs were later found in the milk of healthy cows, and in butter, cheese, and beef. They were also discovered in sheep, pigs and chickens which were fed the contaminated feed. Farm families in Michigan began to complain of numbness in fingers and toes, headaches, fatigue and joint pain.

A survey later by the Mt. Sinai School of Medicine showed that practically everyone in Michigan had significant amounts of PBBs in their blood and body tissues. The health of Michigan residents were compared to a comparable but unexposed control group of Wisconsin residents. It was found that the Michigan people were in poorer health. Neurological symptoms were reported much more frequently than in Wisconsin. Michigan residents complained much more about extreme fatigue. Many of them reported sleeping 18 hours a day. Arthritis-like symptoms were reported in young Michigan men. It was also found that skin problems were more common in Michigan than in the control group. However, new residents did not show any PBBs in their blood.

It was found that during the years 1970 through 1974, the U.S. major manufacturer of PBBs, the Michigan Chemical Corporation, poured the factory's wastewater into the nearby Pine River. Fish concentrated the chemicals to dangerous levels in their tissues. The company disposed of 269,000 pounds of waste containing 60 to 70 percent PBBs in a landfill. Studies have shown the chemicals to be leaching out into surrounding groundwater. This may be a future threat to drinking water.

TCDD (Dioxin)

An explosion in a chemical factory on July 10, 1976 sent a cloud of poison gases over the suburban town of Séveso near Milan. Dioxin or TCDD was one of the chemicals that settled on the houses and gardens of several thousand persons in the area. Dioxin was a highly toxic by product in the manufacture of herbicides and disinfectant soaps.

Domestic and wild animals began to die, and children were brought to the hospital with chemical burns. Three weeks later the area was evacuated. Some of the children began to develop a distinctive skin rash. Medical surveys began to detect signs of damage in the blood, liver, nerves, and vision of some of the exposed people. A number of women underwent abortions because of concern of the possibility of birth defects. Dioxin lingers indefinitely in the body.

14.7 Industrial Waste Disposal

An enormous source of pollutants is the industrial waste disposal from chemical factories. By 1979 30-40 million tons of hazardous waste was produced in the United States and most of it was put in landfills or dumped in water.

Halogenated hydrocarbons including PCBs and PBBs have been found in the wastes of industries that manufacture plastics, textiles, petrochemicals, automobiles, pharmaceuticals and electronic components.

The wastes can wind up in waterways and in landfills. Metal drums of waste material buried years ago eventually leak into the surrounding ground threatening water supplies. PCBs, PBBs, TCDDs and other poisons have been found in the wastewaters of industries that manufacture artificial textiles, pharmaceuticals, petrochemicals, and other 20th century materials.

One technique to prevent such threats to the environment is to use a closed recycling system in factories like the molten metal technology described in Chapter 3. To be successful, closed recycling systems must be economically viable and should prevent any harmful pollutant wastes from leaving the factory area. Many chemical companies including giants like du Pont are investigating such technologies. Perhaps in the future economics combined with government regulations will solve the problem of waste pollutants that can poison the earth.

Plastic Disposal Problems and Solutions

Some plastics are impervious to heating and are insoluble. Plastic wastes are not biodegradable and they fill up landfills. Plastics have increasingly replaced metals in cars and appliances and getting rid of these materials later becomes a problem.

Plastics used in soda bottles are thermoplastic which can be melted down and used again. However, thermoset plastics cannot be melted and reused. At the end of the 20th century both types were used in cars in large amounts. The question of what to do with the large amounts of plastic waste has loomed large.

One possible solution is a process called pyrolysis that breaks down plastics into burnable oil and gas. Pyrolysis breaks materials down by heating them at high temperatures in the absence of air. Because no oxygen is present, plastics left over from cars do not burn, as they would in an incinerator. Instead, the heat, about 1,300 degrees Fahrenheit, breaks the complex hydrocarbons down into simpler molecules producing a liquid resembling heating oil and a combustible gas, along with some ash. The G.M. Technical Center in Warren, Michigan has run trials on pyrolysis for dealing with waste plastics in cars. However, if the waste compounds include vinyl compounds containing sulfur, pollution in the form of hydrochloric and sulfuric acid will result.

14.8 Hormone Mimics and Blockers

It was discovered in the last decade of the 20th century that a number of synthetic compounds could disrupt the endocrine systems of animals and humans. One example is the production of estrogen. Estrogen is a powerful female hormone that is produced by vertebrates. This binds to an estrogen receptor in body cells like a key inserted in a lock. Certain synthetic chemicals mimic estrogen and bind to the estrogen receptor. This can cause a disruption in the body resulting in cancers in males linked to estrogen such as testicular and prostate cancer. Another problem is low sperm count which can cause a loss of fertility.

There are two classes of chemicals that disrupt the endocrine system. Hormone mimics bind to receptors and induce a response. An example is Diethylstilbestrol (DES) which used to be prescribed for pregnant women. DES has caused a rare vaginal cancer in the daughters of mothers who took it. The second class, hormone blockers, do not induce a response but block natural hormones from attaching to the receptors. An example of a hormone blocker is Vinclozolin which is used to kill fungus on fruit. It blocks androgen receptors and prevents the testosterone message from coming through. This results in the production of hermaphrodites instead of males.

Hormones are chemical messengers that regulate important body functions such as reproduction. Hormone receptors act as a lock and the hormones act as a key inserted in the lock and the body responds. Hormone mimics act as a key in the receptor lock and the body may respond in very inappropriate ways. Hormone blockers act like a key jammed into the receptor lock. Natural hormones are prevented from opening the receptor lock.

Very small amounts of artificial chemicals can cause the hormonal problems. These are found throughout human society at the end of the 20th century. Over 50 chemicals have already been discovered. An example is Bisphenol-A found in the inner plastic linings of many metal cans used to package foods. This synthetic chemical also leaches from polycarbonate water jugs. PCBs and dioxin have been found to disturb hormonal functions.

Some of the scientists who have independently discovered the problem often through serendipity, are Dr. Theo Colborn of the Conservation Foundation in Washington, DC,

Dr. Ana Soto and Dr. Carlos Sonnensheim of Tufts Medical School, Boston, and Dr. David Feldman of Stanford University School of Medicine in Palo Alto, California.

Their exploits and others are described in a book, *Our Stolen Future,* by Theo Colborn, et al., published in 1996. This book like Rachel Carson's *Silent Spring* rings an alarm bell for the consequences of using some synthetic chemicals. This is an example of a technology producing unexpected and serious consequences. Much further investigation is needed into the 21st century to determine the extent of the problems and possible technologies to alleviate them.

BIBLIOGRAPHY

Colborn, Theo, John P. Meyers and Dianne Dumanski — *Our Stolen Future, A Scientific Detective Story,* Dutton, 1996

Forester, Tom, Editor — *The Materials Revolution,* The MIT Press, 1982

Gorman, James — *Hazards to Your Health: The Problem of Environmental Disease,* The New York Academy of Sciences, 1979

Meikle, Jeffrey I. — *American Plastic,* Rutgers University Press, 1995

Schneider, Mary Jane — *Persistent Poisons: Chemical Pollutants in the Environment,* The New York Academy of Sciences, 1979

Sterret, Frances S., Editor — *Environmental Sciences,* New York Academy of Sciences, 1987

Van Cleef, Jabez — "Powder Technology," American Scientist, Vol. 79, July-August 1991

CHAPTER 15

EXTRACTING ENERGY FROM THE ATOM

One of the most important technologies developed in the 20th century was the use of atomic energy. Its repercussions in terms of terrorism, war, energy and waste products will continue into the 21st century and beyond. It started out with a simple equation which first saw the light of day on September 27, 1905. This led to the greatest man-made explosion in history during World War II. It also led to a new source of energy with both benefits and very serious problems. The equation has had a greater impact on civilization than any other comparable equation.

15.1 THE MAKING OF THE EQUATION THAT SHOOK THE WORLD

The man who originated the equation was Albert Einstein, one of the greatest scientists who ever lived. He was born to Jewish parents, Hermann Einstein and Pauline Koch Einstein, on March 14, 1879, in the city of Ulm in the Kingdom of Wurttemberg, a part of the German empire since 1871. In 1880 Einstein's family moved to Munich where at the age of six he entered public school, the Volksschule.

Many years later the story went the rounds that Albert was a very poor student. Psychologists went on the radio and told parents not to worry if their children did poorly in mathematics because Einstein had also performed badly. They based this on the modern German marking system of 1 to 5 where 1 was the highest grade and 5 was the lowest grade. Einstein's report card showed a lot of 5s so it was assumed by a number of researchers that he was a bad student and that there was hope for most poor students to become an Einstein. However, during Einstein's time in school, the lowest grade was 1 and the grades went up with the numbers. Later, long after Einstein graduated, the marking system was reversed. Einstein was always one of the smartest in all his classes.

In October 1888 Einstein moved from Volksschule to the Luitpold Gymnasium which was to be his school until the age of fifteen. In all those years he earned the highest or next to the highest marks in mathematics and Latin. However, he disliked the authoritarian teachers and rote learning that characterized German schools of that day. In addition, he disliked gymnastics and sports which were a part of the curriculum.

Albert felt isolated and made few friends at school. He studied many things at home, including music and mathematics. At the age of twelve he was given a book on Euclidean geometry which he later called the holy geometry book. From age 12 to age 16 he studied differential and integral calculus by himself.

In 1895 he left Germany and went to the cantonal school in Aarau in the German speaking part of Switzerland where he studied for the Matura, the high school diploma. For the first time in his life, Albert enjoyed school. He loved the liberal spirit of the school and the teachers who did not rely on external authority. This was a stark contrast to the German Luitpold Gymnasium. On January 28, 1896 Einstein gave up his German citizenship. In the fall of 1896 he received his Matura and on October 29 he registered as a resident of Zurich and became a student at the Eidgenossische Technische Hochschule (ETH).

Upon satisfactory completion of the four-year curriculum at ETH he would qualify as a Fachlehrer, a specialized teacher, in mathematics and physics at the high school. However, he was unhappy with his physics professor, Heinrich Friedrich Weber. Professor Weber was supposed to have told his young student to become anything but a physicist. Einstein relied a great deal on self-study. He received final grades of 5, out of a maximum of 6, in physics, experimental physics and astronomy.

In August 1900 Einstein graduated and became qualified as a Fachlehrer, together with three other students. Those three were immediately appointed assistants at ETH. Einstein received no offer of a job for which he blamed Weber. When Weber died in 1912 Einstein refused to go to his funeral. He even wrote to a friend that Weber's death was good for the ETH. While at the school Einstein saved a small part of his monthly allowance to pay for his Swiss naturalization papers which he obtained on February 21, 1901. He remained a citizen of Switzerland for the rest of his life, although in 1940 he also became a citizen of the United States.

On June 16, 1902, after some temporary employment as a high school teacher, he was appointed a technical expert third class at the Swiss Patent Office in Bern on a trial basis. He enjoyed his work at the Patent Office. Although he took his work seriously and often found it interesting, he still had time and energy to work on his own projects in physics. On September 16, 1904 his provisional appointment was made permanent. He continued his own work on physics which flowered in an incredible burst of creativity in 1905.

On March 17, 1905 he completed a paper on the light quantum hypothesis and the photoelectric effect. He won the 1921 Nobel Prize in Physics for this. On April 30, 1905 he completed his Ph.D. thesis "On a New Determination of Molecular Dimensions." On May 14, 1905 his paper on Brownian Motion was received by the *Annalen der Physik.* On June 30, 1905 his first paper on Special Relativity was received by the same journal. This was followed on September 27 by a short second paper on Special Relativity. It contains the equation $E = mc^2$. It was also published in the *Annalen der Physik.* Einstein concluded this 1905 burst of creativity on December 19th with a second paper on Brownian Motion also published in the *Annalen der Physik.*

The equation $E = mc^2$ came from Einstein's work on Special Relativity. The equation implies that E, the energy equivalence of mass, m, is enormous since c, the speed of light is 300,000 meters per second or 3×10^8. Squaring 3×10^8 results in an even greater number, 9×10^{16}. Many scientists, including Lord Rutherford, the great English physicist believed as late as the 1930's that this enormous energy could not be released by humanity.

15.2 RACING TOWARD THE ATOMIC FISSION BOMB

THE BACKGROUND

In 1932 James Chadwick, a British physicist at Cambridge Laboratory, discovered the neutron, a neutral fundamental particle that later became essential in releasing the energy from the atom in accordance with $E = mc^2$.

In 1934 Enrico Fermi and his co-workers in Rome bombarded Uranium with neutrons and found an unexplained radioactive substance. Fermi speculated that it was a new transuranic element. A German chemist Ida Noddack and others suggested that the atom had fragmented into several lighter elements. However, this was not taken up by other scientists.

Shortly afterwards Otto Hahn and Lise Meitner bombarded Uranium with neutrons at the Kaiser Wilhelm Institute in Berlin. They were soon joined by Fritz Strassmann. A number of scientists in different countries in the next two years bombarded Uranium with neutrons. These included Irene Curie in Paris, Norman Feather and Egon Bretscher at Cambridge in England and Philip Ableson at Lawrence's Radiation Laboratory in Berkeley, California.

Lise Meitner fled Germany in 1938 to escape the Nazis. Hahn and Strassmann decided to study the substances resulting from the bombardment of Uranium by neutrons. They found that the products resembled Barium which was an unexpected result. Hahn wrote a letter describing the results to Meitner who was in Sweden. Meitner's nephew, Otto Frisch, had come home from the Niels Bohr's Institute in Copenhagen to spend the Christmas holidays with her. Meitner showed her nephew Hahn's letter and in the light of both Bohr's recent work and Einstein's energy-mass equation, they set about to analyze what Hahn had produced by neutron bombardment of Uranium. On the day before Christmas 1938 Lise Meitner and Otto Frisch used Einstein's equation to calculate the enormous energy released by the splitting of a Uranium nucleus bombarded by a neutron. They were convinced that the neutrons had split the nucleus of the Uranium atom and that some of the matter had been converted into energy in accordance with Einstein's equation. Frisch returned to Copenhagen but he continued his discussions with his aunt by telephone. This led in January 1939 to a joint authorship of a note to the British scientific journal *Nature* on their explanation of the division of a Uranium nucleus. In this paper Frisch called the process "fission." Later Frisch performed an experiment at Bohr's Institute where he detected fragments from the split Uranium nucleus in an ionization chamber. This confirmed the new phenomenon and Frisch wrote another note to *Nature* describing his experiments and what they meant.

On January 26, 1939 Bohr officially announced the discovery at the Annual Theoretical Physics Conference at George Washington University in Washington, DC. This was soon confirmed by laboratories in Europe and the United States. Throughout that year fission research was conducted in many places. In February 1939 a Paris team of researchers had found that neutrons were released in the fission of Uranium nuclei. This opened up the possibility of a chain reaction with the release of a tremendous amount of energy.

By this time Enrico Fermi had been chased out of Italy by the fascists and was at Columbia University in New York City. Here he found that fission was more likely to occur when the neutrons moved slowly. A moderator such as heavy water would slow down the neutrons and help produce a chain reaction that could produce a tremendous explosion. This meant a bomb that would dwarf the largest explosive existing at that time. In March 1939 Fermi proposed an atomic fission bomb to a group of Naval Officers. He explained the use of a nuclear chain reaction which would result in a crater more than a mile in diameter. The Naval Officers did not see the need for the government to sponsor the work but they donated $1,500 to Columbia University to help Fermi's research.

In Germany scientists informed the Hitler government of the military potential of nuclear fission. The German government procured Uranium and undertook fission. This alarmed many scientists in the rest of the world who understood the meaning of an atomic fission bomb. However, there were still a number of problems that had to be solved.

Before going further into the history of the atom bomb, it is necessary to explain the isotopes or different forms of Uranium. The two isotopes of interest are ${}_{92}U^{235}$ and ${}_{92}U^{238}$. The first number is the number of protons, the positively charged particles in the nucleus. This is called the atomic number. The second number, 235, of the first isotope is the sum of the number of protons and neutrons (the neutral particles in the nucleus). In this case 235 is the sum of 92 protons and 143 neutrons. Similarly, ${}_{92}U^{238}$ has 92 protons and 146 neutrons. For simplicity the two isotopes of Uranium will be referred to here as U^{235} and U^{238}.

One of the main problems was that the predominant isotope in Uranium found in nature, U^{238}, did not lend itself to fission. Two days before France and Britain entered the war, Bohr and John A. Wheeler of Princeton University published a paper on fission. This showed that the rare isotope U^{235} was much more likely to undergo fission. U^{235} became an object of study in many research laboratories.

Frisch, now trapped in Birmingham, England, by the war, studied the separation of the rare U^{235} from U^{238} using a method of thermal diffusion developed by the German chemists Klaus Clusius and Gerhard Dickel to separate Chlorine isotopes. However, the difference in nuclear mass of the two isotopes was very small making the separation of U^{235} from U^{238} very difficult. In Germany, Werner Heisenberg had worked out a theory for energy production by nuclear fission of U^{235}. Work in the Soviet Union was also being done on the assumption that U^{235} was the most likely isotope to fission.

German émigrés in England in 1940 feared that Hitler would develop an atom bomb first. Rudolf Peierls and Frisch calculated the critical mass of U^{235} that would produce a chain reaction necessary to cause a massive explosion. Their calculations showed that about a kilogram was all that was needed. The two German Jewish refugees wrote a memorandum making the first thorough scientific analysis of the feasibility and destructive power of a U^{235} fission bomb. They suggested methods for separating the isotopes of uranium to obtain the critical U^{235}. They also evaluated the destructive effects of the bomb. The memorandum was sent to Mark L. Oliphant, a prominent British physicist, who forwarded it on to George P. Thompson who was in charge of the British government's program of Uranium research.

Thompson, on April 10, 1940, convened the first official British scientific committee on atomic bomb research. The code name for the group was MAUD and it was linked to the Ministry of Aircraft Production. Although Frisch and Peierls were at first excluded from MAUD because technically they were enemy aliens, they and other refugee scientists were included in a technical subcommittee. Frisch and Peierls made a detailed theoretical analysis of various schemes of separating U^{235} from U^{238}. They concluded that gaseous diffusion was the most promising method. Their conclusions were recorded in a series of papers for the MAUD technical subcommittee. Later this method was used in the United States in conjunction with others to obtain U^{235} for the first atomic fission bomb.

THE MAN WHO NEVER UNPACKED HIS SUITCASE

One of the most unusual people involved in the early development of the atomic fission bomb was Leo Szilard. He belonged to a group of remarkable Hungarian Jewish refugee physicists. These included Edward Teller, the so-called father of the hydrogen bomb, John von Neumann who made many contributions to the development of the plutonium implosion bomb and later to computer science, and Eugene P. Wigner who also contributed to the atomic bomb development.

Leo Szilard was one of the most profoundly original minds of the 20th century. He contributed significantly to statistical mechanics, nuclear physics, nuclear engineering, genetics, molecular biology, and political science.

Szilard was born in Budapest on February 11, 1898. He received his Doctorate in Physics from the University of Berlin in 1922. In 1929 he published a famous paper establishing the connection between entropy and information. This became the foundation later of modern cybernetic theory.

During his time at the University of Berlin, Szilard together with Albert Einstein patented an electromagnetic pump for liquid refrigerants. This now serves as the basis for the circulation of liquid metal coolants in nuclear reactors.

In 1933 Szilard was forced to flee Germany by the Nazis. He fled to England but he did not unpack his suitcases because of a feeling of a lack of permanence. While in England, he conceived the idea of a nuclear chain reaction which would yield enormous amounts of energy. In 1938 he went to the United States where he learned of the discovery of fission in Germany by Hahn and Strassman. He still did not unpack his suitcases.

Szilard became convinced that the Germans had the knowledge and capability to convert the discovery of fission into an atomic bomb. He had Edward Teller drive him in 1939 to Long Island where Einstein was vacationing. Szilard drew up a letter to President Franklin Roosevelt for Einstein's signature. The letter stressed that the United States must obtain the atom bomb before Germany did. The letter was transmitted to Roosevelt in late 1939. In response, the President established the Advisory Committee on Uranium from which the Manhattan Project to make the bomb evolved.

15.3 The American Fission Bomb Program

Preliminaries

The Advisory Committee on Uranium in the United States granted funds to buy materials for Fermi at Columbia University. In the early spring of 1940 Szilard had obtained four tons of graphite to be used as a moderator in a U^{235} reactor. Fermi and Szilard by mid-May of 1940 developed the idea of arranging U^{235} blocks in a three dimensional lattice embedded in a pure graphite moderator to produce a chain reaction. Fermi planned to publish these ideas but Szilard convinced him not to do so. The Germans never got this information and they continued to work with heavy water as a moderator. However, a British raid on the Vemork hydroelectric station in occupied Norway destroyed Germany's source of heavy water.

In June 1940 Vannevar Bush, President of the Carnegie Institution, persuaded President Roosevelt to appoint him head of a new organization, the National Defense Research Committee (NDRC). The NDRC enlisted scientists from university research programs. Later in June 1941, Bush became the Director of the Office of Scientific Research and Development (OSRD). The NDRC then became a unit of the OSRD and was led by James Conant.

In June 1940 something occurred at Berkeley which had a profound effect later on the American atomic bomb development. Philip Abelson and Edwin M. McMillan announced the discovery of element 93, the first element beyond Uranium. Abelson and McMillan named their discovery Neptunium. A number of scientists theorized that Neptunium beta decays to produce an element of higher atomic number. A team at Berkeley headed by Glenn T. Seaborg isolated the new element with atomic number 94 in February 1941 and it was named Plutonium. A number of scientists at Berkeley irradiated Plutonium with neutrons and detected fission. This opened up the possibility of a Plutonium bomb.

In July 1941 the MAUD committee in England issued a classified report detailing how a U^{235} bomb could be made. Information on this report was relayed to Vannevar Bush, the Director of OSRD in the United States. Bush also received a report by Seaborg showing that Plutonium could be used in an atomic fission bomb. Bush and James B. Conant drafted plans for an American bomb program. In October 1941 Bush presented these plans to President Roosevelt who authorized a full-scale effort to build an atomic fission bomb.

On December 7, 1941 the Japanese attacked Pearl Harbor and the United States was at war with Germany as well as Japan. This energized the atomic bomb program in the United States. By December 15, 1941 Bush assigned different broad research responsibilities to three distinguished scientists. Arthur Holly Compton was given the job of studying chain reactions. Plutonium research was assigned to Ernest Lawrence, and Uranium isotope separation was the domain of Harold Urey.

Compton assigned the job of building an atomic pile to Fermi. It was decided to build one to prove the concept of a chain reaction which was crucial to the development of an atomic bomb. Neutrons would split U-235 resulting in more neutrons which in turn split additional Uranium atoms. Each collision would convert mass to energy in accordance with Einstein's formula. While theory predicted a chain reaction, it had to be physically proved before work on the bomb could proceed.

Fermi knew that he had to find a way to slow down the neutrons in an atomic pile. Neutrons emitted in fission are quite energetic, which makes them unlikely to cause another fission but are more likely to be captured by non-fissionable U-238. For a successful chain reaction, Fermi knew that the enriched Uranium containing U^{235} must be embedded in a medium that can absorb excess energy from the fast neutrons.

Fermi and Szilard considered several materials to be used as a moderator in an atomic pile. They decided that graphite, a form of carbon, would be the best of available bulk materials. Graphite slows neutrons and it does not absorb many of them. The graphite had to be very pure because even trace amounts of a strongly neutron-absorbing impurity would spoil its effectiveness in an atomic pile. The existing graphite was not pure enough. Szilard talked to executives of various graphite manufacturers about further purifying their product.

Fermi and Szilard in 1941 decided to build a prototype of an atomic pile at Columbia University to test the idea before actually building a working model. Szilard borrowed five hundred pounds of Uranium oxide from the Eldorado Radium Company. Both Fermi and Szilard in 1941 knew that they did not have enough pure material for a self-sustaining chain reaction but they wanted to build a test pile that would point the way. Fermi took measurements and did calculations to find out how much Uranium and graphite was needed and how to arrange the materials.

Fermi's team started by building a stack of graphite bricks measuring about eight feet on a side and eleven feet tall. Cans of Uranium were spaced evenly in a lattice arrangement. Fermi hired members of Columbia's football team to stack the heavy graphite bricks. A radium-beryllium source at the base of the stack provided a steady stream of neutrons.

Fermi made careful measurements which showed that with pure material a working atomic pile was possible. Szilard started to obtain pure graphite from manufacturers who learned how to make it. In August 1941 purer Uranium dioxide was obtained which made a working atomic pile possible. On December 7, 1941 the Japanese attacked the American fleet at Pearl Harbor, Hawaii. Germany declared war and Bush decided to build a working atomic pile at the University of Chicago. The pile was to have the shape of a flattened ball twenty-five feet wide and twenty-feet high. Four hundred tons of purified graphite, forty tons of Uranium oxide, and six tons of Uranium metal were to be used.

Fermi wanted to build the atomic pile in Chicago's Argonne Forest, a number of miles from the University. In October 1942 the construction workers at Argonne went on strike. Fermi asked permission to construct an atomic pile under the west stands of Stagg Field on the University campus. After some worry about an accident in a highly populated area, Fermi received a go ahead. The pile consisted of a lattice of Uranium and graphite with control rods of Cadmium.

The pure graphite logs were cut to about four inches wide and several feet long. These were stacked in layers that formed squared off circles, small in area at first, then larger to create the flattened sphere that would best use the available material. Wooden scaffolding supported the successive layers and made the growing pile look like a cube from the outside.

The Uranium oxide was pressed to form 22,000 egg-shaped slugs. These slugs were fitted into holes drilled into the graphite. Layers of graphite with Uranium alternated with graphite-only layers which had slots milled out to create channels for the Cadmium rods that would control the neutron population. Cadmium is a strong absorber of neutrons. There were a total of fifty-seven layers of graphite, half of them seeded with Uranium. The whole contraption rested on wooden scaffolding.

There was some concern about the safety of the experiment when the atomic pile went critical. Some of the scientists raised the possibility of the pile going out of control and taking a good part of Chicago with it. In order to prevent this, George Weil was assigned the task of positioning a Cadmium control rod. In addition, three men were on top of the pile with buckets of a Cadmium-salt solution to douse the reaction should it start to go out of control. As a last safety measure another man stood by with an ax, ready to chop loose a weighted Cadmium safety rod in case the automatic rod failed.

Weil got his orders from Fermi to start with his control rod about halfway out. Then Fermi had Weil withdraw the control rod six inches at a time. Fermi made measurements of the neutron intensity and the rate of increase of neutrons. At one point an automatic safety rod went into its slot. The neutron intensity had exceeded the point for which it was set.

Fermi put all the control rods into their slots and locked in place. Then he called for a lunch break. Everything seemed to be going according to plan. Fermi decided to go for criticality after lunch. He ordered all but the manually controlled rod removed. He had Weil remove his safety rod and the intensity grew and grew exponentially at a slow controlled rate. Fermi checked the calculations with a slide rule. Finally, Fermi put away his slide rule and declared, "The reaction is self-sustaining." Eleven minutes later the safety rod was reinserted and the neutron intensity fell off rapidly. The experiment was an outstanding success.

Forty-nine people who were present toasted the atomic pile with paper cups filled with Chianti that Eugene Wigner had brought in a straw-encased bottle. Although the power from that pile reached only half a watt, it had proven the theory that led to the atomic bomb and then atomic power plants. The Fermi atomic pile success was transmitted in a message, "The Italian navigator has landed." It was December 2, 1942.

THE MANHATTAN PROJECT

In order to coordinate the work of numerous laboratories in different parts of the country, President Roosevelt placed the Army Corps of Engineers in over-all charge of the development of the atomic bomb. Army Chief of Staff chose Brigadier General Wilhelm D. Styer to follow nuclear developments. On June 18, 1942, Styer ordered Colonel James Marshall to form a Corps of Engineers District responsible for atomic bomb research. Marshall established the Headquarters of the new Corps of Engineers District in the Borough of Manhattan in New York City. The District was called the Manhattan Engineer District and soon it was referred to as the Manhattan Project.

General Marshall and Secretary of War Henry Stimson decided that an aggressive leader of the Manhattan Project was needed. General Styer recommended Colonel Leslie Groves, Deputy Chief of Construction to be in charge of the Manhattan Project. On September 17, 1942 Groves became the Director of the Manhattan Project. He was promoted to Brigadier General on September 23, 1942. Groves had a reputation of handling very large projects and the Manhattan Project was to prove to be the largest engineering project ever undertaken.

General Groves appointed J. Robert Oppenheimer as Director of an atomic bomb laboratory. Oppenheimer was a distinguished physicist at the University of California at Berkeley. Oppenheimer in October 1942 suggested the site of the Los Alamos Ranch School in New Mexico. Oppenheimer had spent summer vacations on his family's nearby ranch in Los Alamos. Groves after initial doubts approved the site and the new laboratory was constructed rapidly. By the end of April 1943 most of the laboratory was constructed.

There were two materials that Los Alamos planned to use in its atomic bombs. One was Pu^{239}, an isotope of Plutonium and the other was U^{235}, an isotope of Uranium. Both materials fissioned readily when bombarded with neutrons, yielding further neutrons in sufficient numbers to sustain in principle an explosive chain reaction.

THE SEPARATION OF URANIUM ISOTOPES

Naturally occurring Uranium contains only 0.7% of the fissionable isotope U^{235}, the balance being essentially U^{238}. Uranium with higher concentrations of U^{235} is called enriched Uranium. U^{235} is readily fissionable when targeted with neutrons. In Britain Frisch and Peierls made a detailed theoretical analysis of a variety of various schemes to separate U^{235} from U^{238}. They concluded that gaseous diffusion was the best method. In Los Alamos it was decided to use a barrier made of sintered nickel powder to separate the two isotopes. A large plant was built in Oak Ridge, Tennessee.

There were also other methods of separation such as centrifuges, electromagnetic and thermal diffusion. For the first weapon U^{235} was produced by using a thermal diffusion plant whose output fed material to a small aseous diffusion plant which fed material to an electromagnetic plant for the major stages of enrichment.

THE GUN ASSEMBLY CONCEPT

Calculations had shown that a critical mass of fissile material had to be obtained in order to produce an explosion. The first idea was to use a gun assembly to drive particles of U^{235} together. The chain reaction is initiated by hurling together in a tube and with tremendous force, two samples of highly enriched material. The gun-assembly method was used in Little Boy, the Uranium bomb that was the first atomic weapon dropped in Japan.

THE IMPLOSION CONCEPT

Theoretical studies at Los Alamos under the Manhattan Project showed that a smaller amount of Plutonium was needed for an atomic bomb compared to Uranium. However, it was soon found that the gun assembly would not work for Plutonium. The very high rate of spontaneous fission of Plutonium made the gun-assembly not feasible. Spontaneous fission was likely to cause predetonation; the explosion would begin before the active material was fully compressed and then fizzle out.

The subcritical masses had to be brought together more rapidly than was possible with the gun assembly. This could be done by implosion which consisted of placing conventional explosive material around the Plutonium. The blast from the conventional explosives would be directed inward, squeezing the Plutonium until it reached critical mass and exploded.

In an implosion device the fissile material is physically compressed by the force of a shock wave created with conventional explosives. Then, at just the right instant, neutrons are released, initiating the ultrafast fission chain reaction. Thus the main elements of an implosion device are a firing system, an explosive assembly, and the core. The firing system consists of vacuum-tube-based, high-energy discharge devices called krytons that can release enough energy to detonate the conventional explosive. The explosive assembly includes "lenses" which precisely focus the spherical, imploding shock wave on the fissile core. A neutron initiator is within the core. This initiator was polonium-beryllium which emits neutrons to start the chain reaction. The Plutonium bomb with its implosion design was named Fat Man.

Converting an Atomic Device into a Bomb

In order to make an actual atomic bomb an infrastructure of laboratories, machine shops, and special equipment had to be designed and built. The work of the scientists had to be transformed into a reliable service weapon. The officer selected by the Navy at the request of Vannevar Bush to do the job was Commander William S. Parsons, a brilliant Ordinance Officer, who had taken a significant part in the development and actual battle testing of the proximity fuse for Naval antiaircraft artillery. In May 1943 he took over his new duties to build the actual atomic bombs that would shorten the war.

In mid-June 1943 Commander Parsons arrived at Los Alamos and took charge of the Ordinance Division which was to build two types of bombs, one using a six foot gun bomb with a muzzle velocity of 1000 feet per second, using Uranium 235 and a second bomb using the implosion concept with Plutonium. Commander Parsons formed a group of scientists, military officers, and enlisted technicians to try out full-scale models of both the gun-assembly, called Little Boy, and an implosion type called Fat Man. He used a modified B-29 to drop-test mock bombs. The tests revealed flaws that would have been fatal in an actual drop. Fuses failed, models yawed and rolled violently. Some models got caught and hung up in the plane. One model fell onto the closed bomb-bay doors. Parsons set to work with his staff and made the necessary corrections.

Parsons as head of Ordinance was in charge of the completion of all nonnuclear components including fail-safe electronics to prevent premature explosion, proximity fuses to detonate the bombs at the best altitude above ground. In addition, as head of project Alberta, he organized a complex technical operation to deliver everything needed overseas including personnel to assemble and deliver the bomb. Parsons sent Commander Frederick L. Ashworth to the Pacific to choose a suitable site for the overseas operation. Ashworth picked Tinian and Parsons from July 2 to July 13, 1945 visited the Island. Then Parsons headed back to New Mexico to watch the test of the implosion bomb called Trinity.

TRINITY

Oppenheimer decided to make an atomic explosion which would test simultaneously both the atomic bomb and the implosion concept. A Plutonium bomb incorporating implosion was mounted on a tower in Alamogordo, New Mexico. The atomic bomb test was called Trinity. The explosion took place at 05:29, July 16, 1944. Captain Parsons observed the explosion from a plane 25 miles away at an altitude of 24,000 feet. A blinding burst of illumination filled the sky followed by a ball of fire and a mushroom cloud that grew to a height of 40,000 feet.

Szilard and other Manhattan Project scientists spent the last month of the war trying to convince President Truman to use the first atomic bomb in a non-lethal demonstration to the Japanese of its destructive power. They failed to persuade Truman and plans were made to drop the bomb on Japan.

THE DELIVERY OF ATOMIC BOMBS OVER JAPAN

One week after Trinity, Captain Parsons went back to the island of Tinian with films of the Trinity test. He took charge of the final preparations for the atomic bombing of Japan. On July 26 the United States Cruiser Indianapolis delivered Little Boy's gun assembly and the U^{235} projectile to Tinian. Three days later the last of the U^{235} target inserts for Little Boy arrived at Tinian in three separate C-54 transport cargo planes. Right after that a Japanese sub sunk the Indianapolis with a loss of 880 men.

In the meantime a special 509th composite group consisting of 15 modified B-29s had arrived at Tinian commanded by Colonel Paul W. Tibbets. The group had previously drop tested dummy models of both types of atomic bombs and had practiced breakaway maneuvers to avoid the destructive power of the bomb immediately after the drop.

Two bombs, Little Boy, the Uranium bomb, and Fat Man, the Plutonium bomb, were sent to Tinian Island in the Pacific. Most of the bomb parts were sent by ship but the Plutonium hemispheres arrived by air. The first piece of fissionable material for Little Boy arrived on July 26, 1945 aboard the U.S. Cruiser Indianapolis.

On July 31, 1945 the 509th Group completed rehearsal for the first mission over Hiroshima. Little Boy was assembled and readied for loading in Colonel Tibbet's B-29 the Enola Gay, named for his mother. Parsons decided that in order to avoid a nuclear explosion at take off with an overloaded aircraft, he would complete the final bomb assembly in the air. However, bad weather over Japan delayed the first mission.

On August 6th the weather cleared over Japan and at 2:45 A.M. Tibbets took off in the Enola Gay accompanied by Parsons. Shortly after take off Parsons aided by Lt. Morris R. Jeppson completed the final assembly around 3:15 A.M. At 7:30 A.M. Parsons returned to the bomb bay and replaced the green safety plugs with red plugs, arming the bomb. At 9:15 A.M. at an altitude of 32,700 feet Little Boy was released over Hiroshima destroying the Japanese city. The results were relayed to President Truman aboard the USS Augusta, returning from Potsdam. Truman released the news to the world and called on the Japanese to surrender. They did not and Fat Man, the Plutonium implosion bomb was made ready.

On August 9, 1945 Fat Man was loaded in a B-29, Bock's Car, at Tinian with a designated primary target, the City of Kokura. Bad weather obscured the primary target and the secondary target, Nagasaki, was bombed. There were no other atomic bombs available. One of the scientists on Tinian, Luis Alvarez, was concerned that the Japanese government might guess that the U.S. had only two atomic bombs and continue the war. He wrote a letter to a former colleague at Berkeley, Professor Ryokichi Sagane of the Physics Department at the University of Tokyo. In this note Alvarez pointed out that the U.S. could build as many atomic bombs as needed to end the war. The letter and two carbon copies were attached to the three parachute gauges dropped to monitor Fat Man's performance. The letters were found and turned over to the Japanese military but were never passed on to Professor Sagane. The day after the bombing the Japanese surrendered. The formal surrender ceremony took place on September 2, 1945 on the USS Missouri.

15.4 The American Hydrogen Bomb

The man who is often referred to as the "Father of the hydrogen bomb" is Edward Teller. He was a Hungarian Jewish physicist who came to the United States in 1935. Teller worked with Fermi on the first self-sustaining nuclear chain reaction at the University of Chicago. He then worked with J. Robert Oppenheimer on theoretical studies on the atomic bomb at the University of California at Berkeley. When Oppenheimer set up the Los Alamos atomic laboratory in 1943, Teller was one of the first scientists to be recruited. While he worked on the fission bomb, he gradually became more interested in a much more powerful hydrogen fusion bomb. When the war ended Teller wanted the U.S. government to develop this type of bomb but public opinion after Hiroshima was against this.

In 1949 the Soviet Union with the aid of espionage exploded an atomic fission bomb. A British atomic scientist who had worked at Alamos, Klaus Fuchs confessed that he had been spying for the Soviet Union since 1942. Fuchs had known of Teller's theoretical work on the hydrogen fusion bomb and had passed the information on to the Soviets. In response to this revelation and the Cold War, President Truman authorized the development of the hydrogen fusion bomb.

Teller went to work but he was not successful until Stanislaw M. Ulam came up with the idea of using the mechanical shock of an atomic bomb to compress an inner fusion core and make it explode. The resulting high density would make the core's thermonuclear fuel fuse in a tremendous explosion. Teller suggested that radiation, rather than mechanical shock, from the atomic bomb be used to compress the fusion material. A device using these ideas was tested at Eniwetok atoll in the Pacific on November 1, 1952. It yielded the equivalent of 10 million tons of TNT.

In this device an uncontrolled self-sustaining, thermonuclear fusion reaction is carried out in heavy hydrogen (deuterium or tritium) to produce an explosion. In a fusion reaction the collision of two nuclei produces reaction products together with a release of energy according to Einstein's formula, $E = mc^2$. The difference in mass results in an enormous release of energy. This thermonuclear reaction requires a tremendous temperature supplied by a fission explosion.

15.5 Obtaining Electrical Power from Atomic Fission

After World War II, in the 1950's, fission atomic reactors were built to generate electricity in the United States, the Soviet Union, England, and Germany.

A nuclear reactor provides the assembly of materials to sustain and control the neutron chain reaction, to transport the heat produced from the fission reactions, and to provide the necessary safety features to cope with the radiation and radioactive materials produced by its operations.

The generation of electrical energy by a nuclear power plant makes use of heat to produce steam or to heat gases to drive turbogenerators. The nuclear power plant is similar to the conventional coal-fired plant, except that the nuclear reactor is substituted for the conventional boiler as the source of heat. The rating of a reactor is usually given in megawatts of electricity or megawatts-thermal. The net output of electricity is about one-third of the thermal output of a fission reactor.

Moderators

Fission neutrons are released at high energies and are called fast neutrons. Neutrons slow down through collisions with materials of low atomic weight called moderators. Examples of moderators are heavy water, light water, graphite, and beryllium. Heavy water has the heavier isotope of hydrogen while light or ordinary water has ordinary hydrogen. Most commercial fission reactors use a moderator for the conversion of fast neutrons to thermal neutrons which move at much lower speeds.

FUEL

Two main types of fuel are used in commercial fission reactors, U^{235}, an isotope of Uranium and Pu^{239}, an isotope of Plutonium. Light water cooled nuclear power reactors use Uranium oxide as a fuel, with an enrichment of several percent U^{235}. Most of this fuel is in the form of cylindrical rods that are fabricated by compacting and sintering cylindrical pellets. The pellets are assembled into metal tubes which are then sealed.

Pu^{239} is produced by neutron capture in U^{238}. It is a by-product in power reactors using Uranium. Commercial Uranium spent fuel can be processed to form Pu^{239}. Plutonium is normally used in conjunction with Uranium 238. A breeder reactor consumes Plutonium fuel and at the same time breeds Plutonium from Uranium 238.

COOLANT

The heat generated in a reactor is removed by a primary coolant flowing through it. Most power reactors use water as a coolant. In a boiling water reactor the water boils directly in the reactor core to make steam that is piped to the turbine. In a pressurized water reactor the coolant water is kept under increased pressure to prevent boiling. It transfers heat to a separate stream of feed water in a steam generator. For both water coolants the water serves as the moderator as well as the coolant.

Liquid metals such as sodium are used in some reactors. However, sodium reacts violently if mixed with water. This requires care in the design and manufacture of sodium to water steam generators and backup systems to cope with occasional leaks.

15.6 HANDLING A DANGEROUS LEGACY, ATOMIC WASTE

The atomic age of the second half of the 20th century left tremendous amounts of residue, some of whose dangerous radiation will last thousands of years. There are three general categories of atomic waste, high level, low level, and transuranic (TRU) which is between the high level and low level wastes.

HIGH LEVEL WASTE

High level waste consists of the spent fuel of civilian and military reactors, plus the liquids produced when the military fuel is processed for atomic weapons. When the "spent" 12-foot long fuel rods are removed from nuclear fission reactors, they still emit a terrific amount of heat and radiation. The radiation is considered dangerous for at least 10,000 years.

To store high level wastes the United States built a facility at Yucca Mountain in Nevada, 90 miles northwest of Las Vegas. The repository is 600 feet below the base of a 1,000-foot mountain. Critics of the facility, including the state of Nevada, charge that shifting land could breach the repository. Over time dangerous radiation could escape through leaks.

The opening of the Yucca Mountain nuclear waste facility was originally scheduled for 1998. It has been pushed back to 2010 after many bitter and prolonged arguments about its safety. In the meantime, 3,000 tons of used nuclear fuel are produced every year. In 1991, 20,000 metric tons of spent fuel were stored at about 70 nuclear plant sites. By the year 2000 this could be raised to 47,000 metric tons. Thus there is an urgent need for the Yucca Mountain nuclear waste facility to open. However, there are serious questions about its safety.

The state of Nevada is subject to earthquakes which could cause the repository to be broken. The area around Yucca Mountain is home to volcanoes, one of which, Lathrop Wells, erupted perhaps as recently as 10,000 years ago. In 1995 some scientists at Los Alamos suggested that storage at Yucca Mountain could result, after thousands of years, in a nuclear explosion. This has been disputed by other scientists.

In the meanwhile the poisonous high level waste from atomic power plants is piling up. The "spent" fuel rods are first placed in the cooling pools of power plants. As the pools fill up, the rods are moved to concrete casks which are held on site. The utilities have complained that eventually they are going to run out of space.

The workers at the nuclear power site are given an exposure of about 300 millirems per year. The specified limit of exposure for the workers is 5 rems and 0.1 rems for the public living near these facilities. While these levels are arrived at by scientists to ensure a low risk of cancer, no one knows for sure. Eventually the high level wastes that are piling up at reactor sites will have to be buried somewhere, perhaps at Yucca Mountain in the early 21st century.

Another method of temporarily handling high level waste is the reprocessing of "spent" Uranium from nuclear power plants into Plutonium. France reprocesses spent Uranium fuel from several countries as well as its own. In November 1992 a Japanese freighter laden with 1.7 tons of Plutonium from reprocessed "spent" Uranium left Cherbourg escorted by French warships. The Japanese claimed that the Plutonium was for breeder reactors. Plutonium is a very dangerous substance. It is possible that terrorists and other irresponsible groups could get their hands on a small amount of Plutonium and make a nuclear bomb.

Transuranic Waste

Transuranic waste is the Plutonium contaminated soil, clothing and similar material from the United States bomb-making program. All TRU elements have long half-lives. The half-life of Plutonium is 24,360 years. Most of the transuranic waste is stored in barrels waiting for permanent storage.

The United States has a Waste Isolation Pilot Program (WIPP) near Carlsbad, New Mexico. This is a salt mine but environmental groups have fought to keep it from opening. TRU is mostly low-volume, low activity material but its extremely long life make it very dangerous. The Department of Energy of the United States hopes one day to fill caverns in WIPP carved from subterranean salt beds with 300,000 drums of Plutonium contaminated waste. It could operate for 30 years before it closes. The drums will be sealed in salt 2,000 feet underground.

There is a worry that many years in the future someone could drill into the site and a gusher of radioactive brine could spew out. To prevent this, experts were asked to devise a system of warning markers that would last ten thousand years and still have meaning over the millennia. Two teams, an "A" team and a "B" team were formed.

The "A" team favored menacing stone monoliths or earthworks to warn off trespassers. The "B team suggested building a giant earthen berm in the shape of a warning symbol such as the trefoil emblem used to indicate radioactivity. They also suggested engraving in stone the periodic table of the elements to show which wastes are stored.

Low Level Waste

Low level waste includes short-lived radioactive materials, such as medical radioactive isotopes. It also includes ancillary refuse such as filters, clothes, etc. from commercial nuclear power plants. Some low-level waste contains long-lived elements of high-level waste but in very low concentrations.

Most low level disposal does not require deep storage. A 1980 United States law says landfills are sufficient and delegates the job to the states. States have joined together in regional compacts. In California that state and three other states formed a group to bury their waste in Ward Valley, a desert area west of Needles, California. Opponents of the dump point to the threat to the Colorado River 20 miles away.

15.7 The Nuclear Sword of Damocles

According to an ancient Greek legend, Damocles was a nobleman of the city of Syracuse in Italy. He sat at a banquet beneath a sword hung by a single hair. The threat of nuclear technology spreading to terrorists backed by certain nation states is the modern equivalent of Damocles sitting at the banquet of nuclear energy.

By 1968 there were five nuclear-weapon states, China, France, Russia, the United Kingdom, and the United States. At the end of 1993, 162 of the 191 countries in the world had signed the Nuclear Non-Proliferating Treaty which stated that no states other than the original five should have nuclear weapons.

However, about every five years, a new nation has acquired a nuclear weapons capability. These include Israel, India, South Africa, and possibly Iraq and North Korea. South Africa which made some nuclear weapons secretly has renounced its nuclear weapons status. Some countries, like South Korea, Switzerland and Sweden started a nuclear weapons program but abandoned it.

The Nuclear Non-Proliferation Treaty commits all the signers to accept safeguards to prevent the diversion of peaceful nuclear energy to nuclear weapons. The International Atomic Energy Agency (IAEA) negotiates the safeguards with the different countries. The safeguards provide for monitoring the flow and storage of all nuclear materials. An annual report is issued by the Director of the IAEA.

There are two general methods of making atomic fission bombs. The first is the separation of U^{235} from U^{238}. The former can sustain a fast fission chain which can be used to either generate electric power or produce a nuclear weapon. U^{235}makes up about 0.71 percent of natural Uranium with the remainder being U^{238}. The generation of electric power requires a concentration of 3 to 4 percent U^{235} which is called low-enriched Uranium. Nuclear weapons require more than 90 percent U^{235} which is referred to as highly enriched Uranium.

There are now four basic methods of separating U^{235} from U^{238}. They all depend on the small difference in mass between the two isotopes. The numbers 235 and 238 represent the relative masses of the two forms of Uranium. The first method is gaseous diffusion where the lighter U^{235} diffuses more rapidly than U^{238}. The second is the use of a centrifuge where the lighter isotope moves towards the center of the column. The third method is electromagnetic separation. This was used by Iraq to try to make an atomic weapon. After the Gulf War, United Nations inspectors destroyed Iraq's dies for making electromagnetic isotope separation equipment at the Tuwaitha reactor. The fourth method is laser separation where the energy of the excited state depends on nuclear mass. This is probably too sophisticated for less technologically advanced nations.

The other general method of producing an atomic fission bomb is from Plutonium. Pu^{239} is made by neutron capture in U^{238} in nuclear reactors. About 70 tons of Plutonium is left in the spent fuel of the world's power reactors. The Plutonium can be separated or reprocessed and used to make weapons. Most of the civilian power-reactor Plutonium is not the best for weapons material but it could be used as such. However, several countries have used large research reactors to produce high grade Plutonium for weapons. In these research reactors the Plutonium is also made in reactors from U^{238} but is withdrawn sooner so that it contains much less of the radioactive higher isotopes of Plutonium. Only about eight pounds of weapon grade Plutonium could destroy most of a city.

In the last decade of the 20th century, terrorists have used trucks carrying conventional chemical explosives to destroy or partially destroy buildings and kill a

Figure 15-1 Tokamak Fusion Test Reactor at Princeton University
Courtesy Princeton Plasma Physics Laboratory

relatively small number of people. Two examples are the New York City World Trade building and the Alfred P. Murray federal building in Oklahoma City. Nuclear bombs carried in a truck would destroy a large part of a city with hundreds of thousands of casualties. This is the nuclear sword of Damocles at the end of the 20th century.

15.8 FUSION POWER FOR THE FUTURE

In the 1950s atomic fission nuclear reactors started out as the answer to the generation of electricity without pollution. By 1991 in the United States 21.8% of the electricity generated was from nuclear fission reactors. It was the second behind coal which generated 54.8%. In France 75% of electricity was generated from fission nuclear reactors in 1991.

However, two events occurred that demonstrated serious problems. One was in the United States at the Three Mile Reactor in Pennsylvania where a failure of the cooling system resulted in a partial meltdown. Fortunately very little radiation escaped but the reactor was closed down. In the Ukraine at Chernobyl there was a meltdown and radiation in large amounts escaped with a resulting loss of life. The radiation was detected in many parts of Europe carried by the winds. The safety factor and the waste disposal problem caused scientists and engineers to look into fusion as a possible way to generate electricity without the problems of atomic fission generators.

THE TOKAMAK

The TOKAMAK (a Russian acronym for a doughnut shape) was proposed by Russian scientists as a way of containing the fusion fuel. Temperatures must reach 30 times that of the sun to obtain fusion. At that temperature no material could hold the fuel. It was proposed that a doughnut shaped magnetic field could serve that purpose. By 1995 three experimental large TOKAMAK experimental fusion test reactors were studying the scientific feasibility of fusion energy.

They were the Joint European Torus at Culham Laboratory in the United Kingdom, the Japan Torus 60 in Naka, Japan and the Tokomak Fusion Test Reactor (TFTR) at Princeton University, New Jersey, Figure 15-1.

In the first TOKAMAK experiments the fuel was deuterium an isotope of hydrogen. Ordinary hydrogen has one proton only in the nucleus. Deuterium has one proton and a neutron in its nucleus. Deuterium is a component of heavy water.

Two atoms of deuterium fuse together to form helium (He) with a mass less than that of the deuterium. The missing mass is converted into energy in accordance with Einstein's formula, $E = mc^2$. In one form of this reaction, D + D yields He^3+ 3.3 MeV + neutron, where MeV stands for million electron volts, a form of energy. He^3 is an isotope of helium.

In 1994 a new fuel was tried in Princeton to increase the energy output of the TFTR. This new fuel was deuterium plus tritium (T). Tritium is another isotope of hydrogen with two neutrons. This reaction is:

D + T yields He^4 + neutron + 17.6 MeV

The major problem with the above two reactions is that the neutrons produced would make the walls of the reactor radioactive. Every six months the reactor would have to be shut down and torn apart in order to remove thousands of pounds of weakened walls, made reactive by the neutrons produced in the fusion reactions that generate the energy.

There is another fuel, deuterium and He^3, an isotope of helium which theoretically would eliminate most of the radioactivity. The reaction is:

D + He^3 yields He^4 + Proton + 18.3 MeV where He^4 is another isotope of helium.
This reaction does not have the radioactive problems of the previous fuels.

A number of physics professors have favored the use of He^3 in fusion generators of the future.

One of the problems with He^3 is the supply of this material. Sources of He^3 are helium mines and the decay of tritium and mines on the moon. There is enough He^3on earth to try it out. As for the moon supply it has been suggested that it would pay to export He^3 from the moon. One shuttle load of this material from the moon would supply enough energy for the United States for one year. At any rate fusion will not be practical until the 21st century. The terrible problems of atomic fission power of the 20th century must be taken seriously in the development of fusion power.

INERTIAL-CONFINEMENT FUSION

Another fusion method for the 21st century is Inertial-Confinement Fusion (ICF). In one form of ICF, powerful lasers heat up a tiny spherical capsule filled with a vapor made of deuterium and tritium. A solid frozen fuel core of deuterium and tritium surrounds the capsule. The vapor in the capsule is ignited by the laser. This compresses the fuel core of deuterium and tritium, creating helium four, neutrons and energy. In an actual large scale ICF, five to ten capsules per second must be ignited every second.

A National Ignition Facility (NIF) using 192 neodymium-glass laser beams has been proposed for the United States. If fully funded, it is expected to lead to ignition by 2005. This would be an experimental facility to demonstrate the process.

However, the problem of deuterium and tritium fuel described previously for the TFTR remains and must be resolved in the 21st century.

BIBLIOGRAPHY

Calder, Nigel — *Einstein's Universe,* The Viking Press, 1979

Einstein, Albert — *Relativity, The Special and the General Theory,*

Bonanza Books, 1962

Fermi, Laura — *Atoms in the Family: My Life with Enrico Fermi,* University of Chicago Press, 1994

Hodeson, Lilian, Paul W. Heriksen, Roger A. Meade, and Catherine Westfall — *Critical Assembly, A Technical History of Los Alamos During the Oppenheimer Years, 1943-1945.* Press Syndicate of the University of Cambridge, 1993

Holloway, David — *Stalin and the Bomb, The Soviet Union and Atomic Energy, 1939-1956.* Yale University Press, 1994

Moore, Ruth — *Neils Bohr, The Man, His Science, and the World They Changed.* Alfred A. Knoff, 1966

Pais, Abraham — *The Science and the Life of Albert Einstein.* Oxford University Press, 1982

Segre, Emilio — *Enrico Fermi, Physicist.* University of Chicago Press, 1972

Sime, Ruth Lewin — *Lise Meitner: A Life in Physics.* University of California Press, 1996

Szilard, Gertrud W. & Spencer R. Weart, Editors — *Leo Szilard: His Version of the Facts. Selected Recollections & Correspondence.* MIT Press, 1978

CHAPTER 16

EPILOGUE

Paraphrasing Einstein and others, technology in the 20th century changed everything except the way we think and feel about others who may be a little different from us.

In the 1930's there arose charlatans who twisted religion and history into hurricanes of hatred that swept away millions of innocent men, women and children. Technology in the form of radio and film was used to spew out a continuous stream of propaganda. This created a climate for the ineffable horrors of the Holocaust.

In the 1990's the same sort of hatreds were whipped up to destroy innocent men, women and children. In the former Yugoslavia the technology of television was used to disseminate propaganda which set off three religious groups of the same ethnic background against each other. The results of the propaganda were the rape and murder of thousands of civilians and the destruction of what had been peaceful integrated neighborhoods.

At about the same time the technology of radio was used to spread hatred between two ethnic groups in Africa in Burundi and Rwanda. This resulted in the massacre of hundreds of thousands of people.

Technology in the 21st century can used for extreme good or extreme evil. In order for technology to be used for benign purposes in a multicultural world, it is absolutely necessary to teach true tolerance for all religions and races at an early age. This includes teaching the Holocaust and the history of slavery in elementary schools. This should be reinforced in high schools and colleges. Tolerance and technology must go hand in hand in the 21st century and beyond if humanity is to survive.

In an almost biblical sense, humanity may be weighed in the balance by nature sometime in the future. Comets and asteroids are a substantial threat to the survival of humanity.

Comets are composed of ice with rocky nuclei. Their orbits are at the edges of the solar system. They visit Earth's neighborhood only when disturbed by unknown forces. Humanity had a front seat at a spectacular display of comet explosions in July 1994 when 21 fragments of a fractured comet bombarded Jupiter. This produced television images of Earth-sized fireballs.

Asteroids are rocks that fly mostly in a belt between Mars and Jupiter. It was discovered that an asteroid had exploded on June 30, 1908 above the Tunguska region of Siberia with a force of some 12 one megaton hydrogen bombs. It devastated a large area of mostly uninhabited forest land. If it had exploded above a city, most of the inhabitants would have perished.

There have been a number of near misses in the 20th century by asteroids passing close to the Earth. In October 1937, an asteroid, Hermes, passed within 400,00 miles of our planet. This compares to the moon's distance of 239,000 miles.

On March 23, 1989, an asteroid, Asclepius, missed the Earth by six hours. It was large enough to destroy civilization. The asteroid was not detected approaching the Earth but was only detected going away.

Interest in asteroids increased sharply when Luis Alvarez, a Nobel Prize winner, proposed in 1982 that a six-mile-wide asteroid had hit the Earth 65 million years ago. A crater was later found in the Yucatan Peninsula in Mexico which had been created 65 million years ago. The collision caused a huge amount of debris that blocked out the sun and resulted in the end of the ruling dinosaurs. This allowed the tiny nocturnal mammals to develop over millions of years into humanity.

In January 1994, the US Defense Department disclosed that from 1975 to 1992, military satellites had detected 136 atomic sized explosions from meteoroids about 20 miles above the Earth. Experts have concluded that the actual rate of explosions may be as high as 80 a year. This is because most of the explosions were undetected by military satellites.

The United States government became concerned and put a small amount of money into threat investigations. One of these was run by Dr. Tom Gehrels at the University of Arizona. He observed many asteroids going past the Earth at close distances, sometimes between the Moon and the Earth.

In December 1995, a federal program, NEAT (Near-Earth Asteroid Tracking), started to look for dangerous space rocks. The program was funded for less than one million dollars annually. It quickly discovered asteroid that swept close to Earth. These had not been found previously.

In April 1995 a conference was held by the *Explorers Club* and the *United Nations Office for Outer Space* on near-Earth objects (NEOs). The proceedings of the conference assessing the hazards NEOs may pose to planet Earth was published by the New York Academy of Science in 1997.

If humanity controls its internal hatreds, it can unite and use the 20th century technology to detect and change the course of Earth-bound comets and asteroids. There is nothing like a common outside deadly threats for uniting people. However, in order to do this an international agency must be established with three functions. The first serve as a center for information on the inanimate visitors from space and the consequence of such a collisions. This benign propaganda should be distributed all over the world.

The second function is set up a well-funded detection system to detect Earth-bound comets and asteroids. This would use many of the technologies created in the 20th century, including satellites, computers, lasers, etc.

The third function is to develop methods of diverting or destroying the threatening comets and asteroids by using 20th century technology like hydrogen bombs.

A humanity that changes its way of thinking and feeling will earn nature's blessing. The wild stallion and its passenger can then ride safely into the 21st century and beyond.

INDEX

D

E

F

G

H

I

J

K

L

Q

R

S

T